Crossley William Crosby Barlow, George Hartley Bryan

Elementary mathematical astronomy

With examples and examination papers

Crossley William Crosby Barlow, George Hartley Bryan

Elementary mathematical astronomy
With examples and examination papers

ISBN/EAN: 9783337275860

Printed in Europe, USA, Canada, Australia, Japan

Cover: Foto ©berggeist007 / pixelio.de

More available books at **www.hansebooks.com**

The University Tutorial Series.

ELEMENTARY
MATHEMATICAL ASTRONOMY,

WITH

EXAMPLES AND EXAMINATION PAPERS.

BY

C. W. C. BARLOW, M.A., B.Sc.,

GOLD MEDALLIST IN MATHEMATICS AT LONDON M.A.,
SIXTH WRANGLER, AND FIRST CLASS FIRST DIVISION PART II. MATHEMATICAL
TRIPOS, CAMBRIDGE,

AND

G. H. BRYAN, D.Sc., M.A., F.R.S.,

SMITH'S PRIZEMAN, LATE FELLOW OF ST. PETER'S COLLEGE, CAMBRIDGE,
JOINT AUTHOR OF "COORDINATE GEOMETRY, PART I.," "THE TUTORIAL ALGEBRA,
ADVANCED COURSE," ETC.

Third Impression (Second Edition).

LONDON: W. B. CLIVE,

University Tutorial Press
(*University Correspondence College Press*),

13 BOOKSELLERS ROW, STRAND, W.C.
1900.

PREFACE TO THE FIRST EDITION.

For some time past it has been felt that a gap existed between the many excellent popular and non-mathematical works on Astronomy, and the standard treatises on the subject, which involve high mathematics. The present volume has been compiled with the view of filling this gap, and of providing a suitable text-book for such examinations as those for the B.A. and the B.Sc. degrees of the University of London.

It has not been assumed that the reader's knowledge of mathematics extends beyond the more rudimentary portions of Geometry, Algebra, and Trigonometry. A knowledge of elementary Dynamics will, however, be required in reading the last three chapters, but all dynamical investigations have been left till the end of the book, thus separating dynamical from descriptive Astronomy.

The principal properties of the Sphere required in Astronomy have been collected in the Introductory Chapter; and, as it is impossible to understand Kepler's Laws without a slight knowledge of the properties of the Ellipse, the more important of these have been collected in the Appendix for the benefit of students who have not read Conic Sections.

All the more important theorems have been carefully illustrated by worked-out numerical examples, with the view of showing how the various principles can be put to practical application. The authors are of opinion that a far sounder knowledge of Astronomy can be acquired with the help of such examples than by learning the mere bookwork alone.

Feb. 1st, 1892.

PREFACE TO THE SECOND EDITION.

The first edition of *Mathematical Astronomy* having run out of print in less than eight months, we have hardly considered it advisable to make many radical changes in the present edition. We have, however, taken the opportunity of adding several notes at the end, besides answers to the examples, which latter will, we hope, prove of assistance, especially to private students; our readers will also notice that the book has been brought up to date by the inclusion of the most recent discoveries. At the same time we hope we have corrected all the misprints that are inseparable from a first edition. Our best thanks are due to many of our readers for their kind assistance in sending us corrections and suggestions.

Nov. 1st, 1892.

CONTENTS.

CHAPTER VI.

DYNAMICAL ASTRONOMY.

INTRODUCTORY CHAPTER.

ON SPHERICAL GEOMETRY.

Properties of the Sphere which will be referred to in the course of the Text.

(1) **A Sphere** may be defined as a surface all points on which are at the same distance from a certain fixed point. This point is the **Centre**, and the constant distance is the **Radius**.

(2) **The surface formed by the revolution of a semicircle about its diameter is a sphere.** For the centre of the semicircle is kept fixed, and its distance from any point on the surface generated will be equal to the radius of the semicircle.

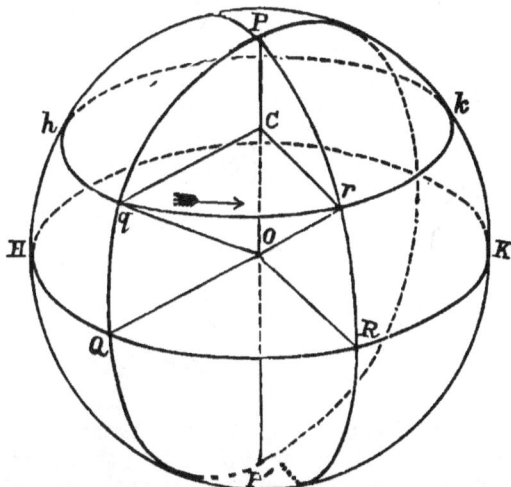

FIG. 1.

(3) Let $PqQP'$ be any position of the revolving semicircle whose diameter PP' is fixed. Let OQ be the radius perpendicular to PP', Cq any other line perpendicular to PP', meeting the semicircle in q. (We may suppose these lines to be marked on a semicircular disc of cardboard.) As the semicircle revolves, the lines OQ, Cq will sweep out planes perpendicular to PP', and the points Q, q will trace out in these planes circles $HQRK$, $hqrk$, of radii OQ, Cq respectively. From this it may readily be seen that **Every plane section of a sphere is a circle.**

Definitions.

(4) A **great circle** of a sphere is the circle in which it is cut by any plane passing through the centre (*e.g.*, *HQRK*, *PqQP'* or *PrRP*). A **small circle** is the circle in which the sphere is cut by any plane *not* passing through the centre (*e.g.*, *hqrk*).

(5) The **axis** of a great or small circle is the diameter of the sphere perpendicular to the plane of the circle. The **poles** of the circle are the extremities of this diameter. (Thus, the line *PP* is the axis, and *P*, *P'* are the poles of the circles *HQK* and *hqk*).

(6) **Secondaries** to a circle of the sphere are great circles passing through its poles. (Thus, *PQP'* and *PRP'* are secondaries of the circles *HQK*, *hqk*).

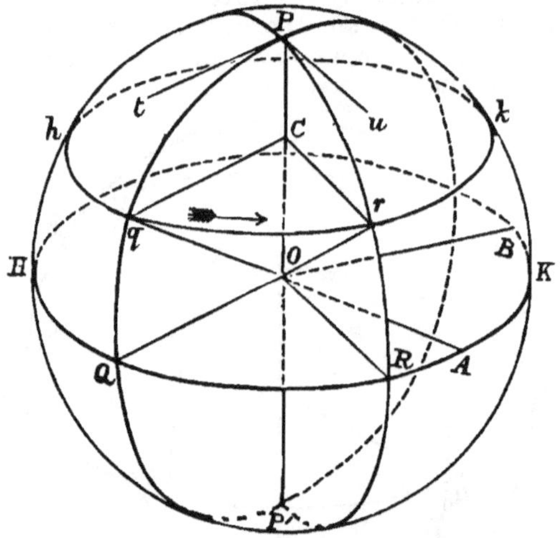

Fig. 2.

(7) **The angular distance between two points on a sphere** is measured by the arc of the great circle joining them, or by the angle which this arc subtends at the centre of the sphere. Thus, the distance between *Q* and *R* is measured either by the arc *QR*, or by the angle *QOR*. Since the circular measure of ∠ *QOR* = (arc *QR*) ÷ (radius of sphere), it is usual to measure arcs of great circles by the angles which they subtend at the centre. This remark does *not* apply to small circles.

(8) **The angle between two great circles** is the angle between their planes. Thus, the angle between the circles *PQ*, *PR* is the angle between the planes *PQP'*, *PRP'*. It is called "the angle *QPR*."

(9) A **spherical triangle** is a portion of the spherical surface bounded by three arcs of great circles. Thus, in Fig. 2, *PQR* is a spherical triangle, but *Pqr* is not a spherical triangle, because *qr* is not an arc of a great circle. We may, however, draw a great circle passing through *q* and *r*, and thus form a spherical triangle *Pqr*.

Properties of Great and Small Circles.

(10) **All points on a small circle are at a constant (angular) distance from the pole.**

For, as the generating semicircle revolves about PP', carrying q along the small circle hk, to r, the arc Pq = arc Pr, and $\angle POq = \angle POr$.

The constant angular distance Pq is called the **spherical, or angular radius** of the small circle. The pole P is analogous to the centre of a circle in plane geometry.

(11) **The spherical radius of a great circle is a quadrant, or, All points on a great circle are distant 90° from its poles.**

For, as Q, by revolving about PP', traces out the great circle $HQRK$, we have $\angle POQ = \angle POR = 90°$, and therefore, PQ, PR are quadrants.

(12) **Secondaries to any circle lie in planes perpendicular to the plane of the circle.**

For PP' is perpendicular to the planes of the circles HQK, hqk, therefore any plane through PP', such as PQP' or PRP', is also perpendicular to them.

(13) **Circles which have the same axis and poles lie in parallel planes.** For the planes HQK, hqk are parallel, both being perpendicular to the axis PP'. Such circles are often called **parallels.**

(14) **If any number of circles have a common diameter, their poles all lie on the great circle to which they are secondaries, and this great circle is a common secondary to the original circles.**

For if OA is the axis of the circle PQP', then OA is perpendicular to POP'. Hence, if the circle PQP' revolves about PP', A traces out. the great circle $HQRK$, of which P, P' are poles. We likewise see that

(15) **If one great circle is a secondary to another, the latter is also a secondary to the former.**

This is otherwise evident, since their planes are perpendicular.

(16) **The angle between two great circles is equal to**
    (i.) The angle between the tangents to them at their points of intersection ;
    (ii.) The arc which they intercept on a great circle to which they are both secondaries ;
    (iii.) The angular distance between their poles.

Let Pt, Pu be the tangents at P to the circles PQ, PR, and let A, B be the poles of the circles. If we suppose the semicircle PQP' to revolve about PP' into the position PRP', the tangent at P will revolve from Pt to Pu, the radius perpendicular to OP will revolve from OQ to OR, and the axis will revolve from OA to OB. All these lines will revolve through an angle equal to the angle between the planes PQP', PRP', and this is the angle QPR between the circles (Def. 8). Hence,

Angle between circles PQ, $PR = \angle tPu = \angle QOR = \angle AOB$.

(17) The arc of a small circle subtending a given angle at the pole is proportional to the sine of the angular radius.

Let qr be the arc of the small circle $hqrk$, subtending $\angle qPr$ at P, and let O be the centre of the circle. Evidently $\angle qCr = \angle QOR$ (since Cq, Cr are parallel to OQ, OR). Hence, the arcs qr, QR are proportional to the radii Cq, OQ,

$$\therefore \quad \frac{\text{arc } qr}{\text{arc } QR} = \frac{Cq}{OQ} = \frac{Cq}{Oq} = \sin POq = \sin Pq.$$

But QR is the arc of a great circle subtending the same angle at the pole P, hence the arc qr is proportional to $\sin Pq$, as was to be shown. Since $qQ = 90°$ Pq, therefore $\sin Pq = \cos qQ$, so that the arc qr is proportional to the cosine of the angular distance of the small circle qr from the parallel great circle QR.

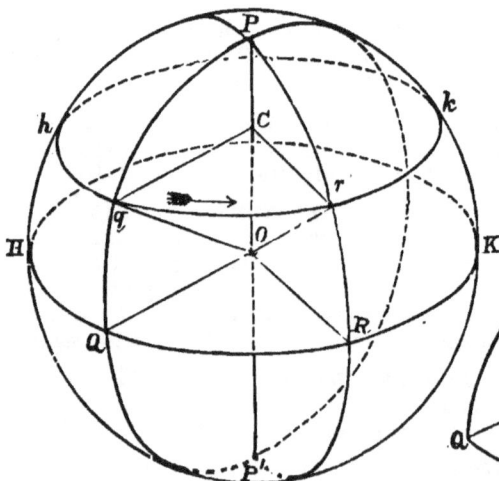

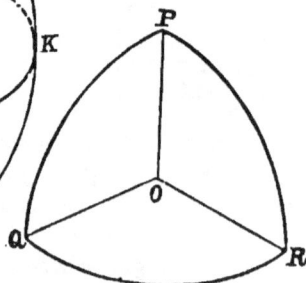

FIG. 3.   FIG. 4.

(18) Comparison of Plane and Spherical Geometry.

It may be laid down as a general rule that great circles and small circles on a sphere are analogous in their respective properties to straight lines and circles in a plane. Thus, to *join* two points on a sphere means to draw the great circle passing through them.

Secondaries to a great circle of the sphere are analogous to perpendiculars on a straight line. The distance of a point from any great circle is the length of the arc of a secondary drawn from the point to the circle. Thus, rR is the distance of the point r from the great circle $HQRK$.

On Spherical Triangles.

(19) Parts of a Spherical Triangle.—A spherical triangle, like a plane triangle, has six parts, viz., its three sides and its three angles. The sides are generally measured by the angles they subtend at the centre of the sphere, so that the six parts are all expressed as *angles*.

Any three given parts suffice to determine a spherical triangle, but there are certain "ambiguous cases" when the problem admits of more than one solution. The formulæ required in solving spherical triangles form the subject of Spherical Trigonometry, and are in every case different from the analogous formulæ in Plane Trigonometry. There is this further difference, that a spherical triangle is completely determined if its three angles are given.

Thus, two spherical triangles will, in general, be equal if they have the following parts equal :—

(i.) Three sides.
(ii.) Two sides and included angle.
(iii.) Two sides and one opposite angle.

(iv.) Three angles.
(v.) Two angles and adjacent side.
(vi.) Two angles and one opposite side.

Cases (iii.) and (vi.) may be ambiguous.

(20) Right-angled Triangles.—If one of the angles is a right angle, two of the remaining five parts will determine the triangle.

(21) Triangle with two right angles.—The properties of a spherical triangle, such as PQR, Fig. 3, in which one vertex P is the pole of the opposite side QR, are worthy of notice. Here two of the sides, PQ, PR, are quadrants, and two angles Q, R are right angles. The third side QR is equal to the opposite angle QPR.

(22) Triangle with three right angles.—If, in addition, the angle QPR is a right angle (Fig. 4), QR will be a quadrant. The triangle PQR will, therefore, have all its angles right angles, and all its sides quadrants, and each vertex will be the pole of the opposite side.

The planes of the great circles forming the sides, are three planes through the centre O mutually at right angles, and they divide the surface of the sphere into eight of these triangles; thus the area of each triangle is one-eighth of the surface of the sphere.

(23) The three angles of a spherical triangle are together greater than two right angles.

[For proof, see any text-book on Spherical Geometry.]

(24) If the sides of a spherical triangle, when expressed as angles, are very small, so that its linear dimensions are very small compared with the radius of the sphere, the triangle is very approximately a plane triangle.

Thus, although the Earth's surface is spherical, a triangle whose sides are a few yards in length, if traced on the Earth, will not be distinguishable from a plane triangle. If the sides are several miles in length, the triangle will still be very nearly plane.

(25) **Any two sides of a spherical triangle are together greater than the third side.** For if we consider the plane angles which the sides subtend at the centre of the sphere, any two of these are together greater than the third, by Euclid XI., 20.

(26) The following application of (25) is of great use in astronomy, and is analogous to Euclid III., 7, 8.

Let $AHBK$ be any given great or small circle whose pole is P, Z any other given point on the sphere, and let the great circle ZP meet the given circle in the points A, B. Then A, B are the two points on the given circle whose distances from Z are greatest and least respectively.

For let H be any other point on the circle. Join ZH, HP.
Then, in spherical $\triangle ZPH$, $ZP + PH > ZH$. But $PH = PA$;
$$\therefore ZP + PA > ZH,$$
i.e., $$ZA > ZH.$$
Also, if Z is on the opposite side of the circle to P, then
$$ZH + PH > PZ; \therefore ZH + PB > PZ; \therefore ZH > PZ - PB,$$
i.e., $$ZH > ZB.$$

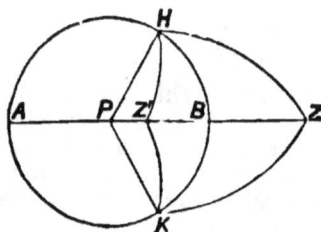

Fig. 5.

If Z' be a point on the same side of the circle as P, then $PZ' + Z'H > PH$. But $PH = PB$. $\therefore PZ' + Z'H > PB$.
$$\therefore Z'H > PB - PZ',$$
i.e., $$Z'H > Z'B, \quad \text{as before.}$$
Hence, A is further from Z, Z', and B is nearer to Z, Z', than any other point on the circle.

(27) If H, K are the two points on the circle equidistant from Z, the spherical triangles ZPH, ZPK have ZP common, $ZH = ZK$ (by hypothesis), and $PH = PK$ [by (10)], hence they are equal in all respects; thus $\angle ZPH = \angle ZPK$, and $\angle PZH = \angle PZK$.
Hence PH, PK are equally inclined to PB, as are also ZH, ZK.
Similar properties hold in the case of the point Z'. These properties are of frequent use.

ASTRONOMY.

CHAPTER I.

THE CELESTIAL SPHERE.

SECTION I.—*Definitions—Systems of Co-ordinates.*

1. **Astronomy** is the science which deals with the celestial bodies. These comprise all the various bodies distributed throughout the universe, such as the Earth (considered as a whole), the Sun, the Planets, the Moon, the comets, the fixed stars, and the nebulæ. It is convenient to divide Astronomy into three different branches.

The first may be called **Descriptive Astronomy.** It is concerned with observing and recording the motions of the various celestial bodies, and with applying the results of such observations to predict their positions at any subsequent time. It includes the determination of the distances, and the measurement of the dimensions of the celestial bodies.

The second, or **Gravitational Astronomy,** is an application of the principles of dynamics to account for the motions of the celestial bodies. It includes the determination of their masses.

The third, called **Physical Astronomy,** is concerned with determining the nature, physical condition, temperature, and chemical constitution of the celestial bodies.

The first branch has occupied the attention of astronomers in all ages. The second owes its origin to the discoveries of Sir Isaac Newton in the seventeenth century; while the third branch has been almost entirely built up in the present century.

In this book we shall treat exclusively of Descriptive and Gravitational Astronomy.

2. **The Celestial Sphere.**—On observing the stars it is not difficult to imagine that they are bright points dotted about on the inside of a hollow spherical dome, whose centre is at the eye of the observer. It is impossible to form any direct conception of the distances of such remote bodies; all we can see is their relative directions. Moreover, most astronomical instruments are constructed to determine only the directions of the celestial bodies. Hence it is important to have a convenient mode of representing directions.

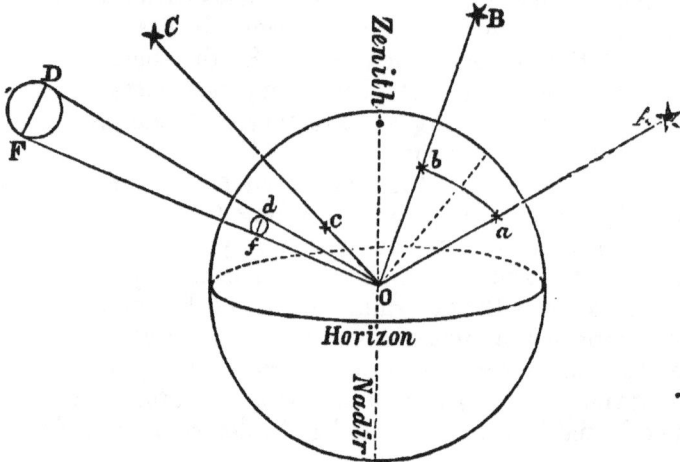

FIG. 6.

The way in which this is done is shown in Figure 6. Let *O* be the position of any observer, *A, B, C*, &c., any stars or other celestial bodies. About *O*, as centre, describe a sphere with any convenient length as radius, and let the lines joining *O* to the stars *A, B, C* meet this sphere in *a, b, c* respectively. Then the points *a, b, c* will represent, on the sphere, the directions of the stars *A, B, C*, for the lines joining these points to *O* will pass through the stars themselves. In this manner we obtain, on the sphere, an exact representation of the appearance of the heavens as seen from *O*. Such a sphere is called the **Celestial Sphere.**

This sphere may be taken as the dome upon which the stars *appear* to lie. But it must be carefully borne in mind that the stars *do not actually lie on a sphere at all*, and that they are only so represented for the sake of convenience.

3. Use of the Globes.—The representation of directions of stars by points on a sphere is well exemplified in the old-fashioned star globes. Such a globe may be used as the observer's celestial sphere; but it must be remembered that the directions of the stars are the lines joining the *centre* to the corresponding points on the sphere; for in every case the observer is supposed to be at the centre of the celestial sphere.

The properties given in the Introduction on Spherical Geometry are applicable to the geometry of the celestial sphere. A knowledge of them will be assumed in what follows.

4. Angular Distances and Angular Magnitudes.— Any plane through the observer will be represented on the celestial sphere by a great circle. The arc of the great circle *ab* (Fig. 6) represents the angle *aOb* or *AOB* which the stars *A*, *B* subtend at *O*. This angle is generally measured in degrees, minutes, and seconds, and is called the **angular distance** between the stars. This angular distance must not be confused with their actual distance *AB*. In the same way, when we are dealing with a body of perceptible dimensions, such as the Sun or Moon (*DF*, Fig. 6), we shall define its **angular diameter** as the angle *DOF*, subtended by a diameter at the observer's eye. This angular diameter is measured by the arc *df* of the celestial sphere, that is, by the diameter of the projection of the body on the celestial sphere. From the figure it is evident that

$$\frac{df}{Od} = \frac{DF}{OD}.$$

Since *DF* is the actual linear diameter of the body, measured in units of length, the last relation shows us that the angular diameter (*df*) of a body varies directly as its linear diameter *DF*, and inversely as *OD*, the distance of the body from the observer's eye.

As the eye can only judge of the dimensions of a body from its angular magnitude, this result is illustrated by the fact that the nearer an object is to the eye the larger it looks, and *vice versâ*. Thus, if the distance of the object be doubled, it will only look half as large, as may be easily verified.

5. The Directions of the Stars are very approximately independent of the Observer's Position on the Earth.

This is simply a consequence of the enormously great distances of all the stars from the Earth. Thus, let x (Fig. 7) denote any star or other celestial body, S, E two different positions of the observer. If the distance SE be only a very small fraction of the distance Sx, the angle ExS will be very small, and this angle measures the difference between the directions of x as seen from E and from S.

In illustration, if we observe a group of objects a mile or two off, and then walk a few feet in any direction, we shall observe no perceptible change in the apparent directions or relative positions of the objects.

FIG. 7.

If Ex' be drawn parallel to Sx, the angle xEx' will be equal to ExS, and will therefore be very small indeed. Hence, Ex will very nearly coincide in direction with Ex'. Thus, considering the vast distances of the stars, we see that

The lines joining a Star to different points of the Earth may be considered as parallel.*

The stars will, therefore, always be represented by the *same* points on a star globe, or celestial sphere, no matter what be the position of the observer. The great use of the celestial sphere in astronomy depends on this fact.

6. Motion of Meteors.—The projection of bodies on the celestial sphere is well illustrated by the apparent motion of a swarm of meteors. Where such a swarm is moving uniformly, all the meteors describe (approximately) parallel straight lines. If we draw planes through these lines and the observer, they will intersect in a common line, namely, the line through the observer parallel to the direction of the common motion of the meteors. The planes will, therefore, cut the celestial sphere in great circles, having this line as their common diameter. These great circles represent the apparent paths of the meteors on the celestial sphere. The paths appear, therefore, to radiate from a common point, namely, one of the extremities of this diameter.

This point is called the **Radiant,** and by observing its position the direction of motion of the meteors is determined.

---

* This is not true in the case of the Moon.

7. Zenith and Nadir.—Horizon.—If, through the observer, a line be drawn in the direction in which gravity acts (*i.e.*, the direction indicated by a plumb-line), it will meet the celestial sphere in two points. One of these is vertically above the observer, and is called the **Zenith;** the other is vertically below the observer, and is called the **Nadir.** (Fig. 6, and Z, N, Fig. 8.)

If the plane through the observer parallel to the surface of a liquid at rest be produced, it will cut the celestial sphere in a great circle. This great circle is called the **Celestial Horizon.** (Fig. 6, and $sEnW$, Fig. 8.)

It is proved in Hydrostatics that the surface of a liquid at rest is a plane perpendicular to the direction of gravity. Hence, the celestial horizon is the great circle whose poles are the zenith and nadir. We might have defined the horizon by this property.

From the above definition, it is evident that, to an observer whose eye is close to the surface of the ocean, the celestial horizon forms the boundary of the visible portion of the celestial sphere. On land, however, the boundary, or **visible horizon** (as it is called), is always more or less irregular, owing to trees, mountains, and other objects.

8. Diurnal Motion of the Stars.—If we observe the sky at different intervals during the night, we shall find that the stars always maintain the same configurations relative to one another, but that their actual situations in the sky, relative to the horizon, are continually changing. Some stars will set in the west, others will rise in the east. One star which is situated in the constellation called the "Little Bear," remains almost fixed. This star is called Polaris, or the Pole Star. All the other stars describe on the celestial sphere small circles (Fig. 8) having a common pole P very near the Pole Star, and the revolutions are performed in the same period of time, namely, about 23 hours 56 minutes of our ordinary time.

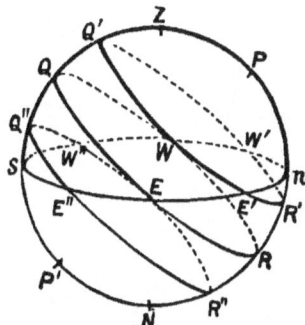

FIG. 8.

9. Celestial Poles, Equator, and Meridian.— The common motion of the stars may most easily be conceived by imagining them to be attached to the surface of a sphere which is made to revolve uniformly about the diameter PP'.

The extremities of this diameter are called the **Celestial Poles.** That pole, P, which is above the horizon in northern latitudes is called the **North Pole,** the other, P', is called the **South Pole.**

The great circle, $EQRW$, having these two points for its poles, is called the **Celestial Equator.** It is, therefore, the circle which would be traced out by the diurnal path of a star distant 90° from either pole.

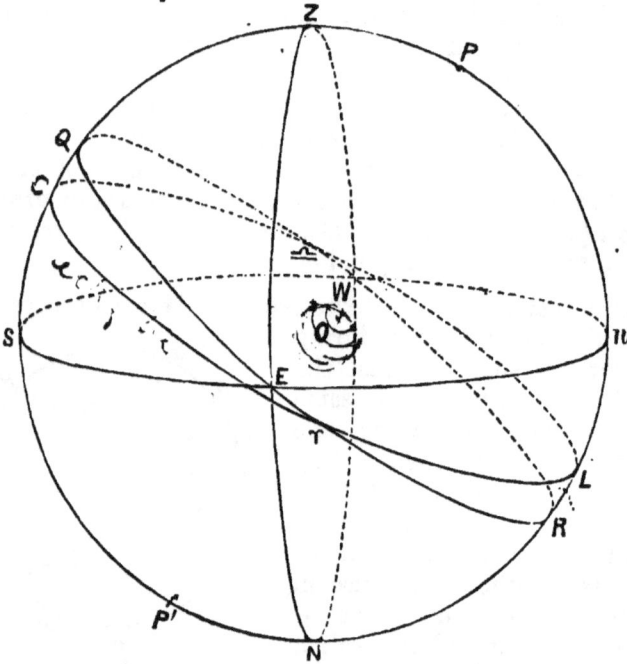

FIG. 9.

The **Meridian** is the great circle ($PZP'N$, Fig. 9) passing through the zenith and nadir and the celestial poles. It cuts both the horizon and equator at right angles [by Spher. Geom. (12), since it passes through their poles].

10. The Cardinal Points.—The **East** and **West Points** (*E*, *W*, Fig. 9) are the points of intersection of the equator and horizon. The **North** and **South Points** (*N*, *S*) are the intersections of the meridian with the horizon.

Verticals.—Secondaries to the horizon, *i.e.*, great circles through the zenith and nadir, are called **Vertical Circles**, or, briefly, **Verticals**. Thus, the meridian is a vertical. The **Prime Vertical** is the vertical circle (*ZENW*) passing through the east and west points.

Since *P* is the pole of the circle *QERW*, and *Z* is the pole of *nEsW*, therefore *E*, *W* are the poles of the meridian *PZP'N*. Hence the horizon, equator, and prime vertical which pass through *E*, *W*, are all secondaries to the meridian ; they therefore all cut the meridian at right angles.

11. Annual Motion of the Sun.—The Ecliptic.— The Sun, while participating in the general diurnal rotation of the heavens, possesses, in addition, an independent motion of its own relative to the stars.

Imagine a star globe worked by clockwork so as to revolve about an axis pointing to the celestial pole in the same periodic time as the stars. On such a moving globe the directions of the stars will always be represented by the same points. During the daytime let the direction of the Sun be marked on the globe, and let this process be repeated every day for a year. We shall thus obtain on the globe a representation of the Sun's path relative to the stars, and it will be found that—

(i.) The Sun moves from west to east, and returns to the same position among the stars in the period called a year ;

(ii.) The relative path on the celestial sphere is a great circle, inclined to the equator at an angle of about $23° 27\frac{1}{2}'$.

This great circle ($C \Upsilon L \simeq$, Fig. 9) is called the **Ecliptic**. We may, therefore, briefly define the ecliptic as the great circle which is the trace, on the celestial sphere, of the Sun's annual path relative to the stars.

The intersections of the ecliptic and equator are called **Equinoctial Points**. One of them is called the **First Point of Aries**; this is the point through which the Sun passes when crossing from south to north of the equator, and it is usually denoted by the symbol Υ. The other is called the **First Point of Libra**, and is denoted by the symbol $\simeq$.

12. Coordinates.—In Analytical Geometry, the position of a point in a plane is defined by two coordinates. In like manner, the position of a point on a sphere may be defined by means of two coordinates. Thus, the position of a place on the Earth is defined by the two coordinates, latitude and longitude. For fixing the positions of celestial bodies, the following different systems of coordinates are used.

13. Altitude or Zenith Distance and Azimuth.—Let Fig. 10 represent the celestial sphere, seen from overhead, and let x be any star. Draw the vertical circle ZxX. Then the position of x may be defined by either of the following pairs of coordinates, which are analogous to the Cartesian and polar coordinates of a point in a plane respectively :—

(a) The arc sX and the arc Xx;
(b) The arc Zx and the angle sZx.

Practically, however, the two systems are equivalent; for, since Z is the pole of sX, $ZX = 90°$, therefore

$$Zx = 90° - xX, \text{ and angle } sZx = \text{arc } sX.$$

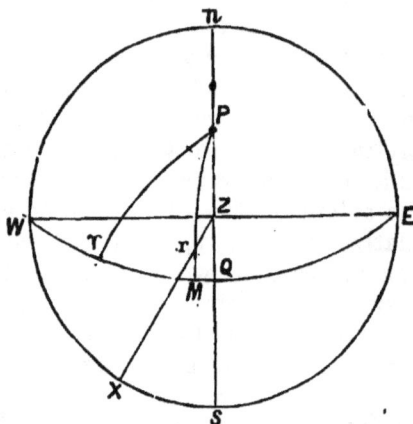

Fig. 10.

The **Altitude** of a star (Xx) is its angular distance from the horizon, measured along a vertical.

The **Zenith Distance** (abbreviation, Z.D.) is its angular distance from the zenith (Zx), or the complement of the altitude.

The **Azimuth** $(sX$ or $sZx)$ is the arc of the horizon intercepted between the south point and the vertical of the star, or the angle which the star's vertical makes with the meridian

*14. Points of the Compass.—In practical applications of Astronomy to navigation, it is usual to measure the azimuth in "points" and "quarter points" of the compass. The dial plate of a mariner's compass is divided into 32 points, by repeatedly bisecting the right angles formed by the directions of the four cardinal points. Thus each point represents an angle of 11¼ degrees. The points are again subdivided into "quarter points" of 2¹³⁄₁₆ degrees. Starting from the north and going round towards the east, the various points are denoted as follows :—

N., N. by E., N.N.E., N.E. by N., N.E., N.E. by E., E.N.E., E. by N.

E., E. by S., E.S.E., S.E. by E., S.E., S.E. by S., S.S.E., S. by E.

S., S. by W. S.S.W., S.W. by S., S.W., S.W. by W., W.S.W , W. by S.

W., W. by N., W.N.W. N.W. by W., N.W., N.W. by N., N.N.W., N. by W.

The quarter points are denoted thus :—E.N.E. ¼ E. means one quarter point to the eastward of E.N.E., that is, 6¼ points, or 70° 18′ 45″, from the north point, taken in an easterly direction. So, too, S.S.W. ¼ W. means 2¼ points, or 28° 7′ 30′, measured from the south point westwards.

15. Polar Distance, or Declination, and Hour Angle.

—From the pole P, draw through x the great circle PxM; this circle is a secondary to the equator EQW.

Then we may take for the coordinates of x the arc Px and the angle sPx. Or we may take the arc xM, which is the complement of Px, and the arc QM, which = angle QPx.

The **North Polar Distance** of a star (abbreviation, N.P.D.) is its angular distance (Px) from the celestial pole.

The **Declination** (abbreviation, Decl.) is the angular distance from the equator (xM), measured along a secondary, and is, therefore, the complement of the N.P.D.

The great circle PxM through the pole and the star is called the star's **Declination Circle**.

The **Hour Angle** of the star (ZPx) is the angle which the star's declination circle makes with the meridian.

The declination may be considered positive or negative, according as the star is to the north or south of the equator, but it is more usual to specify this by the letter N. or S., as the case may be, and this is called the **name** of the declination.

The hour angle is generally measured from the meridian towards the west, and is reckoned from 0° to 360°.

Either the declination and hour angle or the N.P.D. and hour angle may be taken as the two coordinates of a star.

16. Declination and Right Ascension.—The position of a celestial body is, however, more frequently defined by its declination and right ascension.

The declination has been already defined, in § 15, as the angular distance of the star from the equator, measured along a secondary. (*xM*, Fig. 11.)

The **Right Ascension** (R.A.) is the arc of the equator intercepted between the foot of this secondary and the First Point of Aries. Thus, ♈*M*, Fig. 11, is the R.A. of the star *x*.

The R.A. of a star is always measured from ♈ *eastwards* reckoning from 0° to 360°. Thus the star ω *Piscium*, whose declination circle cuts the equator 1° 34′ 18″ *west* of ♈, has the R.A. 360°—1° 34′ 18″, or 358° 25′ 42″.

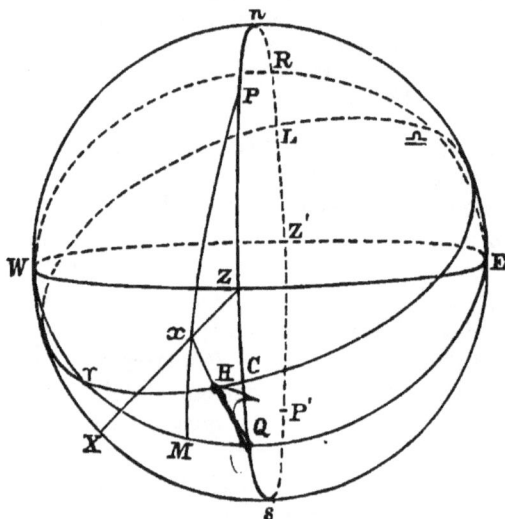

FIG. 11.

17. Celestial Latitude and Longitude.—The position of a celestial body may also be referred to the ecliptic instead of the equator.

The **Celestial Latitude** is the angular distance of the body from the ecliptic, measured along a secondary to the ecliptic. (*Πx*, Fig. 11.)

The **Celestial Longitude** is the arc of the ecliptic intercepted between this secondary and the first point of Aries, measured eastwards from ♈. (♈*Π*, Fig. 11.)

18. Latitude of the Observer.—The celestial latitude and longitude, defined in the last paragraph, must not be confounded with the latitude and longitude of a place on the Earth, as there is no connection whatever between them.

The **Latitude** of a place is the angular distance of its zenith from the equator, measured along the meridian.

Thus, in Fig. 11, ZQ is the latitude of the observer.

Since $PQ = nZ = 90°$; $\therefore ZQ = nP$, or in other words, *The latitude of a place is the altitude of the Celestial Pole.*

The complement of the latitude is called the **Colatitude.**

Hence, in Fig. 11, PZ is the colatitude of the observer, and is *the angular distance of the zenith from the pole.*

In this book the latitude of an observer will generally be denoted by the symbol l, and the colatitude by c.

The longitude of a place will be defined in Chapter III.

19. Obliquity of the Ecliptic.—The inclination of the ecliptic to the equator is called the **Obliquity.** In Fig. 11, $Q \Upsilon C$ is the obliquity. As stated in § 11, this angle is about $23° 27\frac{1}{2}'$. We shall generally denote the obliquity by i.

20. Advantages of the Different Coordinate Systems.—The altitude and azimuth of a celestial body indicate its position relative to objects on the Earth. Owing, however, to the diurnal motion, they are constantly changing.

The N.P.D. and hour angle also serve to determine the star's position relative to the earth, and have this further advantage, that the N.P.D. is constant, while the hour angle increases at a uniform rate.

Since the equator and first point of Aries partake of the common diurnal motion of the stars, the declination and right ascension of a star are constant. These coordinates are, therefore, the most suitable for tabulating the relative positions of the various stars on the celestial sphere.

The celestial latitude and longitude of a celestial body are also unaffected by the diurnal motion. They are most useful in defining the positions of the Sun, Moon, planets, and comets, for the first always moves in the ecliptic, while the paths described by the others are always very near the ecliptic.

21. Recapitulation.—For the sake of convenient reference, we give on the next page a list of all the definitions of this chapter, with references to Figs. 11, 12.

GREAT CIRCLES.	THEIR POLES.
Horizon, $nEsW$.	Zenith, Z; Nadir, Z'.
Equator, $EQWR$.	North Pole, P; South Pole, P'.
Meridian, $ZsZ'n$.	East Point, E; West Point, W.
Prime Vertical, $ZEZ'W$.	North Point, n; South Point, s.

Ecliptic, $\Upsilon C \simeq L$; Equinoctial Points, Υ, $\simeq$, viz. :—First Point of Aries, Υ, and First Point of Libra, $\simeq$; Vertical of Star, ZxX; Declination Circle of Star, PxM.

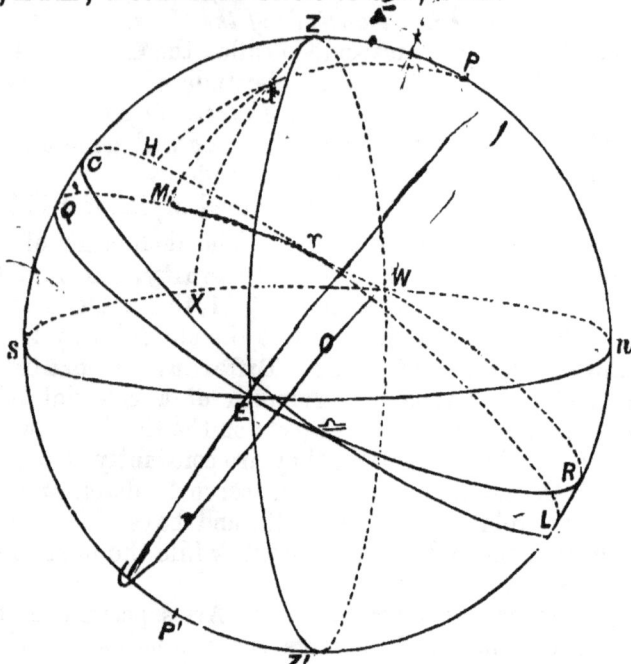

FIG. 12.

COORDINATES.

Altitude, Xx;
or Zenith Distance, Zx. } Azimuth, $sX = sZx$.

North Polar Distance, Px.   Hour Angle, $QM = ZPx$.
Declination, Mx.   Right Ascension, ΥM.
Celestial Latitude, Hx.   Celestial Longitude, ΥH.

OTHER ANGLES. — Obliquity of Ecliptic $(i) = C\Upsilon Q$. Observer's Latitude $(l) = ZQ = nP$. Colatitude $(c) = PZ$.

Notice that the circles on the remote side of the celestial sphere are *dotted*.

SECTION II. *The Diurnal Rotation of the Stars.*

22. Sidereal Day and Sidereal Time.—A Sidereal Day is the period of a complete revolution of the stars about the pole relative to the meridian and horizon. Like the common day it is divided into 24 hours (h.), and these are subdivided into 60 minutes (m.) of 60 seconds (s.) each. The sidereal day commences at "**Sidereal Noon,**" *i.e.*, the instant when the first point of Aries crosses the meridian.

The **Astronomical Clock,** which is the clock used in observatories, indicates sidereal time. The hands should indicate 0h. 0m. 0s. when the first point of Aries crosses the meridian, and the hours are reckoned from 0h. up to 24h., when Υ again comes to the meridian and a new day begins.

From the facts stated in § 8, it appears that the sidereal day is about 4 minutes shorter than the ordinary day. The stars are observed to revolve about the pole at a *perfectly uniform* rate, so that the sidereal day is of invariable length, and the angles described by any star about the pole are proportional to the times of describing them. Thus, the *hour angle* of a star (measured towards the west) is proportional to the interval of sidereal time that has elapsed since the star was on the meridian.

Now, in 24 sidereal hours the star comes round again to the meridian, after a complete revolution, the hour angle having increased from 0° to 360°. Hence the hour angle increases at the rate of 15° per hour. Hence, also, it increases 15' per minute, or 15" per second.

The hour angle of a star is, for this reason, generally measured by the number of hours, minutes, and seconds of sidereal time taken to describe it. It is then said to be **expressed in time.** Thus,

The hour angle of a star, when expressed in time, is the interval of sidereal time that has elapsed since the star was on the meridian.

In particular, since the instant when Υ is on the meridian is the commencement of the sidereal day, we see that

The sidereal time is the hour angle of the first point of Aries when expressed in time.

23. To reduce to angular measure any angle expressed in time.—*Multiply by* 15. *The hours, minutes, and seconds of time will thus be reduced to degrees, minutes, and seconds of angle.*

Conversely, **to reduce to time from angular measure** *we must divide by* 15, *and for degrees, minutes, and seconds, write hours, minutes, and seconds.*

EXAMPLES.—1. To find, in angular measure, the hour angle of a star at 15h. 21m. 50s. of sidereal time after its transit. The process stands thus—

$$
\begin{array}{ccc}
15 & 21 & 50 \\
   &   & 15^{\circ} \\
\hline
230 & 27 & 30
\end{array}
$$

∴ the angular measure of the hour angle is 230° 27′ 30″

2. To find the sidereal time required to describe 230° 27′ 30″ (converse of Ex. 1).

$$
\begin{array}{r|ccc}
15) & 230 & 27 & 30 \\
\hline
 & 15 & 21 & 50
\end{array}
$$

; ∴ required time = **15h. 21m. 50s.**

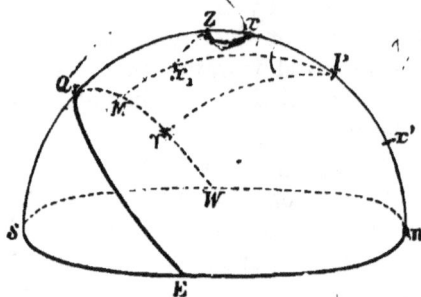

FIG. 13.

24. Transits.—The passage of the star across the meridian is called its **Transit.**

Let x be the position of any star in transit (Fig. 13).

The star's R.A. = ϒQ or ϒPQ = hour angle of ϒ
                   = sidereal time expressed in angle.

Hence, **the right ascension of a star, when expressed in time, is equal to the sidereal time of its transit.**

In practice the R.A. of a star is always expressed in time. Thus, the R.A. of a *Lyræ* is given in the tables as 18h. 33m. 14·8s., and not as 278° 18′ 42″.

Again, let s be the meridian zenith distance Zx, considered positive if the star transits north of the zenith, d the star's north declination Qx, and l the north latitude QZ. We have evidently —

$$Qx = QZ + Zx;$$
$$\therefore \quad d = l + s;$$

or (star's N. decl.)
= (lat. of observer)+(star's meridian Z.D.)

This formula will hold universally if declination, latitude, and zenith distance are considered negative when south.

Hence *the R.A. and decl. of a star may be found by observing its sidereal time of transit and its meridian Z.D., the latitude of the observatory being known.*

Conversely, if the R.A. and decl. of a star are known, we can, by observing its time of transit and meridian Z.D., determine the sidereal time and the latitude of the observatory.

By finding the sidereal time we may set the astronomical clock.

A star whose R.A. and decl. have been tabulated, is called a **known star.**

In Chapter II. we shall describe an instrument called the Transit Circle, which is adapted for observing the times of transit and meridian zenith distances of celestial bodies.

25. General Relation between R.A. and hour angle.—Let x_1 (Fig. 13) be any star not on the meridian. Then

$$\angle QPx_1 = \angle QP\Upsilon - \angle \Upsilon Px_1 = \angle QP\Upsilon - \Upsilon M;$$

hence, if angles are expressed in time,

(star's hour angle) = (sidereal time) − (star's R.A.).

Hence, given the R.A. and decl. of a star, we can find its hour angle and N.P.D. at any given sidereal time, and by this means determine the star's position on the observer's celestial sphere. Or we can construct the star's position thus:—On the equator, in the *westward* direction from Q, measure off $Q\Upsilon$ equal to the sidereal time (reckoning 15° to the hour). From Υ *eastwards*, measure ΥM equal to the star's R.A.; and from M, in the direction of the pole, measure off Mx_1 equal to the star's declination. We thus find the star x_1.

*26. Transformations.—If the R.A. and decl. of a star are given, its celestial latitude and longitude may be found, and *vice versâ*; but the calculations require spherical trigonometry. The process is analogous to changing the direction of the axes through an angle i, in plane coordinate geometry. Again, the Z.D. and azimuth may be calculated from the N.P.D. and hour angle, by solving the triangle ZPx_1. We know the colatitude PZ, Px_1 and $\angle ZPx_1$, and we have to determine Zx_1 and $\angle QZx_1$ ($= 180° - PZx_1$).

In the last article we showed how to find the hour angle in terms of the R.A., or *vice versâ*, the sidereal time being known. Hence we see that, given the coordinates of a star referred to one system, its coordinates referred to any other of the systems can be calculated at any given instant of sidereal time.

27. Culmination and Southing of Stars.—A celestial body is said to **culminate** when its altitude is greatest or least.

Since the fixed stars describe circles about the pole, it readily follows, from Spherical Geometry (26), that a star attains its greatest or least zenith distance when on the meridian, and, therefore, that its culmination is the same as its transit.

This is not strictly the case with the Sun, because, owing to its independent motion, its polar distance is not constant; hence it does not describe strictly a small circle about the pole.

When a star transits S. of the zenith it is said to **south.**

28. Circumpolar Stars.—A **Circumpolar Star** at any place is a star whose polar distance is less than the latitude of the place. Its declination must, therefore, be greater than the colatitude.

On the meridian let Px and Px' be measured, each equal to the N.P.D. of such a star (Fig. 14). Then x and x' will be the positions of the star at its transits. Since $Px' < Pn$, both x' and x will be above n. Hence, during a sidereal day a circumpolar star will transit twice, once above the pole (at x) and once below the pole (at x'), and both transits will be visible. The two transits are distinguished as the **upper** and **lower culminations** respectively, and they succeed one another at intervals of 12 sidereal hours (since $xPx' = 180°$). The altitude of the star is greatest at upper, and least at lower culmination, as may easily be seen from Sph. Geom. (26) by considering the zenith distances. Hence the altitude is never less than nx', and the star is always above the horizon.

Since $nx-nP = Px = Px' = nP-nx'$,

∴ $nP = \frac{1}{2}(nx+nx')$;

that is,

The observer's latitude is half the sum of the altitudes of a circumpolar star at upper and lower culminations.

Also, $Px = \frac{1}{2}(nx-nx')$;

that is,

The Star's N.P.D. is half the difference of its two meridian altitudes.

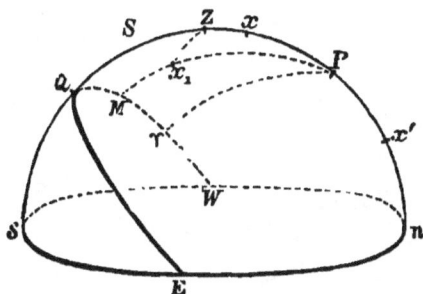

FIG. 14.

These results will require modification if the upper culmination takes place south of the zenith as at S. The meridian altitude will then be measured by sS, and not nS. Here, $nS = 180°-sS$, and we shall, therefore, have to replace the altitude at upper culmination by its *supplement*.

South Circumpolar Stars.—If the *south* polar distance of a star is less than the north latitude of the observer, the star will always remain *below* the horizon, and will, therefore, be invisible. Such a star is called a **South Circumpolar Star.**

EXAMPLE.—The constellation of the Southern Cross (*Crux*) is invisible in Europe, for its declination is 62° 30′ S; therefore its south polar distance is 27° 30′, and it will, therefore, not be visible in north latitudes higher than 27° 30′.

29. Rising, Southing, and Setting of Stars.—If the N. and S. polar distances of a star are both greater than the latitude, it will transit alternately above and below the horizon. This shows that the star will be invisible during a certain portion of its diurnal course. Astronomically, the star is said to **rise** and **set** when it crosses the celestial horizon.

Let b, b' be the positions of any star when rising and setting respectively.

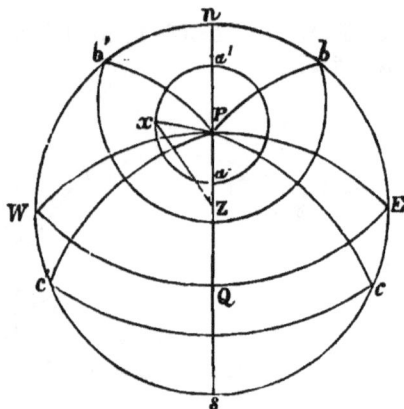

FIG. 15.

In the spherical triangles Pnb, Pnb',

$$Pb = Pb' \text{ (each being the star's N.P.D.)},$$
$$\text{right } \angle Pnb = \text{right } \angle Pnb',$$

and Pn is common.

Hence the triangles are equal in all respects; therefore

$$\angle nPb = \angle nPb',$$

and the supplements of these angles are also equal, that is,

$$\angle sPb = \angle sPb'.$$

But the angle sPb, when reduced to time, measures the interval of time taken by the star to get from b to the meridian, and sPb' measures the time taken from the meridian to b'. Hence,

The interval of time between rising and southing is equal to the interval between southing and setting.

Thus, if t, t' are the times of rising and setting, and T the time of transit, we have $T-t = t'-T.$

$$\therefore T = \tfrac{1}{2}(t+t'), \text{ or}$$

The time of transit is the arithmetic mean between the times of rising and setting.

In order to facilitate the calculations, tables have been constructed giving the values of $T-t$ for different latitudes and declinations.

If the observer's latitude Pn and the star's polar distance Pb are known, it is possible (by Spherical Trigonometry) to solve the right-angled triangle Pbn, and to calculate the angle nPb, and therefore also the angle bPs. This angle, when divided by 15, gives the time $T-t$. Moreover, the sidereal time of transit T is known, being equal to the star's R.A. Hence the sidereal times of rising and setting can be found.

If the star is on the equator, it will rise at E and set at W. Since EQW is a semicircle, exactly half the diurnal path will be above the horizon, and the interval between rising and setting will be 12 sidereal hours. If the star is to the north of the equator, it will rise at some point b between E and n, so that

$$\angle bPs > \angle EPs,$$

i.e., $$\angle bPs > 90°,$$

and the star will be above the horizon for more than 12 hours. Similarly, if the star is south of the equator, it will rise at a point c between E and s, and will be above the horizon for less than 12 hours.

From the equality of the triangles bPn, $b'Pn$ (Fig. 15), we also see that

$$nb = nb', \text{ and } sb = sb'.$$

Hence the diameter (ns) of the celestial sphere, joining the north and south points, bisects the arc (bb') between the directions of a star at rising and setting.

This gives us an easy method of roughly determining, by observation, the directions of the cardinal points; but, owing to the usual irregularities in the visible horizon, the method is not very exact.

SECTION III.—*The Sun's Annual Motion in the Ecliptic—The Moon's Motion—Practical Applications.*

30. The Sun's Motion in Longitude, Right Ascension and Declination.—In § 11, we briefly described the Sun's apparent motion in the heavens relative to the fixed stars. We defined a **Year** as the period of a complete revolution, starting from and returning to any fixed point on the celestial sphere. The **Ecliptic** was defined as the great circle traced out by the Sun's path, and its points of intersection with the Equator were termed the **First Point of Aries** and **First Point of Libra**, or together, the **Equinoctial Points.**

We shall now trace, by the aid of Fig. 16, the variations in the Sun's coordinates during the course of a year, starting with March 21st, when the Sun is in the first point of Aries. We shall, as usual, denote the obliquity by i, so that $i = 23° 27\frac{1}{2}'$ nearly.

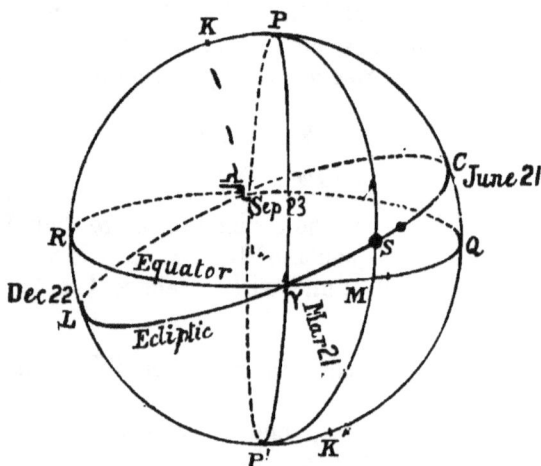

FIG. 16.

On **March 21st** the Sun crosses the equator, passing through the first point of Aries (Υ). This is the **Vernal Equinox**, and it is evident from the figure that

Sun's longitude = 0, R.A. = 0, Decl. = 0.

From March 21st to June 21st the Sun's declination is north, and is increasing.

On **June 21st** the Sun has described an arc of 90° from Υ on the ecliptic, and is at C (Fig. 16). This is called the **Summer Solstice.** If we draw the declination circle PCQ, the spherical triangle ΥCQ is of the kind described in Sph. Geom. (21), and CP is a secondary to the ecliptic. Hence (Sph. Geom. 26) the Sun's polar distance CP is a minimum and therefore its decl. a maximum.

Also $\Upsilon Q = 90°$ and $CQ = \angle C \Upsilon Q = i$. Hence

Sun's longitude = 90°, R.A. = 90° = 6h.,
N. Decl. = i, (a maximum).

From June 21 to September 23 the Sun's declination is still north, but is decreasing.

On **September 23rd** the Sun has described 180°, and is at the first point of Libra ($\triangleq$), the other extremity of the common diameter of the ecliptic and equator. This is the **Autumnal Equinox,** and we have

Sun's long. = 180°, R.A. = 180° = 12h., Decl. = 0.

From Sept. 23 to Dec. 22 the Sun is south of the equator, and its south declination is increasing.

On **December 22nd** the Sun has described 270° from Υ, and is at L (Fig. 16). This is called the **Winter Solstice.** We have $\triangleq L = 90°$, and the triangle $\triangleq RL$ has two right angles at R, L (Sph. Geom. 21). The Sun's polar distance LP is a maximum (Sph. Geom. 26), and

$$\triangleq R = \triangleq L = 90°, \ LR = \angle L \triangleq R = i. \text{ Hence}$$
Sun's longitude = 270°, R.A. = 270° = 18h.,
S. Decl. = i, (a maximum).

From December 22 to March 21 the Sun's declination is still south, but is decreasing.

Finally, on March 21, when the Sun has performed a complete circuit of the ecliptic, we have •

Sun's long. = 360°, R.A. = 360° = 24h., Decl. = 0.

The longitude and R.A. are again reckoned as zero, and they, together with the declination, undergo the same cycle of changes in the following year.

31. Sun's Variable Motion in R.A.—We observe that the Sun's right ascension is equal to its longitude four times in the year, viz., at the two equinoxes and the two solstices. *At other times this is not the case.*

For example, between the vernal equinox and summer solstice we have $\Upsilon M < \Upsilon S$, ∴ Sun's R.A. < longitude.

Hence, even if the Sun's motion in longitude be supposed uniform, its R.A. will not increase quite uniformly. There is a further cause of the want of uniformity, namely, that the Sun's motion in longitude is not quite uniform; but this need not be considered in the present chapter.

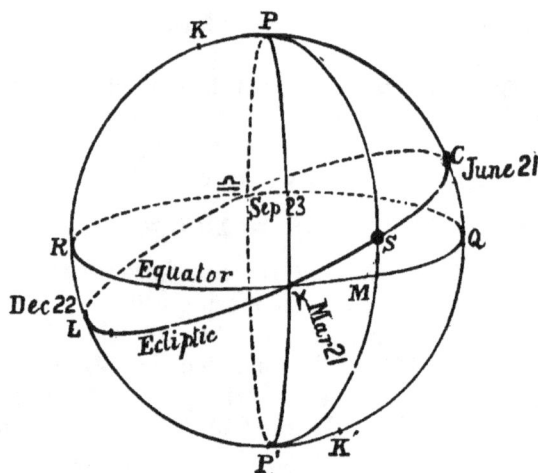

FIG. 17.

32. Direct and Retrograde Motions.—The direction of the Sun's annual revolution relative to the stars, *i.e.*, motion from west through south to east, is called **direct.** The opposite direction, that of the diurnal apparent motions of the stars or revolution from east to west, is called **retrograde.**

The revolutions of all bodies forming the solar system, with the exception of some comets and one or two small satellites, are direct.

We shall see in Chapter III. that the apparent *retrograde* diurnal motion may be accounted for by the *direct* rotation of the Earth about its polar axis.

33. Equinoctial and Solstitial Points—Colures.—

From § 30 it appears that the **Summer** and **Winter Solstices** may be defined as the times of the year when the Sun attains its greatest north and south declinations respectively. The corresponding positions of the Sun in the ecliptic (C, L, Fig. 17) are called the **Solstitial Points.** In the same way the **Equinoctial Points** (Υ, $\triangle$) are the positions of the Sun at the **Vernal and Autumnal Equinoxes** when its declination is zero.

The declination circle $P\Upsilon P'\triangle$, passing through the equinoctial points, is called the **Equinoctial Colure.** The declination circle $PCP'L$, passing through the solstitial points, is called the **Solstitial Colure.** The latter passes through the poles of the ecliptic (K, K').

34. To find the Sun's Right Ascension and Declination.—

In the "Nautical Almanack,"* the Sun's R.A. and declination at noon are tabulated for every day of the year. Their hourly variations are also given in an adjoining column. To find their values at any time of the day, we only have to multiply the hourly variation by the number of hours that have elapsed since the preceding noon, and add to the value at that noon.

EXAMPLE.—To find the Sun's R.A. and decl. on September 4, 1891 at 5h. 18m. in the afternoon. We find from the Almanack for 1891 under September 4 :—

Sun's R.A. at noon = 10h. 52m. 15s., hourly variation 9·04s.
N. Decl. at noon = 7° 12′ 12″   „   „   55·4″

(1)   R.A. at noon = 10h. 52m. 15s.
Increase in 5h. = 9·04s. × 5 =   45·2
    „   18m.   =   2·7
                                     ———————
∴ R.A. at 5h. 18m. = 10h. 53m. 3s.

(2) From the Almanack, decl. is less on September 5, and is therefore *decreasing*.
N. Decl at noon = 7° 12′ 12″
Decrease in 5h. = 55·4″ × 5 =   4′ 37″ } To be
    „   18m.   =   17″ } subtracted.
                                 ———————
∴ N. Decl. at 5h. 18m. = 7° 7′ 18″

———————————————————

* Also in " Whitaker's Almanack," which may be consulted with advantage.

35. Rough Determination of the Sun's R.A.—We can, without the "Nautical Almanack," find to within a degree or two, the Sun's R.A. on any given date, as follows :—

A year contains 365¼ days. In this period the Sun's R.A. increases by 360°. Hence its average rate of increase is very nearly 30° per month, or 1° per day.

Knowing the Sun's R.A. at the nearest equinox or solstice, we add 1° for every day later, or subtract 1° for every day before that epoch. If the R.A. is required in *time*, we allow for the increase at the rate of 2h. per month, or 4m. per day.

EXAMPLES.—1. To find the Sun's R.A. on January 1st. On December 22nd the R.A. = 18h. Hence on January 1st, which is ten days later, the Sun's R.A. = 18h. 40m.

2. To find on what date the Sun's R.A. is 10h. 36m. On September 23rd the R.A. is 12h. Also 12h. – 10h. 36m. = 84m., and the R.A. increases 84m. in 21 days. Hence the required date is 21 days before September 23, *i.e.*, September 2nd.

36. Solar Time.—**Apparent Noon** is the time of the Sun's upper transit across the meridian, that is, in north latitudes, the time when the Sun *souths*. **Apparent Midnight** is the time of the Sun's transit across the meridian below the pole (and usually below the horizon).

An **Apparent Solar Day** is the interval between two consecutive apparent noons, or two consecutive midnights.

Like the sidereal day, the solar day is divided into 24 hours, which are again divided into 60 minutes of 60 seconds each. For ordinary purposes the day is divided into two portions: the morning, lasting from midnight to noon; the evening, from noon till midnight; and in each portion times are reckoned from 0h. (usually called 12h.) up to 12h. For *astronomical* purposes we shall find it more convenient to measure the solar time by the number of solar hours that have elapsed since the preceding *noon*. Thus, 6.30 A.M. on January 2nd will be reckoned, astronomically, as 18h. 30m. on January 1st. On the other hand, 12.53 P.M. will be reckoned as 0h. 53m., being 53 minutes past noon.

During a solar day the Sun's hour angle increases from 0° to 360°. It therefore increases at the rate of 15° per hour. Hence

The apparent solar time = the Sun's hour angle expressed in time.

At noon the Sun is on the meridian. The sidereal time, being the hour angle of Υ, is the same as the Sun's R.A., *i.e.*,

Sidereal time of apparent noon = Sun's R.A. at noon.

At any other time, the difference between the sidereal and solar times, being the difference between the hour angles of Υ and the Sun, is equal to the Sun's R.A. Hence, as in § 25, we have

(Sidereal time) — (apparent solar time) = Sun's R.A.

If a and $a+x$ are the right ascensions of the Sun at two consecutive noons, then, since a whole day has elapsed between the transits, the total sidereal interval is 24h. $+x$, and exceeds a sidereal day by the amount x. But the interval is a solar day.

Hence, **the solar day is longer than the sidereal day, and the difference is equal to the sun's daily motion in R.A.***

37. Morning and Evening Stars.—Sunrise and Sunset.—When a star rises shortly before the Sun, and in the same part of the horizon, it is called a **Morning Star.** Such a star is then only visible for a short time before sunrise. When a star sets shortly after the Sun, and in the same part of the horizon, it is called an **Evening Star.** It is then only visible just after sunset.

It will be readily seen from a figure, that a star will be a morning star if its decl. is nearly the same as the Sun's, while its R.A. is rather less. Similarly, a star will be an evening star if its decl. is nearly the same as the Sun's, but its R.A. somewhat greater. Thus, as the Sun's R.A. increases, the stars which are evening stars will become too near the Sun to to be visible, and will subsequently reappear as morning stars.

The times of sunrise and sunset are calculated in the manner described in § 29. The hour angles of the Sun, when crossing the eastern and western horizons, determine the intervals of solar time between sunrise, apparent noon, and sunset. The two intervals are equal, if the Sun's decl. be supposed constant from sunrise to sunset—a result very approximately true, since the change of decl. is always very small.

---

* Owing to the sun's variable motion in R.A., the apparent solar day is not quite of constant length. In the present chapter, however, it may be regarded as approximately constant.

38. The Gnomon.—Determination of Obliquity of Ecliptic. — The Greek astronomers observed the Sun's motion by means of the Gnomon, an instrument consisting essentially of a vertical rod standing in the centre of a horizontal floor. The direction of the shadow cast by the Sun determined the Sun's azimuth, while the length of the shadow, divided by the height of the rod, gave the tangent of the Sun's zenith distance. To find the meridian line, a circle was described about the rod as centre, and the directions of the shadow were noted when its extremity just touched the circle before and after noon. The sun's Z.D.'s at these two instants being equal, their azimuths were evidently (Sph. Geom. 27) equal and opposite, and the bisector of the angle between the two directions was therefore the meridian line.

The Sun's meridian zenith distances were then observed both at the summer solstice, when the Sun's N. decl. is i and meridian Z.D. least, and at the winter solstice, when the Sun's S. decl. is i and meridian Z.D. greatest. Let these Z.D.'s be z_1 and z_2 respectively, and let l be the latitude of the place of observation. From § 24, we readily see that

$$z_1 = l - i, \quad z_2 = l + i,$$
$$\therefore \ l = \tfrac{1}{2}(z_2 + z_1), \quad i = \tfrac{1}{2}(z_2 - z_1);$$

thus determining both the latitude and the obliquity.

39. The Zodiac.—The position of the ecliptic was defined by the ancients by means of the constellations of the Zodiac, which are twelve groups of stars, distributed at about equal distances round a belt or zone, and extending about 8° on each side of the ecliptic. The Sun and planets were observed to remain always within this belt. The vernal and autumnal equinoctial points were formerly situated in the constellations of Aries and Libra, whence they were called the First Point of Aries and the First Point of Libra. Their positions are *very slowly* varying, but the old names are still retained. Thus, the "First Point of Aries" is now situated in the constellation Pisces.

The early astronomers probably determined the Sun's annual path by observing the morning and evening stars. After a year the same morning and evening stars would be observed, and it would be concluded that the Sun performed a complete revolution in the year.

40. Motion of the Moon.—The Moon describes among the stars a great circle of the celestial sphere, inclined to the ecliptic at an angle of about 5°. The motion is direct, and the period of a complete " sidereal " revolution is about 27⅓ days.

In this time the Moon's celestial longitude increases by 360°. When the Moon has the same longitude as the Sun, it is said to be **New Moon,** and the period between consecutive new Moons is called a **Lunation.** When the Moon has described 360° from new Moon, it will again be at the same point among the stars; but the Sun will have moved forward, so that the Moon will have a little further to go before it catches up the Sun again. Hence the lunation will be rather longer than the period of a sidereal revolution, being about 29½ days.

The **Age of the Moon** is the number of days which have elapsed since the preceding new Moon. Since the Moon separates 360° from the Sun in 29½ days, it will separate at the rate of about 12°, or more accurately 12⅕°, per day, or 30' per hour. This enables us to calculate roughly the Moon's angular distance from the Sun, when the age of the Moon is given, and conversely, to determine the Moon's age when its angular distance is given.

EXAMPLE.—On September 23, 1891, the Moon is 20 days old. To find roughly its angular distance from the Sun and its longitude on that day.

(1) In one day the Moon separates 12⅕° from the Sun; therefore, in 20 days it will have separated 20 × 12⅕, or 244°, and this is the required angular distance from the Sun.

(2) On September 23 the Sun's longitude is 180°; therefore the Moon's longitude is 180° + 244° = 424° = 360° + 64°, or 64°.

This method only gives very rough results; for the Moon's motion is far from uniform, and the variations seem very irregular.

Moreover, the plane of the Moon's orbit is not fixed, but its intersections with the ecliptic (called the **Nodes**) have a retrograde motion of 19° per year. Hence, for rough purposes, it is better to neglect the small inclination of the Moon's orbit, and to consider the Moon in the ecliptic. If greater accuracy be required, the Moon's decl. and R.A. may be found from the Nautical Almanack.

ASTRON.   D

41. Astronomical Diagrams and Practical Applications.—We can now solve many problems connected with the motion of the celestial bodies, such as determining the direction in which a given star will be seen from a given place, at a given time, on a given date, or finding the time of day at which a given star souths at a given time of year.

We have, on the celestial sphere, certain circles, such as the meridian, horizon, and prime vertical, also certain points, such as the zenith and cardinal points, whose positions relative to terrestrial objects always remain the same. Besides these, we have the poles and equator, which remain fixed, with reference *both* to terrestrial objects and to the fixed stars. We have also certain points, such as the equinoctial points, and certain circles, such as the ecliptic, which partake of the diurnal motion of the stars, performing a retrograde revolution about the pole once in a sidereal day. Lastly, we have the Sun, which moves in the ecliptic, performing one retrograde revolution relative to the meridian in a solar day, or one direct revolution relative to the stars in a year, and whose hour angle measures solar time.

In drawing a diagram of the celestial sphere, the positions of the meridian, horizon, zenith, and cardinal points should first be represented, usually in the positions shown in Fig. 18. Knowing the latitude nP of the place, we find the pole P. The points Q, R, where the equator cuts the meridian, are found by making $PQ = PR = 90°$; and the points Q, R, with E, W, enable us to draw the equator.

We now have to find the equinoctial points. How to do this depends on the data of the problem. Thus we may have given—

(i.) The sidereal time;

(ii.) The hour angle of a star of known R.A. and decl.;

(iii.) The time of (solar) day and time of year.

In case (i.), the sidereal time multiplied by 15 gives, in degrees, the hour angle ($Q\Upsilon$) of the first point of Aries. Measuring this angle from the meridian westwards, we find Aries, and take Libra opposite to it. Any star of known decl. and R.A. can be now found by taking on the equator $\Upsilon M =$ star's R.A., and taking on MP, $Mx =$ star's decl.

The ecliptic may be drawn passing through Aries and Libra, and inclined to the equator at an angle of about $23\frac{1}{2}°$ (just over ¼ right angle). As we go round from west to east, or in the *direct* sense, the ecliptic passes from south to north of the equator at Aries ; this shows on which side to represent the ecliptic. Knowing the time of year, we now find the Sun (roughly) by supposing it to travel to or from the nearest equinox or solstice about 1° per day from west to east. Finally, if the Moon's age be given, we find the Moon by measuring $12\frac{1}{4}°$ per day, or 30′ per hour eastwards from the Sun.

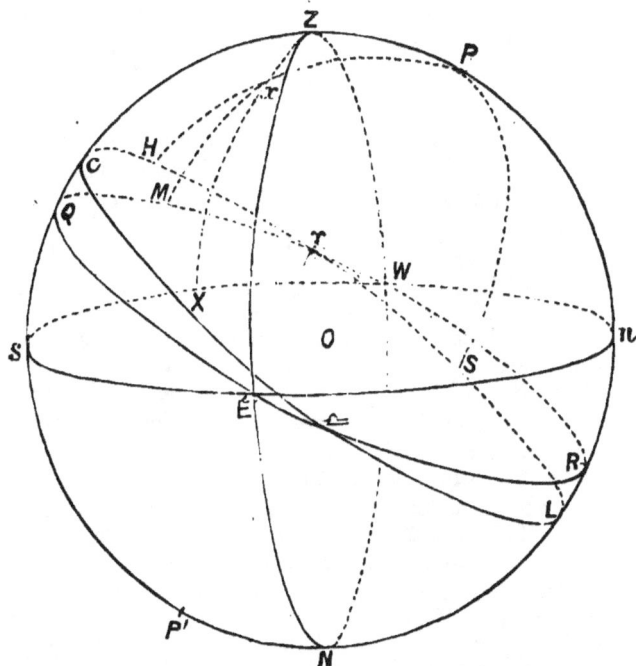

FIG. 18.

In case (ii.), we either know the hour angle, QM or QPM of a known star (x), or, what is the same thing, the sidereal interval since its transit; or, in particular, it is given that the star is on the meridian. Each of these data determines M, the foot of the star's declination circle. From M we measure $M\Upsilon$ westwards equal to the star's R.A. This finds Aries.

In case (iii.), the solar time multiplied by 15 gives the
Sun's hour angle QPS in degrees. From the time of year
we can find the Sun's R.A., ΥPS. From these we find
$QP\Upsilon$ and obtain the position of Aries just as in case (ii.)

It will be convenient to remember that azimuth and hour
angle are measured from the meridian westwards, while
right ascension and celestial longitude are measured from the
first point of Aries eastwards. Thus, since the Sun's diurnal
motion is retrograde, and its annual motion direct, the Sun's
azimuth, hour angle, R.A., and longitude are all increasing.

Most problems of this class depend for their solution chiefly
on the consideration of arcs measured along the equator, or
(what amounts to the same) angles measured at the pole.

In another class of problems depending on the relation between the latitude, a star's decl. and meridian altitude (§ 24),
we have to deal with arcs measured along the meridian.
These two classes include nearly all problems on the celestial
sphere which do not require spherical trigonometry.

EXAMPLES.

1. To represent, in a diagram, the positions of the Sun and Moon,
and the star ζ *Herculis* as seen by an observer in London on Aug. 19,
1891, at 8 p.m., the following data being given :—Latitude of London
= 51°, Moon's age at noon on Aug. 19 = 14 days 19 hours, Moon's
latitude = 2° S., R.A. of ζ *Herculis* = 16h. 37m., decl. = 31° 48' N.

The construction must be performed in the following order :—

(i.) Draw the observer's celestial sphere, putting in the meridian,
horizon, zenith Z, and four cardinal points n, E, s, W.

(ii.) Indicate the position of the pole and equator. The observer's
latitude is 51°. Make, therefore, $nP = 51°$. P will be the pole. Take
$PQ = PR = 90°$, and thus draw the equator, $QERW$.

(iii.) Find the declination circle passing through the Sun. The
time of day is 8 p.m. Therefore the Sun's hour angle is 8 × 15°, or
120°. On the equator measure $QK = 120°$ westwards from the
meridian. Then the Sun $\odot$ will lie on the declination circle PK.
Since $QW = 90°$, we may find K by taking $WK = 30° = \frac{1}{3} WR$.

(iv.) Find the first points of Aries and Libra. The date of observation is August 19. Now, on September 23 the Sun is at $\simeq$. Also
from August 19 to September 23 is 1 month 4 days. In this
interval the Sun travels about 34° from west to east. Hence the
Sun is 34° west of $\simeq$. And we must measure $K\simeq = 34°$ eastwards
from S, and thus find $\simeq$.

The first point of Aries (Υ) is the opposite point on the equator.

(v.) We may now draw the ecliptic $C \Upsilon L \triangle$ passing through the first points of Aries and Libra, and inclined to the equator at an angle of about 23½° (*i.e.*, slightly over ¼ of a right angle). The Sun is above the equator on August 19; hence the ecliptic cuts PK above K. This shows on which side of the equator the ecliptic is to be drawn; we might otherwise settle this point by remembering that the ecliptic rises above the equator to the *east* of Υ.

The intersection of the ecliptic with PK determines ⊙, the position of the Sun.

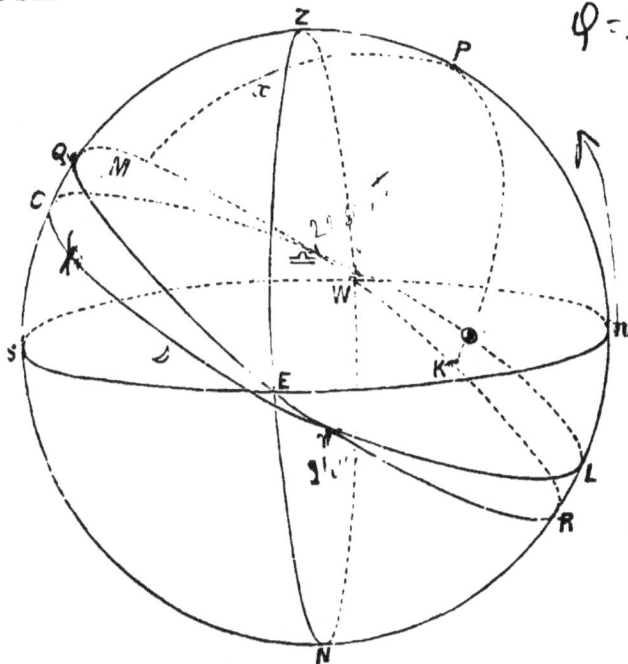

FIG. 19.

(vi.) Having found Υ, we can now find ζ *Herculis*. Its right ascension is 16h. 37m., in time, = 249° 15′ in angular measure. On the equator measure off $\Upsilon M = 249° 15′$ in the direction west to east (*i.e.*, the direction of direct motion) from Υ; we must, therefore, take $\triangle M = 69° 15′$. On the declination circle MP, measure off $Mx = 31° 48′$ towards P. Then x is the required position of ζ *Herculis*.

(vii.) Find the Moon. At 8 p.m. the Moon's age is 14d. 19h + 8h. = 15d. 3h. Hence, the Moon has separated from the Sun by about 185° in the direction west to east. Measure off ⊙ ☽ = 185° from west to east, and put in ☽ about 2° below the ecliptic. The Moon's position is thus found.

2. To find (roughly) at what time of year the Star *a Cygni* (R.A. = 20h. 38m., decl. = 44° 53′ N.) souths at 7 p.m.

Let *a* be the position of the star on the meridian (Fig 20). At 7 p.m. the Sun's western hour angle (*QS* or *QPS*) = 7h. = 105°.

Also ♈*RQ*, the Star's R.A. = 20h. 38m. Hence ♈*RS*, the Sun's R.A. = 20h. 38m. − 7h. = 13h. 38m. ; or, in angular measure, Sun's R.A. = 204° 30′. Now, on September 23, Sun's R.A. = 180°, and it increases at about 1° per day. Hence the Sun's R.A. will be 204° about 24 days later, *i.e.*, about October 17th.

3. At noon on the longest day (June 21) a vertical rod casts on a horizontal plane a shadow whose length is equal to the height of the rod. To find the latitude of the place and the Sun's altitude at midnight.

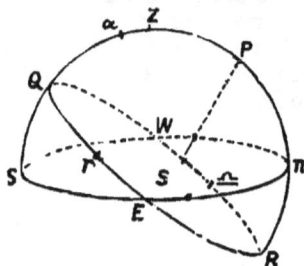

FIG. 20.

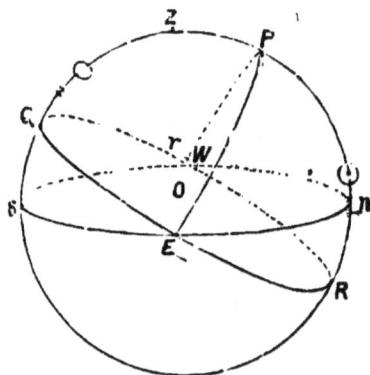

FIG. 21.

From the data, the Sun's Z.D. at noon, *Z*⊙, evidently = 45°. Also, if *QR* be the equator, ⊙*Q* = Sun's decl. = *i* = 23° 27′ (approx.);

∴ latitude of place = *ZQ* = 45° + 23° 27′ = 68° 27′.

If ⊙′ be the Sun's position at midnight,

$$P\odot' = P\odot = 90° - 23° 27' = 66° 33'.$$

But   *Pn* = lat. = 68° 27′.

∴ ⊙′*n* = 68° 27′ − 66° 33′ = 1° 54′ ;

and the Sun will be above the horizon at an alt. of **1° 54′** at midnight.

EXAMPLES.—I.

1. Why are the following definitions alone insufficient?—The zenith and nadir are the poles of the horizon. The horizon is the great circle of the celestial sphere whose plane is perpendicular to the line joining the zenith and nadir.

2. The R.A. of an equatorial star is 270°; determine approximately the times at which this star rises and sets on the 21st June. In what quarter of the heavens should we look for the star at midnight?

3. Explain how to determine the position of the ecliptic relatively to an observer at a given hour on a given day. Indicate the position of the ecliptic relatively to an observer at Cambridge at 10 p.m. at the autumnal equinox. (Lat. of Cambridge = 52° 12′ 51·6″.)

4. Prove geometrically that the least of the angles subtended at an observer by a given star and different points of the horizon is that which measures the star's altitude.

5. Show that in latitude 52° 13′ N. no circumpolar star when southing can be within 75° 34′ of the horizon.

6. Represent in a figure the position of the ecliptic at sunrise on March 21st as seen by an observer in latitude 45°. Also in latitude 67½°.

7. If the ecliptic were visible in the first part of the preceding question, describe the variations which would take place during the day in the positions of its points of intersection with the horizon.

8. Determine when the star whose declination is 30° N. and whose R.A. is 356° will cross the meridian at midnight.

9. The declination and R.A. of a given star are 22° N. and 6h. 20m. respectively. At what period of the year will it be (i.) a morning, (ii.) an evening star? In what part of the sky would you then look for it?

10. Find the Sun's R.A. (roughly) on January 25th, and thus determine about what time *Aldebaran* (R.A. 4h. 29m.) will cross the meridian that night.

11. Where and at what time of the year would you look for *Fomalhaut?* (R.A. 22h. 51m., decl. 30°. 16′ S.)

12. At the summer solstice the meridian altitude of the Sun is 75°. What is the latitude of the place? What will be the meridian altitude of the Sun at the equinoxes and at the winter solstice?

EXAMINATION PAPER.—I.

1. Explain how the directions of stars can be represented by means of points on a sphere. Explain why the configurations of the constellations do not depend on the position of the observer, and why the angular distance of two different bodies on the celestial sphere gives no idea of the actual distance between them.

2. Define the terms—*horizon, meridian, zenith, nadir, equator, ecliptic, vertical, prime vertical,* and represent their positions in a figure.

3. Explain the use of coordinates in fixing the position of a body on the celestial sphere, and define the terms—*altitude, azimuth, polar distance, hour angle, right ascension, declination, longitude, latitude.* Which of these coordinates always remain constant for the same star?

4. Define the *obliquity of the ecliptic* and the *latitude of the observer.* Give (roughly) the value of the obliquity, and of the latitude of London. Indicate in a diagram of the celestial sphere twelve different arcs and angles which are equal to the latitude of the observer.

5. What is meant by a *sidereal day* and a *sidereal hour?* How could you find the length of a sidereal day without using a telescope? Why is sidereal time of such great use in connection with astronomical observations?

6. Show that the declination and right ascension of a celestial body can be determined by meridian observations alone.

7. What is meant by a *circumpolar star?* What is the limit of declination for stars which are circumpolar in latitude 60° N.? Indicate in a diagram the belt of the celestial sphere containing all the stars which rise and set.

8. Define the terms—*year, equinoxes, solstices, equinoctial and solstitial points, equinoctial and solstitial colures.* What are the dates of the equinoxes and solstices, and what are the corresponding values of the Sun's declination, longitude, and right ascension? Find the Sun's greatest and least meridian altitudes at London.

9. Why is it that the interval between two transits of the Sun or Moon is rather greater than a sidereal day? Show how the Sun's R.A. may be found (roughly) on any given date, and find it on July 2nd, expressed in hours, minutes, and seconds.

10. Indicate (roughly) in a diagram the positions of the following stars as seen in latitude 51° on July 2nd at 10 p.m.:—*Capella* (R.A. 5h. 8m. 38s., decl. 45° 53′ 10″ N.), *a Lyræ* (R.A. 18h. 33m. 14s., decl. 38° 40′ 57″ N.), *a Scorpii* (R.A. 16h. 22m. 43s., decl. 26° 11′ 22″ S.), *a Ursæ Majoris* (R.A. 10h. 57m. 0s., decl. 62° 20′ 22″ N.)

CHAPTER II.

THE OBSERVATORY.

Section I.—*Instruments adapted for Meridian Observations.*

42. One of the most important problems of practical astronomy is to determine, by observation, the right ascension and declination of a celestial body. We have seen in Chapter I. that these coordinates not only suffice to fix the position of a star relative to neighbouring stars, but they also enable us to find the direction in which the star may be seen from a given place at a given time of day on a given date (§ 41). Moreover, it is evident that by determining every day the declination and right ascension of the Sun, the Moon, or a planet, the paths of these bodies relative to the stars can be mapped out on the celestial sphere and their motions investigated.

In Section II. of the preceding chapter we showed that the right ascension and declination of a star can be determined by observations made when the star is on the meridian. We proved the following results:—

The star's R.A. measured in time is equal to the time of transit indicated by a sidereal clock (§ 24).

The star's north decl. d can be found from z its meridian zenith distance, and l the latitude of the observatory by the

formula $$d = l + z,$$

where if the decl. is south d is negative, and if the star transits south of the zenith z is negative (§ 24).

Lastly, l can be found by observing the altitudes of a circumpolar star at its two culminations, and is therefore known (§ 28).

Hence the most essential requisites of an observatory must include (i.) a clock to measure sidereal time, (ii.) a telescope so fitted as to be always pointed in the meridian, provided with graduated circles to measure its inclination to the vertical, and with certain marks to fix the position of a star in its field of view.

43. The Astronomical Clock is a clock regulated to indicate sidereal time. It should be set to mark 0h. 0m. 0s. at the time when the first point of Aries crosses the meridian. It will therefore gain about 4 minutes per day on an ordinary clock, or a whole day in the course of a year (§§ 22, 36).

The clock is provided with a seconds hand, and the pendulum beats once every second, producing audible "ticks"; hence an observer can estimate times by counting the ticks, whilst he is watching a star through a telescope.

. The pendulum is a **compensating pendulum,** or one whose period of oscillation is unaffected by changes of temperature. The form most commonly used is **Graham's Mercurial Pendulum,** in which the bob carries two glass cylinders containing mercury (Fig. 22). If the temperature be raised, the effect of the increase in length of the pendulum rod is compensated for by the mercury expanding and rising in the cylinders. The same result is also effected in Harrison's Gridiron Pendulum, described in Wallace Stewart's *Text-Book of Heat*, page 37.

The clock is sometimes regulated by placing small shot in a cup attached to the pendulum.

FIG. 22.

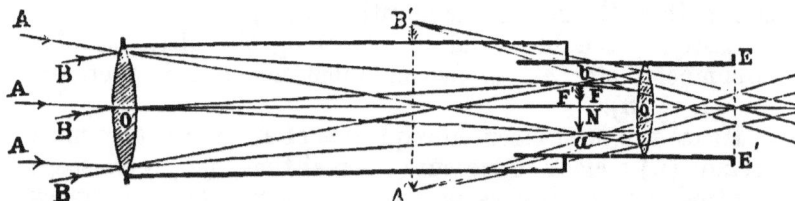

FIG. 23.

44. The **Astronomical Telescope** (Fig. 23) consists essentially of two convex lenses, or systems of lenses, O and O', fixed at opposite ends of a metal tube, and called the **object-glass** and **eye-piece** respectively. The former lens receives the rays of light from the stars or other distant objects, and forms an *inverted* "**image**" (ab) of the objects. The centre O of the round object-glass is called its "optical centre," and the image is produced as follows:—Let AAA be u pencil of rays from a distant star. By traversing the object-glass these rays are *refracted* or bent towards the middle ray AO, which alone is unchanged in direction. The rays all converge to a common point or "focus" at a point a in AO produced, and, if received by the eye after passing a, they would appear to emanate from a luminous point or "image" of the star at a.

Similarly, the rays BBB, coming from another distant star, will converge to a focus at a point b in BO produced, and will give the effect of an "image" of the star at b. All these images (a, b) lie in a certain plane FN, called the *focal plane* of the object-glass, and they form a kind of picture or image of such stars as are in the field of view.

The eye-piece O' acts as a kind of magnifying glass, and enlarges the image ab just as if it were a small object placed in the focal plane FN. The figure shows how a second image $A'B'$ is formed by the direction of the pencils of light after refraction through O'. This is the final image seen on looking through the telescope. The eye must be placed in the plane EE, so as to receive the pencils from A', B'.

If, now, a framework of fine wires or spider's threads (Fig. 25) be stretched across the tube in the focal plane FN, these wires, together with the image (ab), will be equally magnified by the eye-piece. They will thus be seen in focus simultaneously with the stars, and the field of view will appear crossed by a series of perfectly distinct lines, which will enable us to fix any star's position, and thus determine its exact direction in space. Suppose, for example, that we have two wires crossing one another at the point F', and the telescope is so adjusted that the image of a star coincides with F', then we know that the star lies in the line joining F'' to the optical centre O of the object-glass.

45. The Transit Circle (Figs. 24, 26) is the instrument used for determining both right ascension and declination. It consists of a telescope, ST, attached perpendicularly to a light, rigid axis, $WPPE$, hollow in the interior. The extremities of this axis are made in the form of cylindrical pivots, E, W, which are capable of revolving freely in two fixed forks, called **Y**'s, from their shape. These Y's rest on piers of solid stone, built on the firmest possible foundations, and they are carefully fixed, so as always to keep the axis exactly horizontal and pointing due east and west.

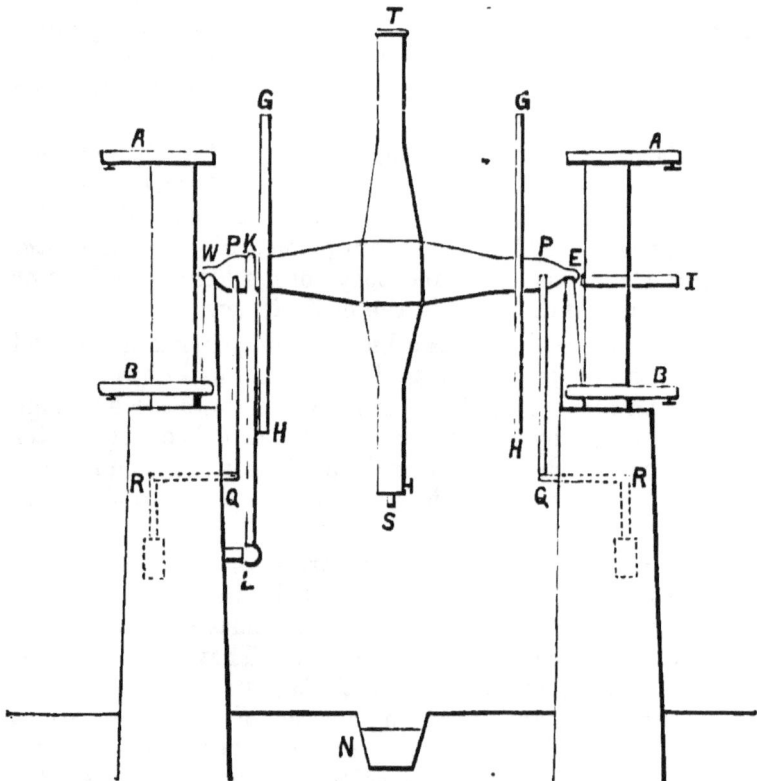

FIG. 24.

In order to diminish the effect of friction in wearing away the pivots, the axis is also partially supported at P, P upon friction rollers (not represented in the figure) attached to a system of levers (Q, Q) and counterpoises (R, R) placed within the piers. These support about four-fifths of the weight of the telescope, leaving sufficient pressure on the Y's to ensure their keeping the axis fixed.

Within the telescope tube, in the focal plane of the object-glass (§ 44), is fixed a framework of cross wires, presenting the appearance shown in Fig. 25. Five, or sometimes seven, wires appear vertical, and two appear horizontal. Of the latter, one bisects the field of view; the other is movable up and down by means of a screw, whose head is divided by graduation marks which indicate the position of the wire.

The line joining the optical centre of the object-glass to the point of intersection of the *middle* vertical wire with the fixed horizontal wire is called the **Line of Collimation.** The wires should be so adjusted that the line of collimation is perpendicular to the axis about which the telescope turns. For this purpose the framework carrying the wires can be moved horizontally, by means of a screw, into the right position. If the Y's have been accurately fixed, then, as the telescope turns,

FIG. 25.

the line of collimation will always lie in the plane of the meridian. Hence, when a star transits we shall, on looking through the telescope, see it pass across the middle vertical wire.

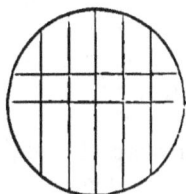

Attached to the axis of the telescope, and turning with it, are two wheels, or **graduated circles,** GH, having their circumferences divided into degrees, and further subdivided by fine lines at (usually) intervals of 5'. By means of these graduations the inclination of the line of collimation to the vertical is read off by aid of several fixed compound microscopes, A, I, B, pointed towards the circle. One of these microscopes (I), called the **Pointer** or **Index,** is of low magnifying power, and shows by inspection the number of degrees and subdivisions in the mark of the circle, which is opposite a wire bisecting its field of view. The pointer should read zero when the line of collimation points to the zenith, and the graduations increase as the telescope is turned northwards.

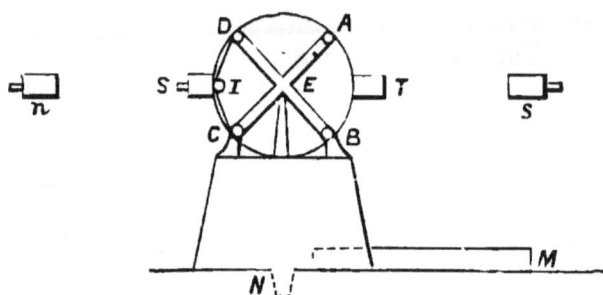

FIG. 26.

46. Reading Microscopes.—In addition to the pointer there are four (sometimes six) other microscopes, called **Reading Microscopes,** arranged symmetrically round each circle, as at $ABCD$ (Fig. 26). These serve to determine the number of minutes and seconds in the inclination of the telescope, by means of the following arrangement. Inside the tube of each microscope in the focal plane of its object-glass* is fixed a graduated scale NL (Fig. 27) in the form of a strip of metal with fine teeth or notches. This scale, and the image of the telescope circle, formed by the object-glass of the microscope, are simultaneously viewed by the eye-glass, and present the appearance shown in Fig. 27.

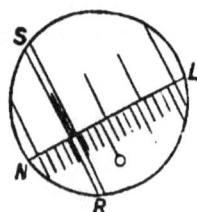

FIG. 27.

A small hole O marks the middle notch, and 5 notches correspond to a division of the telescope circle, hence the number of notches from the hole to the next division of the circle gives the number of minutes to be added to the pointer reading.

---

* A compound microscope, like a telescope, consists of an object-glass, which forms an image of an object, and an eye-piece which enlarges this image. A scale or wires fixed in the plane of the the image will, therefore, be seen in distinct focus, like the wires in the telescope.

To read off the number of seconds, a pair of parallel wires, SR, are attached to a framework, and can be moved across the field of view by means of a screw. One whole turn takes the wires from one notch of the metal scale to the next, i.e., over a space representing 1' on the telescope circle; and the head of the screw is divided into 60 parts, each, therefore, representing 1″. The wires are adjusted so that the graduation on the telescope circle appears midway between them, and the reading of the screw-head then gives the number of seconds. With practice, tenths of a second can be estimated.

The four microscopes of one of the circles are all read, and the best result is obtained by taking the mean of the readings.

47. Clamp and Tangent Screw.—When it is required to rotate the telescope of the transit circle very slowly, this is done by means of the bar represented at LK in Fig. 24. The telescope axis may be firmly clamped to this bar by means of a **clamp** (not represented in the figure), which grips the rim of one of the circles as in a vice. When this has been done, the bar KL, and with it the telescope, may be slowly turned by means of a horizontal screw at L, called the **Tangent Screw,** and provided with a long handle attached to it by a "universal joint." This handle is held by the observer, and he can thus turn the tangent screw without ceasing to watch the stars.

48. Arrangements for Illumination.—As most observations are conducted at night, the wires in the telescope and the graduations of the circles must be illuminated. This is done by a lamp placed exactly in front of one of the pivots, the light from which is concentrated by means of a bull's-eye lens in front and a mirror behind. Part of the rays are reflected, by a complicated arrangement of mirrors and prisms, so as to illuminate the parts of the graduated circle viewed by the microscopes. The rest of the light passes through a plate of red glass down the hollow axis to a ring-shaped mirror, whence it is reflected up to the wires; thus the wires appear as dark lines on a dull red ground. There is also another arrangement for illuminating the wires from in front, if desired, so that they appear bright on a dark ground.

49. Taking a Transit.—Eye and Ear Method.—If

a star is to be observed with the transit circle, its R.A.
and decl. must have been *roughly* estimated beforehand;
hence, its meridian Z.D. [= (star's decl.) − (observer's lat.)]
is known roughly. Before the star is expected to
cross the meridian, the telescope is turned by hand
until the pointer indicates this roughly determined Z.D.;
this adjustment is sufficiently accurate to ensure the
star traversing the field of view. The telescope is then
clamped (§ 47). The observer now "takes a second" from
the astronomical clock, *i.e.*, he observes and writes down the
hour and minute, observes the second, and begins counting
seconds by the clock's ticks. Thus, if he sees the time to be
11h. 23m. 29s., he writes down "11h. 23m.," and at the
subsequent ticks he counts "30—31—32—33—" and so on;
in this way he knows, during the rest of the observation, the
exact time at every clock-beat without looking at the clock.

The star soon approaches the first vertical wire, and passes
it, usually between two successive ticks. With practice, the
observer is able to estimate fractions of a second as follows:—
Suppose the star crosses the wire between the 34th and 35th
tick. The positions of the star are noticed at tick 34 and at
tick 35, and by judging the ratio of their distances from the
wire on the two sides, the observer estimates the time of
crossing the wire by a simple proportion, and writes down
this time, say 34·6. The estimate is difficult to make,
because the two positions of the star are not visible simultaneously, and the star does not stop at them, but moves
continuously; hence to estimate tenths of seconds (as is
usually done) requires much training and practice.

Moreover, the observer must not lose count of the ticks of
the clock, for when he has written down the instant of transit.
over the first wire the star will be nearing the second wire.*

The time of transit over the second vertical wire is now
estimated in the same way, and the process repeated at.
each wire. The average of the times of crossing the five
or seven wires is taken as the time of transit; in this way,

---

* In most instruments the wires are placed at such a distance·
that a star in the equator takes about 13 seconds from one wire,
to the next.

the effect of small errors of observation will be much smaller than if the transit over one wire only were observed.

This method of taking the time of transit is called the **"Eye and Ear Method."**

While observing the transit, the observer turns the telescope by means of the tangent screw, until the horizontal wire bisects the image of the star; during the rest of the observation the star will appear to run along the horizontal wire. After the observation, one of the circles is read by the pointer and the four microscopes. If the circle reads 0° 0′ 0″, when the line of collimation points to the *zenith*, the reading for the star will determine its meridian Z.D., in other cases we must subtract the zenith reading. From the meridian Z.D. the declination can be found.

50. **The Chronograph.**—To obviate the difficulty of observing transits by the eye and ear method, an instrument called the **Chronograph** is now frequently used. A cylindrical barrel, covered with prepared paper, is made to turn slowly and uniformly by clockwork about an axle, on which a screw is cut. In this way the barrel is made to move forward in the direction of its axis, about one-tenth of an inch in every revolution. The observer is furnished with a key or button, which is in electric communication with a pen or marker. At the instant when the star crosses one of the vertical wires, the observer depresses the key, and a mark is made upon the paper of the barrel. The astronomical clock, also, has electric communication with the marker, and marks the paper once every second, the beginning of a new minute being indicated, in some instruments, by the omission of the mark, in others, by a double mark. In this way, a record is made of the times of transit over the wires, the marks being arranged in a spiral, owing to the forward motion of the barrel. The distance of the *beginning* of any transit-mark from the previous second-mark can be measured at leisure with very great accuracy, and the time of transit may thus be readily calculated. Indeed, there is no difficulty in recording, by this method, the transits of two, or even more, near stars which are simultaneously in the field of view of the telescope, for the transit-marks of the different stars can be readily distinguished from one another afterwards.

51. Corrections.—After the transit of a star has been observed, certain corrections have to be allowed for in practice before its true R.A. and decl. are obtained. These corrections, which depend on errors of observation, may be conveniently classified as follows :—

(a) Corrections required for the Right Ascension :
    1. Error and rate of the astronomical clock.
    2. Personal equation of the observer.
    3. Errors of adjustment of the transit circle, including
        (a) Collimation error.
        (b) Level error.
        (c) Deviation error.
        (d) Irregularities in the form of the pivots.
        (e) Corrections for the "verticality" and "wire intervals."

(b) Corrections required in finding the Declination :
    1. Reading for zenith point, or for the nadir, horizontal or polar point.
    2. Errors of imperfect centering of the circles.
    3. Errors of graduation.
    4. Errors of " runs " in the reading microscopes.

Besides these corrections, which we now proceed to describe, there are others of a physical nature, such as refraction, parallax, aberration, the description of which will be given later. A correction is always regarded as positive when it must be added to the *observed* value of a quantity in order to get the *true* value, negative if it has to be subtracted.

(a) CORRECTIONS REQUIRED FOR THE RIGHT ASCENSION.

. **52. Clock Error and Rate.**—A good astronomical clock can generally be regulated so as not to gain or lose more than about 2s. in a sidereal day. But to estimate times with greater accuracy, it is necessary to apply a correction to the time indicated, owing to the clock being either fast or slow.

The **Error** of a clock is the amount by which the clock is *slow* when it indicates 0h. 0m. 0s. Thus, the error must be added to the indicated time in order to obtain the correct time. If the clock is fast, its error is negative.

The **Rate** of the clock is the increase of error during 24 hours. It is, therefore, the amount which the clock *loses* in the 24 hours. If the clock *gains*, the rate is negative.

The rate of a clock is said to be **uniform** or **constant** when the clock loses equal amounts in equal intervals of time. In a good astronomical clock, the rate should remain uniform for several weeks.

53. Correction for Error and Rate.—If the error of a clock and its rate (supposed uniform) are known, the correct time can be readily found from the time shown by the clock. The method will be made clear by the following example :—

EXAMPLE.—If the error of an astronomical clock be 2·52s., and its rate be 0·44s., to find to the nearest hundreth of a second the correct time of a transit, the observed time by the clock being 19h. 23m. 25·44s.

Here in 24h. the clock loses 0·44s.

$\therefore$ in 1h. it loses $\frac{1}{24}$ × 0·44s. = 0·0183s.

Hence, loss in 19h. = 0·0183s. × 19 = 0·348s.,

and loss in 23m. = 0·007s.

At 0h. 0m. 0s. the clock error is = 2·52s. ;

$\therefore$ at 19h. 23m. 25·44s., clock is too slow by 2·52s. + 0·355s. = 2·88s.,

$\therefore$ the correct time = 19h. 23m. 25·44s. + 2·88s.

= 19h. 23m. 28·32s.

54. Determination of Error and Rate of Clock.— The clock error is found by observing the transit of a known star, *i.e.*, a star whose R.A. and. decl. are known.

If the clock were correct, the time of transit (when corrected for all other errors) would be equal to the star's R.A. (see § 24). If this is not the case, we have evidently

(Clock error) = (Star's R.A.)

— (observed time of transit).

This determines the clock error at the time of transit.

To find the rate, the transits of the same star are observed on two consecutive nights.

Let t and $t-x$ be the observed times of transit; then x is the amount the clock has lost in 24 hours, *i.e.*, the rate of the clock. Therefore

(Rate of Clock) = (observed time of 1st transit)

—(observed time of 2nd transit).

Having found the rate of the clock and its error at the time of transit, the error at 0h. 0m. 0s. may be found by subtracting the loss between 0h. 0m. 0s. and the transit.

Stars used in finding clock error are known as **" Clock Stars."**

55. Personal Equation is the error made by any particular observer in estimating the time of a transit.

Of two observers, one may habitually estimate the transit too soon, another may estimate it too late, but experience shows that the error made by each observer in taking times of transit by the same method is approximately constant.

If all observations are made. by the same individual there will be no need to take account of personal equation, because the error made in taking a transit will be compensated by the error made in observing the clock stars to set the clock. If the two operations are performed by different observers, we must allow for the *difference* of their personal equations.

Personal equation may be measured by an apparatus for observing the transit of a fictitious star, *i.e.*, a bright point moved by clockwork; in this case the actual time of its transit is known, and can be compared with the observed time. Personal equation is positive if the observer is too quick, so that the correction must be *added* to the observed time to get the true time, as in § 51.

56. Errors of Adjustment of the Transit Circle.— If the transit circle is in perfect adjustment, the line of collimation of the telescope must always lie in the plane of the meridian. If not, we must correct for the small errors of adjustment. The conditions required for perfect adjustment, together with the corresponding corrections when these conditions are not fulfilled, may be classified as follows :—

(*a*) The line of collimation should be perpendicular to the axis about which the telescope rotates. If not, the corresponding correction is called **Collimation Error.**

(*b*) The axis of rotation must be horizontal. **Level Error.**

(*o*) The axis must point due east and west. **Deviation** (or Azimuthal) **Error.**

(*d*) The pivots resting on the Y's must be truly turned, and form parts of the same circular cylinder. Correction for **shape of pivots.**

(*e*) The vertical wires in the transit must be truly vertical (*i.e.*, parallel to the meridian) and equidistant. **Verticality** and **Thread Intervals.**

*57. Collimation Error.—We have seen (§ 45) that the frame-work carrying the vertical wires in the transit telescope can be adjusted by a screw, so that collimation error can be corrected. Suppose, for simplicity, that no other error is present. Then the line of collimation will always make a constant small angle with the meridian, and this angle will measure the collimation error.

To correct this error, two telescopes, called Collimators, are pointed towards each other, one due north, the other due south of the instrument (n, s, Fig. 26). Both contain adjustable "collimating marks," formed by cross wires in their focal planes. The transit telescope being first pointed vertically, and two apertures in the side of its tube being uncovered, the observer looks through the telescope s, and sees through the apertures into the telescope n. He then brings the wires in s into coincidence with the images of the wires in n; he then knows (from the optical theory of the telescope) that the lines of collimation of n, s are *parallel*. Suppose (e.g.) that they make a small unknown angle x'' W. of S., and E. of N., respectively.

He now looks through the transit telescope into the collimator s. He adjusts the middle vertical wire of the transit to coincide with the image of the cross mark in s, reading the graduated screw by which the adjustment is made. The line of collimation of the transit is now x'' *west* of the meridian. He points the telescope into n, and similarly adjusts the wires : the line of collimation is now x'' *east* of the meridian. He now turns the adjusting screw to a reading *midway* between the two observed readings ; the line of collimation is then in the meridian, and collimation error has been removed.

*58. Level Error is measured by the inclination to the horizon of the axis of rotation of the telecope. It causes the line of collimation to trace out, on the celestial sphere, a great circle inclined to the meridian at an angle equal to the level error.

Level error is found by pointing the telescope (corrected for collimation error) downwards over a trough of mercury (N, Figs. 24, 26, 28).

An eye-piece is provided, called a "collimating eye-piece" (EF, Fig. 28, p. 49), containing a plate of glass M, which reflects the light from a lamp straight down the tube. The mercury will form a reflected image of the telescope, which may be treated just as if it were a real telescope or collimator; the wires in the actual telescope will appear bright, and those in the image will appear dark. By the law of reflection, if the middle wire coincide with its image, the line of collimation will be vertical, and (since there is no collimation error) there will be no level error. If not, the wires are moved by the screw until the vertical wire coincides with its image. The observer reads the angle through which the screw has been turned, and thus measures the level error. The wires are then replaced (otherwise collimation error would be introduced) and level error is corrected by adjusting the Y's (§ 59).

*59. **Deviation Error** is measured by the small angle which the axis of rotation of the telescope makes with the plane of the prime vertical. It causes the line of collimation of an otherwise correctly adjusted transit circle to describe a great circle through the zenith whose inclination to the meridian is equal to the deviation error.

Deviation error can be discovered by observing the times of upper and lower transit of a circumpolar star, such as the pole star. Suppose (*e.g.*) that the telescope axis points slightly south of east; then it is readily seen by a diagram that when the telescope is pointed north of the zenith, the line of collimation will be slightly east of the meridian. Then, at upper transit, if the observed circumpolar star is north of the zenith it will reach the middle wire *before* reaching the meridian. At lower transit it will not reach the wire till *after* passing the meridian. Hence, the time from upper to lower transit will be rather greater than 12h., and the time from lower to upper transit will be rather less than 12h. By observing the difference of the intervals the deviation error can be found.

In many observatories, the Y's of the transit circle can be adjusted by screws, one moving vertically, to correct for level error, the other horizontally, to correct for deviation error.

When these errors are corrected, the cross wires of the collimators are brought into coincidence with the middle wire of the telescope when pointed horizontally.

*60. **The correction for the shape of the pivots** is rather complicated, but, in a good instrument, it should be very small. When the pivots are much worn by friction, they should be re-turned.

The errors may be measured by making a small mark on the end of each pivot, and observing, by means of reading microscopes, the motions of the marks as the instrument is slowly turned round. If the pivots are true, the marks should remain fixed, or describe circles.

*61. **Verticality of the Wires** may be tested by observing one of the collimators, whose cross wires are adjusted as in § 59. If the cross wires always appear to intersect on the middle wire of the transit when the instrument is turned through any small angle, we know that the middle wire is vertical.

*62. *Wire Intervals.*—By "**Equatorial Wire Intervals**" are meant the intervals of time taken by a star on the equator in passing from one vertical wire of the transit to the next.

If the intervals between successive wires are unequal, the mean of the times of transit over the wires will not in general be the same as the time of transit over the middle wire. We may imagine a straight line so drawn across the field of view that the time of transit across it is exactly equal to the mean of the times of transit over the five or seven wires. This line is called the **Mean of the Wires**.

By carefully determining the equatorial wire intervals, the very small interval between the transits over the mean of the wires and over the middle wire can be found.

For a star not in the equator, the wire intervals are proportional to the *secant* of the declination. This follows from Sph. Geom. (17).

(b) CORRECTIONS REQUIRED IN FINDING THE DECLINA-
TION OF A STAR.

63. Zenith Point.—In § 45 we stated that the pointer of the transit circle is usually adjusted to read 0° 0′ when the line of collimation is pointed to the zenith. But it would be very difficult to adjust the microscopes to give a mean reading of exactly 0° 0′ 0″ for the zenith. Hence it is necessary to determine the **zenith point,** or zenith reading, and in calculating the meridian Z.D. of any star, this must be subtracted from the reading for the star.

Let Z and N be the readings when the telescope is pointed to the zenith and nadir, respectively, H and H' the readings for the north and south points of the horizon; then evidently,

$$Z = H - 90° = N - 180° = H' - 270°.$$

Also, if x is the reading for the meridian transit of any star, then star's meridian Z.D. $= x - Z$, if north of the zenith, or, $= 360° - (x - Z)$, if south of the zenith.

64. To find the Nadir Point, use is made of the **Colli-mating Eye Piece,** already mentioned in § 56, and represented in Fig. 28. It consists of two lenses E, F, between which is a plate of glass, M, inclined at an angle of 45° to the axis. This plate illuminates the wires from above by partially reflecting the light from a lamp on them, at the same time allowing them to be seen through the eye-glass, E.

The telescope is pointed downwards over the trough of mercury, N; and the rays of light from any one of the wires, Q, will produce by reflection a distinct image of the wire at q in the focal plane. By turning the telescope with the tangent screw, the fixed horizontal wire may be made to coincide with its image; it will then be vertically over the "optical centre" of the object-glass (§ 44). The line of collimation will, therefore, point to the

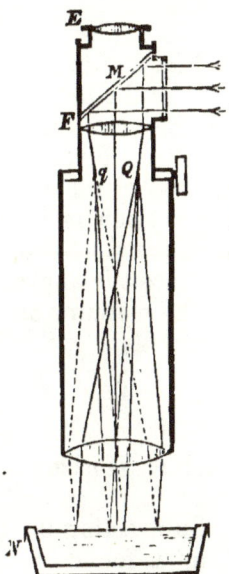

Fig. 28.

nadir, and the nadir reading is given by the pointer and microscopes. Subtracting 180°, we have the the zenith reading.

65.—Determination of Horizontal Point.—Method of Double Observation.

—Both the horizontal reading and the meridian altitude of a star can be determined by observing the star, both directly and by reflection, in a trough of mercury placed in a suitable position (M, Figs. 26, 29).

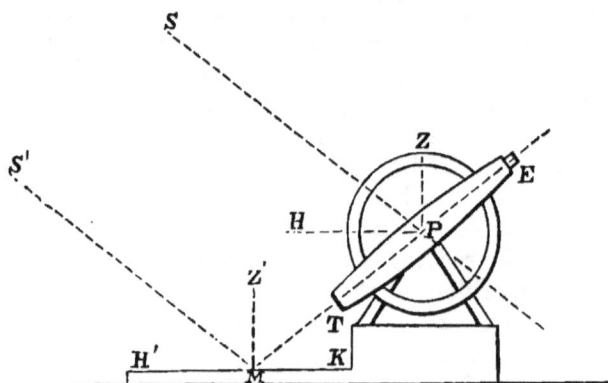

FIG. 29.

Fig. 29 illustrates the method of double observation. Let PZ be the direction of the line of collimation corresponding to the zero reading, PH the horizontal direction, PS and MTP the directions of the star viewed directly and its image viewed by reflection. The reading of the circle for the direct observation is the angle ZPS, the reading for the reflection is the angle ZPM.

Since the angles of reflection and incidence $S'MZ'$, TMZ' at the mercury are equal, and MS', PS are parallel, we have evidently $\angle SPH = S'MH' = TMK = MPH$;

$\therefore$ star's altitude, $SPH = \frac{1}{2} SPM$;

$$= \frac{1}{2} (ZPM - ZPS)$$

= half the difference of the two readings.

Also : Horizontal reading, $ZPH = \frac{1}{2} (ZPM + ZPS)$;

= half the sum of the two readings.

Subtracting 90° from the north horizontal point, the zenith point is found.

*66. In using this method with the transit circle of a fixed observatory, the star will remain sufficiently long in the field of view to allow of both observations being made at the same transit, and the fact of the star not being quite on the meridian will not affect the results perceptibly. But there will *not* be time to read the circles by means of the four microscopes, between the two observations. This difficulty is obviated by proceeding thus:— Before the first observation, point the telescope (by means of the pointer) in such a direction that the reflection of the star in the mercury will cross the field of view during *the* transit; for this purpose the star's meridian altitude must be known *approximately.* Clamp the telescope, and read the microscopes. When the star appears in the field of view, adjust the *moveable* horizontal wire (by means of its graduated screw) till it crosses the star, keeping the telescope fixed. Now unclamp the telescope, and point it to the star direct, turning it with the tangent screw until the *moveable* horizontal wire again crosses the star. After the observation, read the graduated screw of the horizontal wire, and also the pointer and microscopes.

Since the star is bisected by the same wire at each observation, the difference in the readings gives the angle through which the telescope was rotated, and this angle is evidently double the star's altitude. Half the sum of the readings gives what would be the reading if the moveable wire were pointed horizontally. This must be corrected by adding the angular interval between the moveable and fixed wires as determined from the graduated screw, and we then have the reading for the horizon point when the fixed wire is used.

67. Polar Point.—In order to find the declination of a star by means of the transit circle, it is necessary to know the reading when the telescope is pointed to the pole. This may be found, just as in § 28, by observing the upper and lower transits of a circumpolar star. The mean of the two readings gives the polar point.

The N.P.D. of any star is found by taking the difference of the readings for the star and the polar point. The declination is, of course, the complement of the N.P.D.

We may also find declinations thus:—Since angles are measured from the zenith northwards, it is evident (by drawing a figure or otherwise) that the reading for the point of the equator above the horizon is given by

Equatorial point = (Polar point) + 270°.

Since the decl. is the angular distance from the equator, we have

(North Decl.) = (Reading for star) (Equatorial point).

If the star transits north of the zenith, its reading must be increased by 360°.

The latitude of the observatory is given by

Latitude = Altitude of pole

= (North horizontal point) — (Polar point).

*68. Errors of Graduation.—The operation of testing the accuracy
of the graduations on the circles of the transit circle is very long
and laborious. One of the two graduated circles is so attached to
its axis, so that it can be turned through any angle relative to the
telescope. Then, by reading the microscopes belonging to both
circles, every graduation on one circle is compared with every
graduation on the other circle, and any errors of graduation are thus
detected and measured. The effect of such errors is much reduced
by using all the four microscopes, and taking the mean of their
readings.

*69. Errors due to Imperfect Centering of the Circles.—By
taking the mean of the microscope readings, all errors due to imperfect centering are eliminated. In proof, let us suppose that only
two microscopes (A, C, Fig. 26) are used, but that these are opposite
to one another. If the circle is truly centred, with its centre on
the line AC, the two readings will differ by 180°. If, now, the graduated circle is displaced, without being rotated, till its centre is at a
distance h from AC, then the points of the scale, now under AC,
will be at distances h from the points formerly under AC, both being
displaced in the same direction. Hence, since both readings are
measured the same way round the circle, one will be increased
and the other will be decreased by the same angle. The arithmetic
mean of the two readings will, therefore, be unaltered by the displacement of the centre, and will be independent of any small error
due to imperfect centering. The same is, of course, true of the
mean reading for the other pair of microscopes, B, D.

The error in centering may be discovered by taking the difference
of the readings of a pair of opposite microscopes. This difference
should be 180' if the circle is properly centred; if not, the amount
by which it differs from 180° will determine how much the centre of
the circle is to one side or the other of the line joining the centres
of the pair of microscopes.

*70. Error of Runs.—In the reading microscopes, one turn of the
micrometer screw should move the parallel wires over a space corresponding to exactly 1' on the graduated circle, so that the wires
should be brought from one mark of the circle to the next by exactly
five turns of the screw. In practice it will probably be found that
rather more or rather less than five turns will be necessary. In this
case the readings of the teeth and of the micrometer screw-head will
differ slightly from true minutes and seconds of arc on the circle,
and a correction will be required. This error is called **Error of
Runs.**

*71. Collimation, Level and Deviation Errors have no appreciable effect on observations for declination, provided that such
errors are small compared with the star's N.P.D. Hence, they may
be left out of account, except in observations of the Pole Star.

72. General Remarks.—We first described the Transit Circle, and the methods of "taking a transit"; we afterwards described the corrections which must be applied to the results of the observations in finding the right ascension and declination of a star. But in practical work the various errors must be determined *before* any observation can be made. Among these, collimation, level and deviation error, and the nadir point should be found daily, as they may be affected by heat or cold, or by shaking the instrument.

Clock error and rate are also determined daily by observing certain "clock stars." The accuracy of the corrections may be tested by observing various "known stars" of different declinations. If the corrections have been accurately made, the observed right ascensions and declinations should agree with their values as given in astronomical tables.

Before determining clock error and rate by means of a "clock star," the R.A. of *one* such star must be known. Since the R.A. is measured from the first point of Aries, that point must first be found. The method of finding it will be described in Chap. IV.

73. Observations on the Sun, Moon, and Planets.— The positions of the Sun, Moon, and Planets are defined by the coordinates of their **centres**. In finding these, the angular diameters must be taken into account.

In observing the Moon or a planet, the fixed horizontal wire is adjusted to touch the illuminated edge of its disc, and the times at which its edge touches the vertical wires are observed. To find the coordinates of the centre, a correction is made for the angular semi-diameter of the body, which must be determined independently. It must not be forgotten that the image formed by the telescope is **inverted.**

In observing the Sun, the semi-diameter may be found during the observation by adjusting the moveable horizontal wire to touch one edge of the disc, while the fixed wire touches the other edge. The reading of the micrometer screw gives the Sun's angular diameter. In finding the time of transit, the times of contact of the disc on arriving at and leaving each wire are separately observed; their arithmetic mean for any wire is the time of transit of the centre.

SECTION II.—*Instruments adapted for Observations off the Meridian.*

74. The Transit Circle can only be used to observe celestial bodies during the short period before and after their transit that they remain in the field of view. It is, therefore, unsuited for *continuous* observation of a celestial body, such as is required more particularly in Physical Astronomy. For this purpose, a telescope must be mounted in such a way that it can be pointed in any required direction, or moved so as to keep the same body always in the field of view. There are two such forms of mounting, and the telescopes thus mounted are called the Altazimuth and the Equatorial.

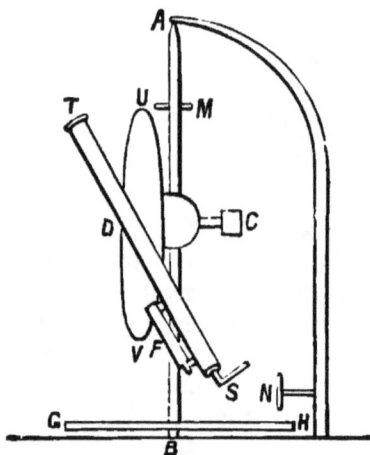

FIG. 30.

75. **The Altazimuth.**—In this instrument, a telescope, *ST*, is supported so that it can turn freely about a horizontal axis, *CD*, sometimes called the **secondary axis.** This secondary axis, with the attached telescope, is capable of turning about a fixed vertical axis, *AB*, sometimes called the **primary axis,** which is supported at its upper and lower ends as shown in the figure.

Both axes are provided with graduated circles, *GH*, *UV*,

attached to, and turning with them. Each circle is read by means of one or more "pointer" microscopes, M and N. There are also clamps, furnished with tangent screws, by means of which the circles may be fixed in any desired position, or rotated slowly if required. At C is a counterpoise, which balances the telescope and the circle UV, and so prevents their weight from bending the axis AB.

By rotating the whole instrument about the vertical axis AB, the telescope can be brought to any required azimuth. If now the circle GH be clamped, the telescope can be turned about CD to any required altitude. The microscope N should indicate zero when the telescope is pointed in the plane of the meridian, and the microscope M should indicate zero when the telescope is horizontal. If now the telescope be pointed so that a star is in the middle of its field of view, the readings of the two microscopes N, M will give the star's azimuth and altitude respectively. The time of observation being also known, the position of the star on the celestial sphere is completely determined, and its R.A. and decl. can be calculated if required. But for observations of this class, the altazimuth is not nearly so reliable as the transit circle.

As the altazimuth possesses two independent motions, while the transit circle possesses only one, the former instrument is liable to a far greater number of errors of adjustment; moreover, its telescope is far less firmly and rigidly supported, and the instrument is therefore more liable to bend.

A large altazimuth in Greenwich Observatory is used for observing the Moon's motion, when it is so near the Sun that it cannot be accurately investigated by meridian observations alone.

A portable telescope, mounted on a tripod stand, such as is commonly used for observing the stars at night, is an altazimuth unprovided with graduated circles.

A **Finder** (F) is usually attached to a large altazimuth, whose field of view is of small angular breadth. This is a small telescope of lower magnifying-power, with a larger field of view, the centre of which is marked by cross wires. To point the large telescope to any celestial body, the altazimuth is so adjusted that the body is seen in the centre of the finder. It will then be in the field of view of the large telescope.

76. The Equatorial (Fig. 31).—If we suppose an alta-zimuth inclined so that its primary axis, instead of being vertical, is pointed in the direction of the pole, we shall have an Equatorial.   In this instrument the framework carrying the telescope turns as a whole about about the primary axis *AB*, which is supported at *A* and *B*, so as to point towards the pole.   Attached perpendicularly to this axis, and turning with it, is a graduated circle, called the **Hour Circle,** which read by a " pointer " microscope *N.*

The framework *AB* carries a secondary axis perpendicular to the primary axis, and the telescope *ST* is attached perpendicularly to this secondary axis, about which it is free to turn.   The axis of the telescope carries another graduated circle called the **Declination Circle** which is read by the " pointer " microscope *M.*

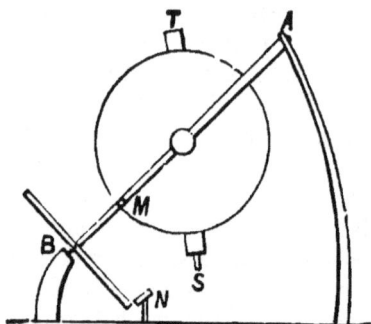

FIG. 31.

The declination circle should read zero when the telescope is pointed in the plane of the equator, and the hour circle should read zero when the telescope is in the plane of the meridian.   If now the telescope is pointed towards any celestial body, the readings of the two microscopes will give, respectively, the declination and hour angle of the body.

When it is required to observe the same body continuously with the equatorial, the declination circle is clamped, and the observer must slowly rotate the hour circle by hand, so as to keep the body observed in the field of view.

In large instruments the hour circle can be attached to a clamp which is worked by clockwork in such a manner that the whole framework turns uniformly round the primary axis *AB* once in a sidereal day. This motion will ensure that the star under observation shall always remain in the centre of the field of view.

The pointer-microscope of the hour circle may be made to revolve with the clamp, and to mark zero when the telescope is pointed towards the first point of Aries; its reading will then give the right ascension of any observed star. But the declination and right ascension cannot be determined with any great degree of accuracy by reading the circles of the equatorial. There are the same difficulties as in the altazimuth; moreover, the primary axis, being inclined to the vertical, is more liable to bend under the weight of the telescope.

The clockwork by which the equatorial is driven could not be regulated by an ordinary pendulum, as this would make the telescope move forward in a series of jerks, one at every beat. For this reason, a conical pendulum revolving uniformly must be used. The reader will find the principle of the conical pendulum explained in most text-books on elementary dynamics; a working example may be seen in the "Watt's Governor" of a steam-engine.

In most modern equatorials, the primary axis is not supported as in Fig. 31, but on a pillar just underneath the secondary axis. The advantage is that the primary axis is less liable to bend than when supported at its two ends *A, B.*

77. Uses of the Equatorial.—Amongst these the following may be mentioned:—

(i.) "Differential" observations, *i.e.*, micrometric observations of the relative distances and positions of two near stars simultaneously visible.

(ii.) Observations of the appearance, structure, and magnitude of the celestial bodies.

(iii.) Stellar photography.

(iv.) Spectroscopic analysis.

78. **Micrometers.**—Any instrument used for measuring the small angular distance between two bodies simultaneously visible in the field of view of a telescope is called a **Micrometer.** Thus the moveable horizontal wire in the transit circle, with its graduated screw, is a micrometer, for if the instrument be so adjusted that the fixed wire crosses one star, while the moveable wire crosses another neighbouring star, the distance between the wires, as read off on the screw head, gives the difference of declination of the stars. The moveable wire in the field of view of the reading microscope is identical in principle with a micrometer.

79. **The Screw and Position Micrometer** (Fig. 32) serves to find both the angular distance between two neighbouring stars and the direction of the line joining them. It contains a framework of wires placed in the focal plane of the telescope. Two of these wires are parallel, and one of them can be separated from the other by turning a screw with a graduated head. A third wire, which we will call the "transverse wire," is fixed in the framework perpendicular to the two former. The whole apparatus, together with the eye piece of the telescope, can be rotated so

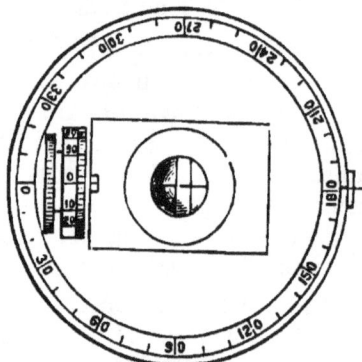

FIG. 32.

that the wires may appear in any required direction across the field of view. A graduated circle, called the **Position Circle,** is attached to the eye-piece, and measures the angle through which it has thus been turned. Besides the wires, the framework contains a transverse strip of metal marked with notches, at distances apart corresponding to complete turns of the micrometer screw, an arrangement similar to that employed in the reading microscope (§ 45).

In observing two stars, the equatorial and micrometer are so adjusted that one of the stars may appear at the intersection of the two fixed wires, while the other appears at the intersection of the fixed and moveable wires.

The number of notches of the scale, together with the reading of the screw-head, determine the distance between the images of the stars in turns and parts of a turn of the screw-head. To find the angular distance between the stars, we only require to multiply by the known angular distance corresponding to one turn of the screw.

The reading of the position circle determines the direction of the small arc joining the stars. The position-circle should read zero if the stars have the same R.A. Then the reading in any other position will determine their **position angle,** *i.e.*, the angle which the line joining the stars makes with a declination circle through one of the stars. •

*80. **Dollond's Heliometer** is another form of micrometer, depending on the principle that if the object-glass of an astronomical telescope be cut across in two, each half will form an image of the whole field of view, in the same way as if the lens were still complete.† In the Heliometer one half of the object-glass can be made to slide along the other by means of a graduated screw.

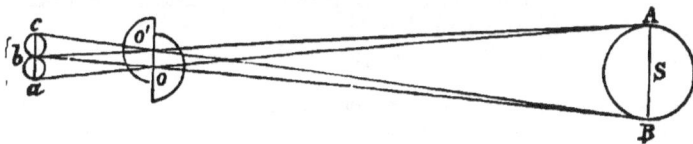

Fɪɢ. 33.

Suppose that we want to measure the angular diameter of the Sun (*S*, Fig. 33). When the halves of the object-glass are together, so that their optical centres coincide, one image of the Sun will be formed. When the two halves are separated, two separate images will be formed in the focal plane of the telescope. and will be seen simultaneously. The half-lenses are separated, till the two images *touch*, as *ab* and *bc*. Let *O*, *O'* be the optical centres of the two halves of the objective. The distance *OO'* is read off on the screw-head; from this reading the Sun's angular diameter may be found.

For at *b*, the point of contact of the images, the half-lens *O* forms an image of the lower limb *B*, and the half-lens *O'* forms an image of the upper limb *A*. Hence, *BOb* and *AO'b* are straight lines, and *ObO'* is the angular diameter *BbA*. But the focal length *Ob* is known Hence, if *OO'* is also known, the angular diameter *ObO'* can be found.

---

† To show this, it is only necessary to cover up half the object-glass of an astronomical telescope. (N.B.—*Not an opera-glass.*)

In measuring the angular distance between two stars, the heliometer is adjusted so that the image of one star formed by one half-
lens *O* coincides with the image of the other star formed by the
other half-lens *O'*. The principle is the same as before.

***81. To find the angular distance corresponding to a revolution
of the micrometer screw**, the simplest plan is to observe the Sun's
diameter, and to compare the reading with its known value. The
latter is given in the Nautical Almanack for every day at noon.

To test the zero reading of the position circle, the equatorial
is pointed to a star near the equator, and fixed, and the micrometer
is turned till the diurnal rotation causes the star to run along the
transverse wire. The circle should then read 90°.

82. Stellar Photography.—For photographic purposes,
the equatorial is driven by clockwork, carrying with it a
sensitized plate, on which an image of the heavens is projected.
In this way a photograph of part of the sky is obtained, and
on such a photograph the distances and relative positions of
the various stars, nebulæ, &c., can be accurately measured.
Moreover, by continuing the exposure sufficiently long, even
the faintest rays of light will produce an impression on the
photographic plate; and it is thus possible to detect stars and
nebulæ which would be invisible to the eye.

*83. Spectrum Analysis.—A description of the spectrum is given
in Wallace Stewart's *Text-Book of Light*, Chap. VIII., and the spectroscope is described in § 91 of the same treatise.

A detailed account of the methods of spectrum analysis would be
out of place in this book, as the subject belongs to the domain of
Physical Astronomy. The general principle is this:—We can, by
means of the spectroscope, analyse the constituent waves of the
light rays which reach us from the Sun and stars. We can compare
these constituents with those emitted or absorbed by the various
chemical elements in a state of vapour. Such comparisons enable
us to infer what chemical elements are present in different celestial
bodies.

84. Other Instruments.—The instruments described in
this chapter are all such as are used in fixed observatories.
Besides these, certain *portable* instruments are used in astronomical observations. Among the latter class the Zenith
Sector will be described in the next chapter, in connection
with the determination of the Earth's form and radius; and
the Sextant and Chronometer will be explained in treating of
the methods of finding latitude and longitude at sea.

EXAMPLES.—II.

1. Describe the Altazimuth. Why is it not so well suited for continuous observations as the equatorial, and, in particular, why is it quite unsuitable for stellar photography?

2. Show that the altitude of a star is greatest when the star is on the meridian.

3. From the result of Question 2, show how the meridian zenith distance of a star might be found by observing its altitude with an altazimuth.

4. How may we most easily set the astronomical clock?

5. Show that the rate of a clock might be found by observations on successive nights with *any* telescope provided with cross wires, and pointed constantly in a fixed direction.

6. Distinguish, with examples, *direct* and *retrograde* angular motion. Is R.A. measured direct or retrograde?

7. Show that in latitude 45° the interval between the time of any star's passing due east and its time of setting is constant.

8. Show that, if a transit circle be not centred truly, the consequent error can be eliminated by taking the mean of the readings of the microscopes.

9. In a double observation made with the transit circle, the readings of the pointer directly and by reflection are 59° 35′ and 125° 20′; the means of the microscope readings are in the two cases 3′ 42″ and 1′ 13″. The moveable wire reads ᴦ 2″, and the reflected star runs along the *fixed* horizontal wire. Find the zenith reading.

10. Explain how it is that photography has revealed the existence of stars which are so faint as to be invisible.

11. Find the decl. of α Ophiuchi from the following observations, made at Greenwich (lat. 51° 28′ 31″ N.) :—Pointer reading 321° 10′, microscope readings, 1′ 2″, 0′ 50″, 0′ 46″, 0′ 58″, the zenith reading being 0° 0′ 16″.

12. Find also the R.A. of α Ophiuchi. Given: Time by sidereal clock = 17h. 29m., the numbers of seconds at the transits over the five wires being 37·4s., 50·2s., 1m. 2·9s., 1m. 15·2s., 1m. 27·4s. Clock error = −10·6s.; personal equation = +0·4s.

EXAMINATION PAPER.—II.

1. Classify the various observations which are taken in astronomical investigations, and state the respective instruments which may be used for those observations.

2. Define the *right ascension* and *declination* of a star, and describe shortly the principles of the methods of finding them.

3. Describe how the time of transit of a star across each of the five or seven wires of a transit instrument is observed, and explain how the time of transit across the meridian is deduced. Define the *equatorial interval* of two wires.

4. Describe the Reading Microscope, and show how the zenith distance of a star may be found by direct observation with the transit circle.

5. Enumerate the errors of a transit instrument, and explain how level error may be measured and corrected.

6. Explain what is meant by *collimation error*, and draw a diagram showing the circle traced out on the celestial sphere by the line of collimation in an instrument which has a small collimation error east of the meridian. Is the correction, to be applied to the times of transit, positive or negative in such a case?

7. Describe the Equatorial, and explain the adjustments and principal uses of the instrument.

8. Describe the Screw and Position Micrometer, and explain how the value of a turn of the screw may be found.

9. What is meant by the *error* and *rate* of a clock, and the *personal equation* of an observer? How are they usually found?

10. On 1st March, 1872, the time of transit of β *Libræ*, at Greenwich, was observed to be 15h. 9m. 6·15s., and on the 3rd March the observed time was 15h. 9m. 4·73s. The tabular R.A. of the star was 15h. 10m. 7·25s. Find the error and rate of the clock on 3rd March.

CHAPTER III.

THE EARTH.

SECTION I.—*Phenomena depending on Change of Position on the Earth.*

85. Early Observations of the Earth's Form.—One of the first facts ascertained by the early Greek astronomers was that the Earth's surface is globular in form. Even Homer (B.C. 850 *circ.*) speaks of the sea as convex, and Aristotle (B.C. 320) gives many reasons for believing the Earth to be a sphere. Among these may be mentioned the appearances presented when a ship disappears from view. If the surface of the ocean were a plane, any person situated above this plane would (if the air were sufficiently clear) see the whole expanse of ocean extending to the furthermost shores, with all the ships sailing on its surface. Instead of this, it is observed that as a ship begins to sail away its lowest part will, after a time, begin to sink below the apparent boundary of the surface of the sea; this sinking will continue till only the masts are visible, and, finally, these will disappear below the convex surface of the water between the ship and the observer.

Another reason is suggested, by observing the stars. If the Earth's surface were a plane, any star situated above the plane would be seen simultaneously from all points of the Earth, except where concealed by mountains or other obstacles, and any star below the plane would be everywhere simultaneously invisible. In reality, stars may be visible from one place which are invisible from another; and all the appearances presented were found by the Greeks to agree with what might be expected on a spherical Earth. Eratosthenes even made a calculation of the Earth's size from the distance between Alexandria and Assouan and their latitudes (§ 91) deduced from the Sun's greatest meridian altitudes. He found the circumference to be 250,000 stadia, or furlongs.

Lastly, the Earth's spherical form will account for the circular form of the Earth's shadow in a lunar eclipse.

86. General Effects of Change of Position.—In § 5, we showed that, owing to the great distance of the stars, they are seen in the same direction whatever be the position of the observer. In confirmation of this fact, it is found by observation that the angular distance between any two stars (after allowing for refraction) is observed to be independent of the place of observation.

But the directions of the zenith and horizon vary with the position of the observer. If we suppose the Earth spherical, the vertical at any point on it will be the radius drawn from the Earth's centre, while the plane of the horizon will be a tangent plane to the Earth's surface; both will depend on the place. This circumstance accounts for the difference in appearance of the heavens as seen simultaneously from different places.

87. Earth's Rotation.—The apparent rotation of the heavens is accounted for by supposing that the stars are at rest, and that the Earth rotates once in a sidereal day, from west to east, about an axis parallel to the direction of the celestial pole. The observer's zenith, horizon and meridian turn about the pole from west to east, relatively to the stars, and this causes the hour angles of the stars to increase by 360° in a sidereal day, in accordance with observation.

It is impossible to decide from observations of the stars alone whether it is the Earth or the stars which rotate, just as when two railway trains are side by side it is very difficult for a passenger in one train, when observing the other, to decide which train is in motion. That the Earth rotates has, however, been conclusively proved by means of experiments, which will be described when we come to treat of dynamical astronomy.

88. Definitions.—The **Terrestrial Poles** are the two points in which the Earth's axis of rotation meets its surface.

The **Terrestrial Equator** is the great circle on the Earth whose plane is perpendicular to the Earth's axis.

A **Terrestrial Meridian** is the section of the Earth's surface by a plane passing through its axis. If we suppose the Earth to be a sphere, a meridian will be a great circle passing through the terrestrial poles.

89. Phenomena depending on Change of Latitude.—

Assuming the Earth to be spherical, let $pOqp'r$ be a meridian section, C being the Earth's centre, p, p' the poles, q, r points on the equator. Then, if an observer is situated on the meridian at O, the direction of his celestial pole P will be found by drawing OP parallel to the Earth's axis $p'Cp$ (§ 87), while his zenith Z will lie in CO produced.

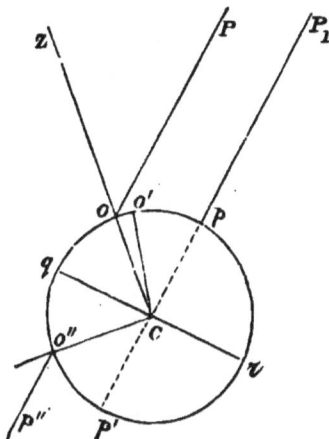

FIG. 34.

Since OP is parallel to CpP_1, therefore,
$$\text{angle } ZOP = OCp,$$
∴ altitude of pole at $O = 90° - ZOP = 90° - OCp = qCO$.
But the latitude of O has been shown to be the altitude of the pole; therefore

The latitude of a place on the Earth is the angle subtended at the Earth's centre by the arc of the meridian drawn from the place to the equator.

Since the angle qCO is proportional to the arc qO,

The latitude of a place is proportional to its distance from the equator.

Suppose the observer to go northwards along the meridian from O to O', then, from what has just been shown, the altitude of the pole increases from $\angle qCO$ to $\angle qCO'$, hence

The increase in the altitude of the pole $(= \angle OCO')$ is proportional to the arc OO', i.e., to the distance travelled northwards.

90. **Southern Latitudes.**—To an observer situated in the southern hemisphere of the Earth, as at O'', the North Pole of the heavens is below, and the South Pole, p'' is above the horizon. The **South Latitude** of the place is measured by the altitude of the South Pole, p'', and is equal to the angle $q C O''$.

At the terrestrial equator, the altitude of the pole is zero; hence the pole is on the horizon. At the terrestrial North Pole p, the altitude of the celestial pole is 90°, therefore the celestial pole coincides with the zenith. Hence, also, an altazimuth, if taken to the North Pole, would there become an equatorial.

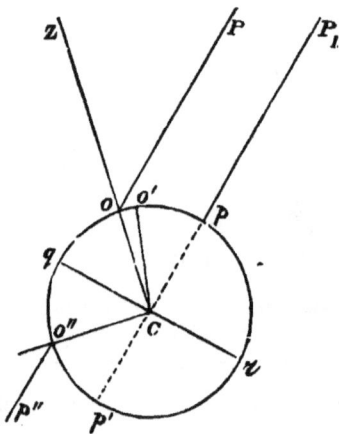

FIG. 35.

At the Earth's North Pole, those stars are only visible which are north of the equator, and they always remain above the horizon. On travelling southwards, other stars, whose declination is south, are seen in the south parts of the celestial sphere, and on reaching the Earth's equator all the stars will be above the horizon at some time or other, but the Pole Star will only just rise above the horizon, near the north point. After passing the equator, the Pole Star and other stars near the North Pole disappear.

91. Radius of the Earth.—The Earth's radius may be found by measuring the distance between two places on the same meridian, and finding their difference of latitude.

Let the places of observation be O, O' (Fig. 35). Let the latitudes qCO, qCO' be l and l' degrees respectively, and let the length $OO' = s$. We have, supposing the Earth spherical,

$$\frac{\text{angle } OCO'}{360°} = \frac{\text{arc } OO'}{\text{circumference of Earth}};$$

$$\therefore \text{Earth's circumference} = s \times \frac{360}{l' - l};$$

$$\text{and Earth's radius} = \frac{\text{circumference}}{2\pi} = \frac{180}{\pi} \frac{s}{l' - l},$$

which determines the Earth's radius in terms of the data.

By observations of this kind the Earth's radius is found to be very nearly 3,960 miles. For many purposes it will be sufficiently approximate to take the radius as 4000 miles. Its circumference is found by multiplying the radius by 2π, and is about 24,900 miles, or, roughly, 25,000 miles.

Conversely, knowing the Earth's radius, we can find the length of the arc of the meridian corresponding to any given difference of latitude.

92. Metre, Nautical Mile, Geographical Mile, Fathom.—The French **Metre** was originally defined as the ten-millionth part of the length of a quadrant of the Earth's meridian.

A **Nautical mile** is defined as the length of a minute of arc of the *meridian*. Thus a quadrant of the meridian contains 90×60, or 5,400 nautical miles, and the Earth's circumference contains 21,600 nautical miles.

A **Fathom** is the thousandth part of a nautical mile. It contains almost exactly six feet.

A **Geographical Mile** is defined as the length of a minute of arc measured on the Earth's *equator*. Taking the Earth as a sphere, the nautical mile and geographical mile are equal.

93. The "Knot."—Use of the Log Line in Navigation.—A nautical mile is sometimes called a knot. But the **Knot** is more correctly the unit of velocity used in navigation, being a velocity of one nautical mile per hour. Thus, a ship sailing 12 knots travels at 12 nautical miles an hour.

The velocity of a ship is measured by means of the **Log Line.** This consists of a "log," or float, attached to a cord which can unwind freely from a small windlass. The log is "heaved" or dropped into the sea, and allowed to remain at rest, the cord being "paid out" as the ship moves away. By measuring the length paid out in a given interval of time (usually half a minute), the velocity of the ship may be found. To facilitate the measurement, the line has knots tied in it at such a distance apart that the number of knots paid out in the interval of time is equal to the number of nautical miles per hour at which the ship is sailing. It is from these that the unit of velocity derives the name of knot.

Now one nautical mile per hour $= \dfrac{1}{120}$ nautical mile per half-minute. Hence, for this interval, the knots should be tied on the line at intervals of $\dfrac{1}{120}$ of a nautical mile apart.

94. From the definitions of §§ 92, 93, it is easy to reduce metres or nautical miles to ordinary feet and miles, and conversely.

EXAMPLES.

1. To find the number of miles in an arc of 1°.

An arc of $1° = \dfrac{\text{circumference of Earth}}{360} = \dfrac{24900}{360}$ miles $= 69\frac{1}{6}$ miles.

2. To find the number of feet in one fathom.

By Ex. 1, 60 nautical miles $= 69\frac{1}{6}$ ordinary miles; *i.e.*, 60,000 fathoms $= 69\frac{1}{6} \times 5280$ feet;

$$\therefore 1 \text{ fathom} = \dfrac{69\frac{1}{6} \times 5280}{60000} \text{ feet} = 6\cdot086 \text{ feet.}$$

3. To express a metre in terms of a yard.

By definition, 40,000,000 metres = Earth's circumference = 24,900 miles;

$$\therefore 1 \text{ metre} = \dfrac{24900 \times 1760}{40,000,000} \text{ yards} = 1\cdot0956 \text{ yards.}$$

95. Terrestrial Longitude.—The Longitude of a place on the Earth is the angle between the terrestrial meridian through that place, and a certain meridian fixed on the Earth, and called the **Prime Meridian.**

Thus, in Fig. 36, if PRP' represents the prime meridian, the longitude of any place q is measured by the angle RPq.

The longitude of q is also measured by RQ, the arc of the equator intercepted between the meridian of the place and the prime meridian.

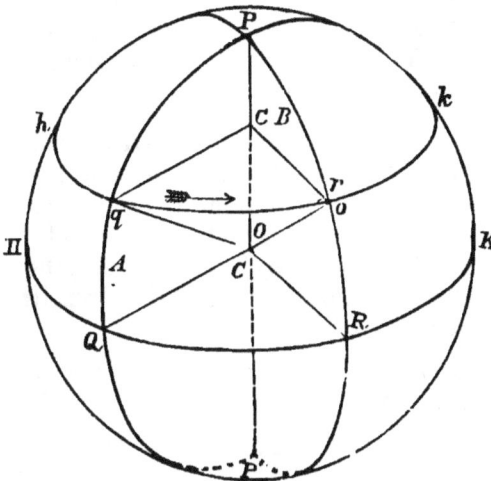

FIG. 36.

Since the latitude of q is measured by the arc Qq, we see that latitude and longitude are two coordinates defining the position of a place on the Earth just as decl. and R.A., or celestial latitude and longitude define the position of a star.*

The choice of a prime meridian is purely a matter of convenience. The meridian of Greenwich Observatory is universally adopted by English-speaking nations. The French use the meridian of Paris, and the University of Bologna has recently proposed the meridian of Jerusalem as the *universal* prime meridian. Longitudes are measured both eastward and westward from the prime meridian, from 0° to 180°, not from 0° to 360°.

---

*Note, however, that terrestrial latitude and longitude, being referred to the equator, correspond more nearly to declination and right ascension than to celestial latitude and longitude.

96. Phenomena depending on Change of Longitude.

(i.) Let q, r (Fig. 37) be two stations in the same latitude, and let the longitude of q be $L°$ west of r, so that $\angle rPq = L°$. As the Earth revolves about its axis at the rate of 360° per sidereal day, or 15° per sidereal hour, the points q, r will be carried forward in the direction of the arrow. After an interval of $\frac{1}{15} L$ sidereal hours, q will have revolved through $L°$ and will arrive at the position originally occupied by r. Hence the appearance of the heavens to an observer at q will be same as it was, $\frac{1}{15} L$ sidereal hours previously, to an observer at r. The stars will rise, south, and set $\frac{1}{15} L$ hours earlier at r than at q.

(ii.) If A, B be two places in *different* latitudes, whose difference of longitude is $L°$, the transits of a star at A and B will take place when the meridian planes PAP' and PBP' (which are evidently also the planes of the celestial meridians of A, B respectively), pass through the direction of the star. Hence, in this case also, the transits will occur $\frac{1}{15} L$ hours earlier at B than at A.

Now an observer at B will set his sidereal clock to indicate 0h. 0m. 0s. when ♈ crosses the meridian of B. When ♈ transits at A, the clock at B will mark $\frac{1}{15} L$ h., but an observer at A will *then* set his clock at 0h. 0m. 0s. Hence, if the two clocks be brought together and compared, the clock from B will be $\frac{1}{15} L$ h. faster than the clock from A. This fact may be expressed briefly by saying that the **"local" sidereal time** at B is $\frac{1}{15} L$ h. faster than the local sidereal time at A.

Since the Earth makes one revolution relative to the Sun in a *solar* day, in like manner the **local solar time** at B will be $\frac{1}{15} L$ *solar* hours faster than the local solar time at A.

Therefore, whether the local times be sidereal or solar, we have **Longitude of A *west* of B = long. of B *east* of A**
$$= 15\{(\text{local time at } B)-(\text{local time at } A)\}.$$
In particular, **Long. west of Greenwich**
$$= 15\{(\text{Greenwich time})-(\text{local time})\}$$
$$= 15 (\text{Greenwich time of local noon}).$$

97. To find the length of any arc of a given parallel of latitude, having given the difference of longitude of its extremities.

[A small circle of the Earth parallel to the equator is called a **Parallel of Latitude.**]

Let qr be the given arc of the parallel $hqrk$, l its latitude, and let qPr, the difference of longitudes of q and r, be $= L°$. Let a be the radius of the Earth.

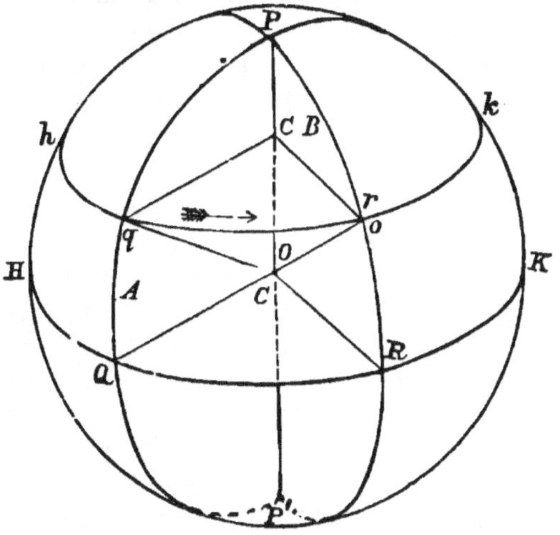

FIG. 37.

If the meridians of q, r meet the terrestrial equator in Q, R, we have, by Sph. Geom. (17),

arc $qr =$ arc $QR \times \sin Pq =$ arc $QR \times \cos l$.

But   arc QR : circumference of Earth $= L° : 360°$;

$$\therefore \text{ arc } QR = 2\pi a L/360 = \frac{1}{180}\,\pi a L\ ;$$

$$\therefore \text{ arc } qr = \frac{\pi a L \cos l}{180}.$$

COROLLARY.—Since $1'$ of arc of the equator measures a geographical mile, it follows that

In latitude l, the arc of a parallel corresponding to $1'$ difference of longitude is $\cos l$ geographical miles.

98. Changes of Latitude and Longitude due to a Ship's Motion.—Suppose a ship, in latitude l, to sail m nautical miles in a direction A degrees west of north. If m is small, we may easily see (by drawing a diagram) that the ship would arrive at the same place by sailing $m \cos A$ nautical miles due north, and then sailing $m \sin A$ nautical miles due west. Hence,

The ship's latitude will increase by $m \cos A$ minutes (§ 92).

Its W. long. will increase by $m \sin A$ sec l minutes (§ 97, cor.).

NOTE.—The *shortest distance* between two points on a sphere is along a great circle. Hence, the shortest distance between two places in the same latitude is less than the arc of the parallel joining them (except at the equator). But the difference is imperceptible when the arc is small.

99. To explain the Gain or Loss of a Day in going round the World.—If a traveller, starting from a place A, go round the world eastward, and if, during the voyage, the Earth revolves n times relative to the Sun, the traveller will have performed one more revolution relative to the Earth in the same direction, and therefore $n + 1$ revolutions relative to the Sun. Hence, to a person remaining at A, the voyage will appear to have taken n days, while to the traveller, $n + 1$ days will appear to have elapsed—in other words, the traveller will, apparently, have "gained a day."

But, as he goes eastward, he will find the local time continually getting faster, and he will have to move the hands of his watch forward 1h. for every 15°, or 4m. for every 1° of longitude. Thus, by the end of the voyage he will have put his watch forward through 24h., and the day apparently gained will be made up of the times apparently lost every time the watch is put forward to local time.

Similarly, a traveller going round the world westward, and starting and arriving back simultaneously with the first traveller, will have made $n - 1$ revolutions relative to the Sun, instead of n. Hence, the journey will appear to have taken $n - 1$ days, and he will apparently have lost a day.

But, during the journey, he will have been continually moving the hands of his watch backwards, so that the 24h. apparently lost will be made up of the times apparently gained each time the watch is put back to local time.

Section II.—*Dip of the Horizon*

100. Definitions.—Let O be an observer situated above the surface of the land or sea. Draw OT, OT' tangents to the surface. Then it is evident, from the figure, that only those portions of the Earth's surface will be visible whose distance from the observer O is less than the length of the tangents OT, OT'.

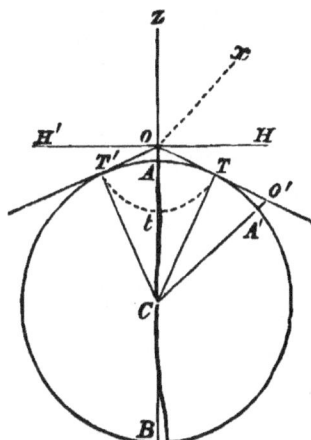

Fig. 38.

The boundary of the portion of the Earth's surface visible from any point is called the **Offing or Visible Horizon.** Hence, if $OACB$ be the Earth's diameter through O, and the Earth be supposed spherical, the offing at O is the small circle TtT', formed by the revolution of T about OB, and having for its pole the point A vertically underneath O. If, however, the Earth be not supposed spherical, the form of the offing will, in general, be more or less oval, instead of circular.

Conversely, since it is observed that the " offing " at sea is very approximately circular, whatever be the position of the observer, it may be inferred that the Earth is approximately spherical.

The **Dip of the Horizon** at O is the inclination to the horizontal plane of a tangent from O to the Earth's surface.

Hence, if HOH' be drawn horizontally (*i.e.*, perpendicular to OC), the dip of the horizon will be the angle HOT.

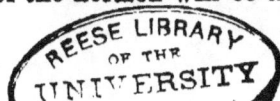

101. **To determine the Distance and Dip of the Visible Horizon at a given height above the Earth.**

Let $h = AO =$ given height of observer;
$a = CA =$ Earth's radius;
$d = OT =$ required distance of horizon;
$D = \angle HOT =$ required dip expressed in *circular measure*;
D'' the number of seconds in the dip D.

(i.) By Euclid III. 36, $OT^2 = OA \cdot OB$;
$$\therefore \quad d^2 = h(2a+h) = 2ah + h^2.$$

This determines d accurately. But in practical applications h is always very small compared with $2a$; therefore h^2 may be neglected in comparison with $2ah$, and we have the approximate formula, $d^2 = 2ah \therefore d = \sqrt{(2ah)}$.

(ii.) Since CTO is a right angle,
$$\therefore \quad \angle OCT = \text{complement of } \angle COT = \angle TOH = D.$$

Therefore, D being expressed in circular measure, we have

$$D = \frac{\text{arc } AT}{\text{radius } CT}.$$

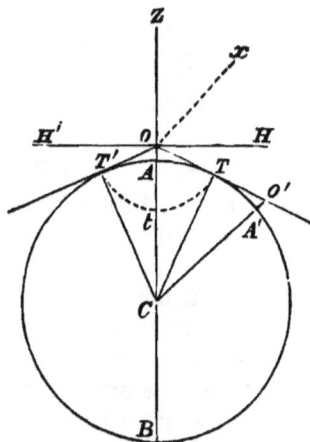

Fig. 39.

Now, in practical cases, where the dip is small, the arc AT will not differ perceptibly in length from the straight line OT. We may, therefore, take arc $AT = d$;

$$\therefore \quad D = \frac{d}{a} = \frac{\sqrt{(2ah)}}{a} = \sqrt{\frac{2h}{a}}.$$

To reduce to seconds, we must multiply by $180 \times 60 \times 60/\pi$, the number of seconds in a unit of circular measurement, and we have

$$D'' = \frac{180 \times 60 \times 60}{\pi} \sqrt{\frac{2h}{a}}.$$

COROLLARY 1.—Let a, h, d be measured in miles, and let h' be the number of feet in the height h.

Then $h' = 5280h$, and taking the Earth's radius a as 3960 miles, we have

$$d = \sqrt{\frac{2 \times 3960 \times h'}{5280}} = \sqrt{\left(\frac{3h'}{2}\right)},$$

a very useful formula.

COROLLARY 2.—Since the offing is a circle whose radius is very approximately equal to OT or d, we have

Area of Earth's surface visible from $O = \pi d^2 = 2\pi a h = \frac{3}{2}\pi h'$ in square miles.

*102. Accurate Determination of Dip.—The use of approximations can be avoided by the exact formula:

$$\tan D = \frac{TO}{CT} = \frac{\sqrt{(2ah + h^2)}}{a} = \sqrt{\frac{h(2a + h)}{a^2}},$$

which is adapted to logarithmic computation.

In this, as in the preceding formulæ, no account has been taken of the effect of refraction due to the atmosphere.

For this reason it is important to determine dip of the horizon by practical observations. An instrument called the Dip Sector is constructed for this purpose.

Tables have also been constructed, giving the dip of the horizon as seen from different heights. They are of great use at sea, where the altitude of a star is usually found by observing its angular distances from the offing.

103. Disappearance of a Ship at Sea.—When a ship has passed the offing, the lower part will be the first to disappear. Let $A'O'$ (Fig. 38) be the position of the ship; let its distance OO' be s, and let $k = A'O'$ be the height above sea level of the lowest portion just visible from O. By the approximate formula we have $OT = \sqrt{(2ah)}$, $O'T = \sqrt{(2ak)}$

$$\therefore \quad s = \sqrt{(2ah)} + \sqrt{(2ak)}.$$

This formula determines the distance s at which an object of given height k disappears below the horizon.

ASTRON.   G

104. Effect of Dip on the Times of Rising and Setting.—To an observer on land, the offing is generally more or less broken by irregularities of the Earth's surface. At sea, however, the offing is well defined, and if the dip of the horizon in seconds be D'', the visible horizon, which bounds the observer's view of the heavens, is represented on the celestial sphere by a small circle parallel to the celestial horizon, and at a distance D'' below it ($n'E's'$, Fig. 40).

Hence the stars appear to rise and set when they are at an angular distance D'' below the celestial horizon. Thus they will rise sooner and set later than they would if there were no dip.

Taking the observer's latitude to be l, let x', x be the positions of a star of declination d, when rising across the visible horizon $n'E's'$ and the celestial horizon nEs

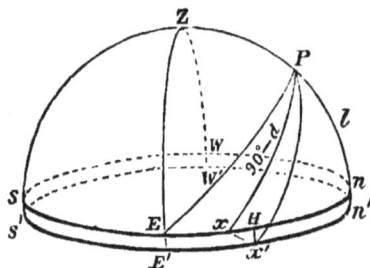

FIG. 40.

respectively. Draw $x'H$ perpendicular to nEs, then $x'H = D''$.

Then, if the star rise t seconds earlier at x' than at x, we have

$$15\ t = \angle x'Px \text{ (in seconds of angle)}$$

$$= \frac{\text{arc } xx'}{\sin xP} = \frac{\text{arc } xx'}{\cos d}. \text{ (Sph. Geom., 17.)}$$

But treating the small triangle $x'xH$ as plane (Sph. Geom., 24), and remembering that $\angle Pxx' = 90°$, we have

$$xx' = \frac{x'H}{\sin x'xH} = \frac{D''}{\cos nxP};$$

$$\therefore\ t = \frac{1}{15}D'' \sec d \cdot \sec nxP.$$

Evidently the acceleration at rising = retardation at setting.

COROLLARY 1.—**To an observer at the Equator,** P coincides with n, $\therefore \angle nxP = 0$,

$\therefore$ the time of rising is accelerated by $\frac{1}{15}D'' \sec d$ **seconds.**

COROLLARY 2.—**If the star is on the equator,** $d = 0$, x coincides with E, and $\angle nEP = nP = l$,

$\therefore$ the acceleration $= \frac{1}{15}D'' \sec l$ **seconds.**

SECTION III.— *Geodetic Measurements—Figure of the Earth.*

105. **Geodesy** is the science connected with the accurate measurement of arcs on the surface of the Earth. Such measurements may be performed with either of the two following objects :—
(i.) The construction of maps.
(ii.) The determination of the Earth's form and magnitude.
Only the second application falls within the scope of this book.

106. **Alfred Russell Wallace's Method of Finding the Earth's Radius.**—An approximate measure of the Earth's radius can be readily found by means of the following simple experiment, due to Mr. A. R. Wallace.

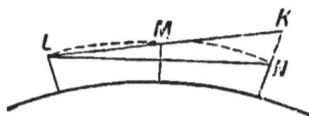

FIG. 41.

Let L, M, N (Fig. 41) be the tops of three posts of the same height set up in a line along the side of a straight canal. Owing to the Earth's curvature the straight line LM will, if produced, pass a little above N. Hence, in order to see L, M in a straight line, an observer at the post N will have to place his eye at a point K, a little above N, and the height KN may be measured. Let KL, KM be also measured.

Since the posts are of equal height, L, M, N will lie on a circle concentric with, and almost coinciding with, the Earth's surface. Let the vertical KN meet this circle again in n. By Euclid III. 36,

$$KL \cdot KM = KN \cdot Kn; \quad \therefore \quad Kn = KL \cdot KM/KN,$$

and Radius of Earth $= \frac{1}{2} Kn$ (very approximately)

$$= \frac{KL \cdot KM}{2KN}.$$

This method cannot be relied on where accuracy is required, for the small height KN is very difficult to measure, and a very slight error in its measurement would affect the final result considerably. Moreover the observations are considerably affected by refraction.

107. Ordinary methods of Finding the Earth's Radius.—Where greater accuracy is required, the radius of the Earth is obtained by measuring the length of an arc of the meridian and determining the difference of latitude of its extremities; the radius may then be calculated as in § 91. The instruments required for the observations include—

(i.) Measuring rods, such as the double bar ;

(ii.) A theodolite, for measuring angles ;

(iii.) A zenith sector.

108. Measurement of a Base Line.—The first step is to measure, with extreme accuracy, the length of the arc joining two selected points, several miles apart, on a level tract of country ; this line is called a **Base Line.** A series of short upright posts are placed at equal distances apart along the base line, and they are adjusted till their tops are seen exactly in the same vertical plane, and are on the same level as shown by a spirit level. Across these posts are laid measuring rods of metal, whose length is very accurately known, and these are also adjusted in a line, and made level by the spirit level. These rods are not allowed to touch, but the small distances between their ends are measured with reading microscopes. In this way, a base line several miles long can be measured correctly to within a small fraction of an inch.

*109. The Double Bar.— If the measuring rods be made of a single metal, their length will vary with the temperature. This disadvantage is, however, sometimes obviated by the use of the double bar (Fig 42).

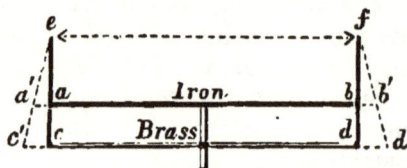

FIG. 42.

It consists of two bars, ab, cd, one of iron, the other of brass. These are joined together in the middle, and to their ends are hinged perpendicular pointers eac, fbd of such length that

$ea : ec = fb : fd$

= coefficient of linear expansion of iron : that of brass,

= about 11 : 18.†

If the temperature be raised, the rods will expand, say to $a'b'$, $c'd'$. But $aa' : cc' = ea : ec$, therefore e, and similarly f, will remain fixed. Hence the distance ef will be unaffected by the changes of temperature.

† Wallace Stewart's *Heat*, Table 22.

110. Triangulation.—When once a base line has been measured, the distance between any two points on the Earth can be determined by the measurement of *angles* alone. For, calling the base line *AB*, let *C* be any object visible from both *A* and *B*. If the angles *CAB*, *CBA* be observed, we can solve the triangle *ABC* and determine the lengths of the sides *CA*, *CB*. Either of these sides, say *CA*, may now be taken as the base of a new triangle, whose vertex is another point, *D*. Thus, by observing the angles of the triangle *ACD* we can determine *DA*, *DC* in terms of the known length of *AC*. Proceeding in this way, we may divide any country into a network of triangles connecting different places of observation *A, B, C, D*, and the distance between any two of the places calculated, as well as the direction of the line joining them. Finally, two stations *C, H* are taken, which lie on the same meridian, and the distance *CH* is calculated; in this way it is possible to measure an arc of the meridian.

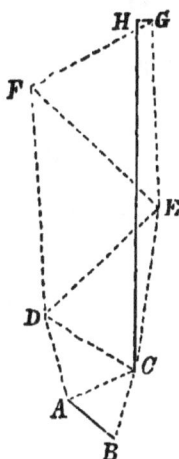

FIG. 43.

111. The Theodolite.—The measurement of the angles is far easier in practice than the measurement of a base line. The instrument used for measuring angles is called a **Theodolite,** and is really a portable form of altazimuth. It is provided with spirit-levels, by means of which the instrument can be adjusted so that the horizontal circle is truly horizontal, and the vertical axis, therefore, truly vertical; the direction of the north point is usually found by means of a compass needle. Most theodolites are only furnished with a small arc of the vertical circle, sufficient for measuring the altitude of one terrestrial object as seen from another.

By reading the horizontal circle of the theodolite, the azimuths of *B, C*, as seen from *A*, are found. By using the difference of azimuth instead of the angle *ABC*, it becomes unnecessary to take account of the height of the various stations above the Earth. For if *A, B, C* are replaced by any other points, *A', B', C'*, at the sea level, and vertically above or below *A, B, C*, the vertical planes joining them will be unaltered in position, and therefore the azimuths will also be unaffected.

112. Having thus found, with great accuracy, the length of the arc joining two stations on the same meridian, it only remains now to observe their difference of latitude.

The Zenith Sector is the most useful instrument for this purpose. It consists essentially of a long telescope ST (Fig. 44), mounted so as to turn about a horizontal axis, A, near its object-glass; this axis is adjusted to point due east and west (as in the transit circle). Attached to the lower end near the eye piece is a graduated arc of a circle GH, whose centre is at A. The line of collimation of the telescope is indicated by cross-wires placed in the field of view. A fine plumb-line, AP, is attached to the axis A, and hangs freely in front of the graduated arc. The plumb-line should mark zero when the line of collimation points to the zenith. When the instrument is pointed to any star, the reading opposite the plumb-line will be the star's zenith distance This reading can be determined with great accuracy by means of a reading microscope.

FIG. 44.

113. A star is selected which transits near the zenith* and its meridian zenith distances are observed at the two stations. Let these be z and z' degrees. Then if l_1 and l_2 are the latitudes of the stations, and d the declination, we have, by § 24,

$$l'-l = (d-z')-(d-z) = z-z'.$$

Hence, if s is the measured length of the arc of the meridian joining the stations, and r the radius of the Earth, § 91 gives

$$r = \frac{180}{\pi} \; \frac{s}{l'-l} = \frac{130}{\pi} \; \frac{s}{z-z'},$$

whence the Earth's radius is found.

---

* This position is chosen because the effects of atmospheric refraction are least in the neighbourhood of the zenith.

114. Exact Figure of the Earth.—If the Earth were an exact sphere, the same value would be found for the radius *r* in whatever latitude the observations were made. But in reality the length of a degree of latitude, and therefore also *r*, is found to be larger when the observation is made near the poles than when made near the equator, and hence it is inferred that the meridian curve is somewhat oval.

Let $PQP'R$ represent the meridian curve, OO' two near places of observation on it. Then, if OK and $O'K$ be drawn normal (*i.e.*, perpendicular) to the Earth's surface at O, O', they will be the directions of the plumb lines of the zenith sectors at O, O'. Hence the observed difference of latitudes or meridian altitudes at O, O' will give the angle OKO'.

Regarding the small arc OO' as an arc of a circle whose centre is K, we shall have approximately,

Circular measure of $OKO' = $ arc $OO' \div OK$,

$$\therefore \ OK = \frac{\text{arc } OO'}{\text{circ. measure of } OKO'} = \frac{180}{\pi} \frac{s}{l' - l},$$

and hence *r*, calculated as in § 113, is the length OK.

The length OK is called the **radius of curvature** of the arc, and K is called the **centre of curvature** ; they are respectively the radius and centre of the circle whose form most nearly coincides with the meridian along the arc OO'.

This radius of curvature OK is not, in general, equal to OC, the distance from the centre of the Earth, owing to the Earth not being quite spherical.

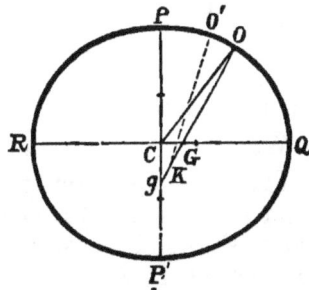

FIG. 45.

As the result of numerous observations, the meridian curve is found to be an **ellipse** (see Appendix), whose greatest and least diameters, called the **major** and **minor axes,** are the Earth's equatorial and polar diameters respectively. The Earth's surface is the figure formed by making the ellipse revolve about its minor axis PCP'. This figure is called an **oblate spheroid.**

115. To find the Equatorial and Polar Radii of Curvature of the meridian curve, supposing it to be an ellipse.—Let $PQP'R$ be the ellipse. Let $2a$, $2b$ be the lengths of its equatorial and polar diameters QCR, PCP'. Let r_1, r_2 be the required radii of curvature at Q and P respectively.

Take any point O on the ellipse, and let the normal at O meet the two axes in G and g respectively.

It is proved in treatises on Conic Sections* that

$$OG : Og = CP^2 : CQ^2 = b^2 : a^2.$$

First take O very near to Q. Then OG will become equal to the radius of curvature r_1; also Og will evidently become ultimately equal to CQ or a.

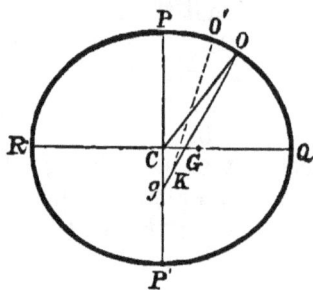

Fig. 46.

Therefore, $r_1 : a = b^2 : a^2$; $\therefore$ $r_1 = b^2/a$.

Next take O very near to P. Then OG will become equal to b and Og to r_2.

Therefore, $b : r_2 = b^2 : a^2$; $\therefore$ $r_2 = a^2/b$.

Thus r_1, r_2 are found in terms of a, b.

Conversely, if r_1 and r_2 are known, a and b may be found; for, by solving, we find $a = \sqrt[3]{(r_2^2 r_1)}$, $b = \sqrt[3]{(r_1^2 r_2)}$.

We notice that since $a > b$, $\therefore$ $r_1 < r_2$.

That the equatorial radius of curvature is less than the polar is also evident from the shape of the curve. This, as the figure shows, is most rounded at Q, R, and flattest or least rounded at P, P'. Hence it will require a smaller circle to fit the shape of the curve at the equator than at the poles.

116. Exact Dimensions of the Earth.—The lengths of the Earth's equatorial and polar semi-diameters, a, b, are

$$a = 3963 \cdot 296 \text{ miles,} \quad b = 3949 \cdot 791 \text{ miles.}$$

Thus, the Earth's equatorial semi-diameter exceeds its polar semi-diameter by $13 \cdot 505$ miles.

* Appendix, Ellipse (9).

The **mean radius** of an oblate spheroid is the radius of a sphere of equal volume, and is equal to $\sqrt[3]{(a^2b)}$. Thus, the Earth's mean radius is approximately 3958·8 miles.

The **ellipticity** or **compression** (c) is the fraction

$$c = \frac{a-b}{a}.$$

For the Earth, $c = \dfrac{1}{293}$ nearly.

The **eccentricity** (e) is given by the relation

$$e^2 = \frac{a^2 - b^2}{a^2}.$$

Hence $\qquad b^2 = a^2(1-e^2) = a^2(1-c)^2$;

$\qquad \therefore\ 1-e^2 = (1-c)^2 = 1-2c+c^2$;

$\qquad \therefore\ e^2 = 2c - c^2 = c(2-c)$.

Since c is small, $2-c = 2$, approx.; $\quad \therefore\ e^2 = 2c$, approx., which gives the Earth's eccentricity $e = ·0826$.

117. Geographical and Geocentric Latitude.—The **Geographical Latitude** of a place is the angle which the normal to the Earth's surface at that place makes with the plane of the equator. It is the latitude defined in § 18, Thus, $\angle QGO$ (Fig. 46) is the geographical latitude of O.

The **Geocentric Latitude** is the angle subtended at the Earth's centre by the arc of the terrestrial meridian between the place and the equator. Thus, $\angle QCO$ is the geocentric latitude of O.

*118. Relations between the Geocentric and Geographical Latitudes.—Let $\angle QGO = l$, $\angle QCO = l'$. Draw ON perp. to CQ.

Then $GN : CN = OG : Og = b^2 : a^2$; $\ \therefore\ NO/CN = (NO/GN) \times (b^2/a^2)$;

$\qquad \therefore\ \tan l' = \tan l \times b^2/a^2 = (1-e^2)\tan l.$

We deduce also $\tan(l-l') = \dfrac{e^2 \sin 2l}{2(1-e^2\sin^2 l)} = \tfrac{1}{2}e^2\sin 2l$ (approx.), since e^2 is small.

EXAMPLES.—III.

1. Show that the locus of points on the Earth's surface at which the Sun rises at the same instant is half a great circle; and state the corresponding property possessed by the other half.

2. Find the least height of a mountain in Corsica in order that it may be visible from the sea-level at Mentone, at a distance of 80 miles.   •

3. At the equator, in longitude $L°$, a given vertical plane declines $a°$ from the north towards the west; find the latitude and longitude of the places to whose horizon the given plane is parallel.

4. Prove that, at either equinox, in latitude l, a mountain whose height is $1/n$ of the Earth's radius will catch the Sun's rays in the morning $\dfrac{12}{\pi \cos i} \sqrt{\dfrac{2}{n}}$ hours before he rises on the plain at the base.

5. Estimate to the nearest minute the value of this expression for a mountain three miles high in latitude 45°.

6. Find the distance of the horizon as seen from the top of a hill 1056 feet high.

7. Find, to the nearest mile, the radius of the Earth, supposing the visual line of a telescope from the top of one post to the top of another post two miles off, cuts a post, half way between, 8 inches below the top, the posts standing at equal heights above the water in a canal.

8. In Question 7, what would be the length of a nautical mile, adopting the usual definition.

9. Supposing the Earth spherical, and of radius r, and neglecting the refraction of the air, show that, if from the top of a mountain of height a above the level of the sea, the summit of another mountain is seen beyond the horizon of the sea, and at an elevation e above the horizon, and if its distance be known to be D, its height is approximately given by

$$a \div eD + D\left(\frac{D}{2r} - \sqrt{\frac{2a}{r}} \right).$$

10. A railway train is moving north-east at 40 miles an hour in latitude 60°; find approximately, in numbers, the rate at which it is changing its longitude.

MISCELLANEOUS QUESTIONS.

1. Explain the different systems of coordinates by which a star's position is fixed in the heavens.

2. Show, by a figure, where a star will be found at 9 p.m. on the 5th of June in latitude 50°N., if the star's right ascension is 12 hours and its declination 5° south.

3. Define *dip*, *azimuth*, *culmination*, *circumpolar*, *zenith*. Why would it be insufficient to define the declination of a star as its distance from the equator measured along a declination circle?

4. Three stars, A, B, C, are on the same meridian at noon, B being on the equator, and A and C equidistant from B on either side. Prove that the intervals between the setting-times of A and B and B and C are equal.

5. Show how to find approximately the Sun's R.A. at a given date. Obtain its approximate value for March 1, August 10, October 23, and January 15.

6. Describe the transit circle.

7. Define a morning and evening star. Show that on the 1st of September a star, whose declination is 0°, and R.A. 11h. 28m., is an evening star, but that it is a morning star three weeks later.

8. Assuming the Earth to be a sphere, show how its radius may be practically measured.

9. Explain clearly the nature and uses of the zenith sector.

10. A, B, C are the tops of the masts of three ships in a line, and are at equal heights above the sea-level, and O is the centre of the Earth. If the distance BC be x miles, and r is the Earth's radius in miles, show that $\angle BAC = \frac{1}{2} \angle BOC$; and hence deduce that

$$\angle BAC = \frac{180 \times 60 \times 60}{\pi} \frac{x}{2r} \text{ seconds.}$$

Find this angle, having given $x = 2$, $r = 3960$, $\pi = 3\frac{1}{7}$.

EXAMINATION PAPER.—III.

1. Assuming the Earth to be a sphere, show that, as we travel from the equator due north, our astronomical latitude (*i.e.*, the altitude of the Pole) will increase.   Taking this increase as 1° for every 69 miles, find the circumference and the radius of the Earth.

2. Define the *metre*, the *nautical mile*, and the *knot*, and calculate their values in feet and feet per second respectively, taking the Earth's radius as 3960 miles.

3. How is the speed of a ship estimated ?   Find, in feet, the distance apart of the knots on a log line, so constructed that the number run out in half a minute measures the ship's velocity in nautical miles per hour.

4. What are the difficulties in measuring an arc of the meridian and how are they met ?

5. Find the Earth's radius in fathoms, and in metres.   Express the nautical mile in French units of length.

6. Obtain formulæ for the distance of the visible horizon from a place whose height is given.   Deduce that, if the height h be measured in inches, the distance in miles will be $\sqrt{\dfrac{h}{8}}$, taking the Earth's radius as 3960 miles.

7. Define the *dip of the horizon*, and show how to find it.   Prove that the number of seconds in the dip is nearly 52 times the distance in miles of the offing.

8. If A, B, and C be the tops of three equal posts arranged in order two miles apart along a straight canal, show that the straight line AB passes 5 feet 4 inches above C, and that AC passes 2 feet 8 inches below B.

9. Find the length of a given parallel of latitude intercepted between two given circles of longitude.

10. Is the Earth an exact sphere ?   Show that a degree of latitude increases in length as we go northward.   Distinguish a *nautical* from a *geographical mile*.

CHAPTER IV.

THE SUN'S APPARENT MOTION IN THE ECLIPTIC.

SECTION I.—*The Seasons.*

119. In Section III. of Chapter I.* we described the Sun's annual motion among the stars, and showed how, in consequence of this motion, the Sun's right ascension increases at an average rate of nearly 1° per day, while his declination fluctuates between the values 23° 27½' north, and 23° 27½' south of the equator. We shall now show how this annual motion, combined with the diurnal rotation about the poles, gives rise to the variations, both in the relative lengths of day and night, and in the Sun's meridian altitude, during the course of the year; how these variations are modified by the observer's position on the Earth; and how they produce the phenomena of summer and winter.

Although both the diurnal and annual apparent motions of the Sun are known to be *really* due to the Earth's motion, it will be convenient in this section to imagine the Earth to be fixed, while the Sun and stars are moving; thus the zenith, pole, horizon, meridian, and equator will be considered fixed, as they actually appear to be to an observer on the Earth.

As the change in the Sun's declination during a single day is very small, the Sun's apparent path in the heavens from morning till night is very approximately a small circle parallel to the equator, and may be regarded as such for purposes of explanation. The effects of the variation in the declination will, however, become very apparent when we compare the Sun's diurnal paths at different seasons of the year.

Throughout this section we shall denote the obliquity of the ecliptic by i, the Sun's declination at any time by d, his zenith distance at noon by z, and the observer's latitude by l.

---

* The student will do well to revise Chapter I., Section III., before proceeding further.

120. Zones of the Earth.—Definitions.—From § 24 it is evident that if the Sun passes through the zenith at noon, d must $= l$.

But d lies between i (north) and i (south).

Therefore l must lie between the limits i N. and i S.

Thus, if the Sun be vertically overhead at some time in the year, the latitude must not be greater than $23° 27\frac{1}{2}'$ N. or S.

Again, from § 28 we see that the Sun, like a circumpolar star, will remain above the horizon during the whole of its revolution provided that $90° - d < l$.

This requires that $l > 90° - i$.

Thus, if the Sun be visible all day long during a certain period of the year, the latitude must be greater than $66° 32\frac{1}{2}'$ N. or S.

These circumstances have led to the following definitions.

The **Tropics** are the two parallels to the Earth's equator in north and south latitude i, or $23° 27\frac{1}{2}'$. The northern tropic is called the **Tropic of Cancer,** the southern the **Tropic of Capricorn.**

The **Arctic** and **Antarctic Circles** are respectively the parallels of north and south latitude $90° - i$, or $66° 32\frac{1}{2}'$.

These four parallels divide the Earth's surface into five regions or **zones.**

The portion between the tropics is called the **Torrid Zone.**

The portion between the tropic of Cancer and the arctic circle is called the **North Temperate Zone.** The portion between the tropic of Capricorn and the antarctic circle is called the **South Temperate Zone.**

The portions north of the arctic circle, and south of the antarctic circle are called the **Frigid Zones,** and are distinguished as the **Arctic** and **Antarctic Zones.**

121. Sun's Diurnal Path at Different Seasons and Places.—We shall now describe the various appearances presented by the Sun's diurnal motion at different times of the year, beginning in each case with the vernal equinox. We shall first suppose the observer at the Earth's equator, and shall then. describe how the phenomena are modified as he travels northward towards the pole.

122. At the Earth's equator, $l = 0$, and the poles of of the celestial sphere are on the horizon (P, P', Fig. 47). Hence, between sunrise and sunset, the Sun has always to revolve about the poles through an angle 180°, and the days and nights are always equal, each being 12 hours long.

On March 21 the Sun is on the celestial equator, and it describes the circle EZW, rising at the east point, passing through the zenith at noon, and setting at the west point.

Between March 21 and Sept. 23, the Sun is north of the celestial equator; it therefore rises north of E., transits north of the zenith Z, and sets north of W. Its N. meridian zenith distance z is always equal to its N. declination d (since by § 24, $z = d - l$ and $l = 0$).

Hence, from March 21 to June 21, z increases from 0 to i N. On June 21, z has its greatest N. value i, and the Sun describes the circle $E'Q'W'$. where $ZQ' = i$.

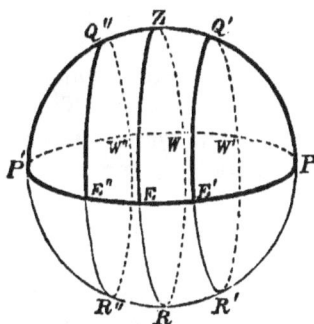

FIG. 47.

From June 21 to Sept. 23, z decreases from i to 0.

On Sept. 23, the Sun again describes the great circle EQW.

Between Sept. 23 and March 21, the Sun is south of the equator, and therefore it transits south of the zenith. We now have $z = d$, both being S.

From Sept. 23 to Dec. 22, the Sun's south Z.D. at noon, z, increases from 0 to i.

On Dec. 22, z has its greatest value i (south) and the Sun describes the circle $E'Q''W''$ where $ZQ'' = i$.

From Dec. 22 to March 21, z diminishes again from i to 0. On March 21, the Sun again describes the circle EQW, and the same cycle of changes is repeated the following year.

123. In the Torrid Zone North of the Equator.

On March 21, the Sun describes the equator EQW (Fig. 48), rising at E and setting at W. Here $\angle ZPE = \angle ZPW = 90°$, and the day and night are each 12h. long. The Sun transits S. of the zenith at Q, where $ZQ = z = l$.

From March 21 to June 21, d increases from 0 to i, and the Sun's diurnal path changes from EQW to $E'Q'W'$.

The hour angles at rising and setting increase from ZPE and ZPW to ZPE' and ZPW', respectively; hence the days increase and the nights decrease in length. The day is longest on June 21, when the hour angle ZPE' is greatest. The increase in the day is proportional to the angle EPE', and is greater the greater the latitude l.

At first the Sun transits S. of the zenith, and $z = l - d$. When $d = l$, $z = 0$, and the Sun is directly overhead at noon. After this, the Sun transits N. of the zenith, and $z = d - l$. On June 21, z attains its maximum N. value $ZQ' = i - l$.

From June 21 to Sept. 23, the phenomena occur in the reverse order. The diurnal path changes gradually back to EQW. The day diminishes to 12h. The Sun, which at first continues to transit N. of the zenith, becomes once more vertical at noon when d again $= l$, and then transits S. of the zenith.

From Sept. 23 to Dec. 22, the Sun's path changes from EQW to $E''Q''W''$.

The eastern hour angle at sunrise decreases to ZPE''; thus the days shorten and the nights lengthen. The day is shortest on Dec. 22.

Also z increases from l to $l + i$.

On Dec. 22, z attains the maximum value $ZQ'' = l + i$, and the Sun is then furthest from the zenith at noon.

From Dec. 22 to March 21, the length of the day increases again to 12 hours, and the Sun's meridian zenith distance decreases to $z = l$.

124. On the Tropic of Cancer, $l = i$. — The variations in the lengths of day and night partake of the same general character as in the Torrid Zone. But the Sun only just reaches the zenith at noon once a year, namely, on the longest day, June 21. At other times the Sun is south of the zenith at noon, and z attains the maximum value $2i$ on December 22.

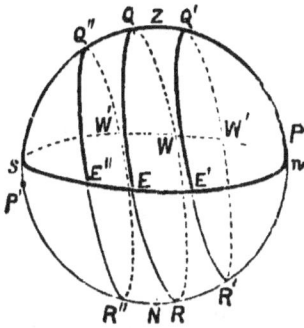

FIG. 48.

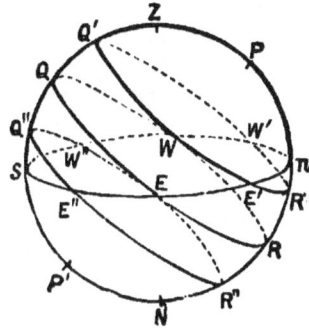

FIG. 49.

125. In the North Temperate Zone $l > i$ but $< 90^{c} - i$.
Here the variations in the lengths of day and night are similar, but more marked, owing to the greater latitude.

On March 21, the Sun describes the equator $EQWR$ (Fig. 49), which is bisected by the horizon; hence the day is 12h. long. The length of the day increases from March 21 to June 21. The day is longest on June 21, when the Sun describes $E'Q'W'R'$, and the hour angles ZPE', ZPW' are greatest.

The days diminish to 12h. on Sept. 23, when the Sun again describes $EQWR$. The day is shortest on Dec. 22, when the Sun describes $E''Q''W''R''$.

From Dec. 22 to March 21, the days increase in length, and on March 21 the day is again 12 hours long.

The difference between the longest and shortest days is the time taken by the Sun to describe the angles $E'PE''$, $W''PW'$, and is therefore

$$= \tfrac{1}{15} \left(\angle E'PE'' + \angle W''PW' \right) = \tfrac{2}{15} \cdot \angle E'PE''.$$

It will be seen that $\angle E'PE''$ is greater in Fig. 49 than in Fig. 48, thus the variations are more marked in the temperate zone than in the torrid zone. The variations increase as the latitude increases.

The Sun never reaches the zenith in the temperate zone, but always transits south of the zenith. The Sun's zenith distance at noon is least on June 21, when $z = ZQ' = l - i$, and is greatest on Dec. 22, when $z = ZQ'' = l + i$. At the equinoxes (March 21 and Sept. 23), $z = ZQ = l$.

126. **On the Arctic Circle,** $l = 90° - i$. Hence on June 21, when the Sun's N.P.D. $= 90° - i$, the Sun at midnight will only just graze the horizon at the north point without actually setting. On Dec. 22 at noon, the Sun's Z.D. $= 90°$, and the Sun will just graze the horizon without actually rising. As in the preceding case, the days increase from Dec. 22 to June 21, and decrease from June 21 to Dec. 22; on March 21 and Sept. 23, the day and night are each 12h. long.

127. **In the Arctic Zone** we have $l > 90° - i$, and the variations are somewhat different (Fig. 50).

On March 21, the Sun describes the circle EQW, and the day is 12h. long.

As d increases, the days increase and the nights decrease, and this continues until $d = 90° - l$. When this happens, the Sun at midnight only grazes the horizon at n.

Subsequently, while $d > 90° - l$, the Sun remains above the horizon during the whole of the day, circling about the pole like a circumpolar star. This period is called the **Perpetual Day.**

During the perpetual day, the Sun's path continues to rise higher in the heavens every twenty-four hours until June 21, when the Sun traces out the circle $R'Q'$. The Sun's least and greatest zenith distances will then be $ZQ' = l - i$, and $ZR' = 180° - i - l$ respectively.

After June 21, the Sun's path will sink lower and lower. When d is again $= 90° - l$ the perpetual day will end. Subsequently, the Sun will be below the horizon during part of each day. The days will then gradually shorten and the nights lengthen.

On Sept. 23, the Sun will again describe the circle EQW, and the day and night will each be 12 hours long.

The days will continue to diminish till the Sun's south declination $d' = 90° - l$. When this happens the Sun at noon will only just graze the horizon at s.

While $d' > 90° - l$, the Sun remains continually below the horizon. This period is called the **Perpetual Night.**

On Dec. 22 the Sun traces out the circle $R''Q''$ below the horizon.

When d' is again $= 90° - l$, the perpetual night will end.

Subsequently, the day will gradually lengthen until March 21, when it will again be 12 hours long.

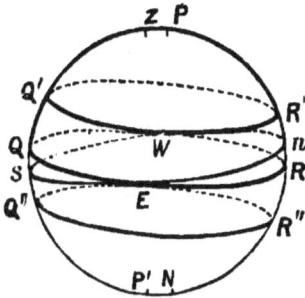

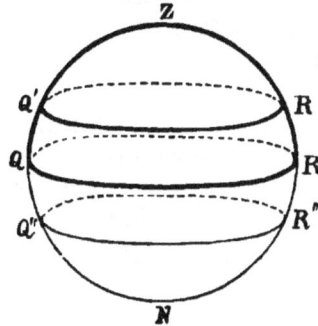

FIG. 50.   FIG. 51.

128. At the North Pole (Fig. 51) the phenomena are much simpler. The celestial equator coincides with the horizon. Hence, from March 21 to Sept. 23, the Sun will be above the horizon, and there will be perpetual day. The Sun's altitude will attain its greatest value i on June 21, when the Sun will trace out the circle $Q'R'$.

From Sept. 23 to March 21 there will be perpetual night. The Sun will be at its greatest depth below the horizon on Dec. 22, when it will trace out the circle $Q''R''$.

129. Phenomena in the Southern Hemisphere.— At a place south of the equator, the variations will partake of the same general character as those in the corresponding north latitude, but the seasons will be reversed. The south pole will be above the horizon, instead of the north pole, and the days will increase in length as the Sun passes to the south of the equator. In fact, if we consider two antipodal points or places at opposite ends of a diameter of the Earth, the day at one place will coincide with the night at the other.

Hence, at any place between the equator and antarctic circle, Dec. 22 will be the longest day, and June 21 the shortest.

Within the antarctic circle there will be perpetual day for a certain period before and after Dec. 22, and perpetual night for a certain period before and after June 21.

The variations in the Sun's north zenith distance at noon will be the same as the variations in the south zenith distance in the corresponding north latitude six months earlier.*

130. **The Seasons.**—Having thus described the variations in the Sun's daily path at different times and places, we shall now show how these variations account for the alternations of heat and cold on the Earth.

Astronomically, the four seasons are defined as the portions into which the year is divided by the equinoxes and the solstices. Thus, in northern latitudes,

Spring commences at the Vernal Equinox (March 21),
Summer ,, ,, Summer Solstice (June 21),
Autumn ,, ,, Autumnal Equinox (Sept. 23),
Winter ,, ,, Winter Solstice (Dec. 22).

It is obvious that the temperature at any place will depend in a great measure upon the length of the day. While the Sun is above the horizon, the Earth is receiving a considerable portion of the heat of his rays, the remaining portion being absorbed by the Earth's atmosphere through which the rays have to pass. When the Sun is below the horizon, the Earth's heat is radiating away into space, although the heated atmosphere retards this radiation to a considerable extent. Thus, on the whole, the Earth is most heated when the days are longest, and conversely.

The variations in the Sun's meridian altitude have a still greater influence on the temperature. When the Sun's rays strike the surface of the Earth nearly perpendicularly, the same pencil of rays will be spread over a smaller portion of the surface than when the rays strike the surface at a considerable angle; hence the quantity of heat received on a square foot of the surface will be greatest when the Sun is most nearly vertical. By this mode of reasoning it is shown in Wallace Stewart's *Text-Book of Light*, § 10, that the intensity of illumination of a surface is proportional to the cosine of the angle of incidence, and the same argument holds good with

* The student will find it instructive to trace out *fully* the variations in S. latitudes corresponding to those described in §§ 122-128. See diagram, p. 421.

regard to radiant heat as well as light. Hence the Sun's heating power when above the horizon is always proportional to the cosine of the Sun's zenith distance or the sine of its altitude. In this proof, however, the absorption of heat by the Earth's atmosphere has been neglected. But when the Sun's rays reach the Earth obliquely, they will have to pass through a greater extent of the Earth's atmosphere, and will, therefore, lose more heat than when they are nearly vertical. This cause will still further increase the effect of variations in the Sun's altitude in producing variations in the temperature.

131. **Between the Tropics** the combination of the two causes above described tends to produce high temperatures, subject only to small variations during the year. The Sun's meridian altitude is always very great, and the variations in the lengths of day and night are small. If the latitude be north, the Sun's heating power is greatest while the Sun transits north of the zenith. During this period the Sun's meridian altitude is least when the days are longest. Thus the effects of the two causes in producing variations in the Sun's heat counteract one another, to a certain extent, and give rise to a period of nearly uniform but intense heat.

In the North Temperate Zone, the Sun is highest at noon when the days are longest, and therefore both causes combine to make the spring and summer seasons warmer than autumn and winter. But the highest average temperatures occur some time *after* the summer solstice, and the lowest temperatures occur after the winter solstice ; for the Earth is gaining heat most rapidly about the summer solstice, and it continues to gain heat, but less rapidly, for some time afterwards. Similarly, the Earth is losing heat most rapidly at the winter solstice, and it continues to lose heat, but less rapidly, for some time afterwards. For this reason, summer is warmer than spring, and winter is colder than autumn. ˙

As we go northwards, the Sun's altitude at noon becomes generally lower throughout the year, and the climate therefore becomes colder. At the same time, the variations in the length of the day become more marked, causing a greater fluctuation of temperature between summer and winter.

Within the Arctic Circle there is a warm period during the perpetual day, but the Sun's altitude is never sufficiently great to cause very intense heat. During the perpetual night the cold is extreme; and the low altitude of the Sun, when above the horizon at intermediate times, gives rise to a very low average temperature during the year.

In the Southern Hemisphere the seasons are reversed; for, in south latitude l, when the Sun's south declination is d, the same amount of heat will be received from the Sun as in north latitude l, when his north declination is d. Hence, the seasons corresponding to our spring, summer, autumn and winter will begin respectively on September 23, December 22, March 21, and June 21, and will be separated from the corresponding seasons in north latitude by six months.

132. **Other Causes affecting the Seasons and Climate.**—It is found (as will be explained in the next section) that the Sun's distance from the Earth is not quite constant during the year. The Sun is nearest the Earth about December 31, and furthest away on July 1 (these are the dates of perigee and apogee respectively). As shown in Wallace Stewart's *Text-Book of Light*, § 9, the intensity of illumination, and therefore also of heating, due to the Sun's rays, varies inversely as the square of the Sun's distance. Hence the Earth receives, on the whole, more heat from the Sun after the winter solstice than after the summer solstice. This cause tends to make the winter milder and the summer cooler in the northern hemisphere, and to make the summer hotter, and the winter colder in the southern hemisphere.

The variations in the Sun's distance are, however, small, and their effect on the seasons is more than counteracted by purely terrestrial causes arising from the unequal distribution of land and water on the Earth. The sea has a much greater capacity for heat than the rocks forming the land; it is not so readily heated or cooled. In the southern hemisphere the sea greatly preponderates, the largest land-surfaces being in the northern hemisphere. Hence, the climate of the southern hemisphere is generally more equable, and the seasons are not so marked as in the northern hemisphere, quite in contradiction to what we should expect from the astronomical causes.

133. Times of Sunrise and Sunset.—The times of sunrise and sunset at Greenwich are given for every day of the year in *Whitaker's* and other almanacks. For any other latitude, the Sun's declination must be found from the almanack, the times of sunrise and sunset can then be found by means of tables of double entry constructed for the purpose (§ 29). These are called "Tables of Semidiurnal and Seminocturnal Arcs.". They give, for different latitudes and declinations, the interval between apparent noon and sunset, *i.e.*, the apparent time of sunset, or half the length of the day. Subtracting this from 12 hours, the apparent time of sunrise is found, and is half the length of the night.

If, as in § 129, we consider two antipodal places *A* and *B*, the planes of their horizons will be parallel, and the Sun will be above the horizon at *A* when he is below the horizon at *B*, and *vice versâ*. Hence, the apparent time of sunrise (measured from noon) in N. latitude *l* will be the apparent time of sunset (measured from midnight) in S. latitude *l* on the same date.

For this reason the tables are usually constructed only for N. latitudes. For S. latitudes they give the time of sunrise instead of sunset.

The times found in this manner will be the *local solar times*. To reduce to Greenwich solar time we must add or subtract 4m. for each degree of longitude, according as the place is W. or E. of Greenwich.

134. To find the length of the perpetual day and night at places within the Arctic or Antarctic Circles.

The perpetual day lasts while the Sun's declination at local midnight is greater than the colatitude (or complement of the latitude), during spring and summer. The perpetual night lasts while the Sun's S. decl. at local noon is greater than the colat. during autumn and winter. The Sun's decl. at Greenwich noon being given for every day of the year, in the Nautical Almanack, it is easy to find, to within a day, the durations of the perpetual day and night in any given latitude greater than $66° 32\frac{1}{2}'$.

135. To find the time the Sun takes to rise or set.—Let D'' be the Sun's angular diameter, measured in seconds. When the Sun begins to rise, his upper limb just touches the horizon, and his centre is at a depth $\frac{1}{2}D''$ below the horizon. When the Sun has just finished rising, his lower limb touches the horizon, and his centre is at an altitude $\frac{1}{2}D''$ above the horizon. During the sunrise, the centre rises through a vertical height D''. The problem is closely similar to that of § 104, where the effect of dip is considered. Hence if t seconds be the time taken in rising, d the declination of the Sun's centre, and x the inclination to the vertical of the Sun's path at rising ($Hx'x$ or nxP, Fig. 40) we have

$$t = \tfrac{1}{15} D'' \sec d \sec x,$$
$$= 4 \sec d \sec x \times (\odot\text{'s angular diameter in minutes}).$$

As in § 104, this gives, for a place on the equator,

$$t = \tfrac{1}{15} D'' \sec d,$$

and at an equinox in latitude l,

$$t = \tfrac{1}{15} D'' \sec l.$$

EXAMPLE.—At an equinox in latitude 60°, the ⊙'s angular diameter being 32′,
the time taken to rise will be $= 4 \times 32 \times \sec 60°$ seconds
$$= 256\text{s.} = 4\text{m. }16\text{s.}$$

136. Note.—It may be mentioned that, owing to atmospheric refraction, the Sun really appears to rise earlier and set later than the times calculated by theory. As the phenomena of refraction will be discussed more fully in Chapter VI., it will be sufficient to mention here that the rays of light from the Sun are bent to such an extent by the Earth's atmosphere that the *whole* of the Sun's disc is visible when it would just be entirely below the horizon if there were no atmosphere.

Moreover, there is daylight, or rather twilight, for some time after the Sun has vanished, so that what is commonly called night does not begin for some time after sunset.

For the same reasons, the perpetual day at a place in the arctic circle is lengthened, and the perpetual night shortened, by several days.

The time taken in rising and setting is, however, practically unaffected.

SECTION II.—*The Ecliptic.*

137. The First Point of Aries.—In determining the right ascensions of stars, the first step must necessarily be to find accurately the position of the first point of Aries, since this point is taken as the origin from which R.A. is measured. In other words, we must first find the R.A. of *one* star. When this is known we can use that star as a "clock star," to determine the sidereal time and clock error; and, these being known, we can *then* find the R.A. of any other star, as explained in Chapter II. But until the position of Υ has been found, the methods of Chapter II. will only enable us to find the *difference* of R.A. of two stars by observing the difference of their times of transit, as indicated by the astronomical clock, and will determine neither the sidereal time nor the clock error, nor the R.A.'s of the stars.

138. First Method.—The position of Υ may be found thus:—At the vernal equinox the Sun's declination changes from south to north, or from negative to positive. Let the Sun's declination be observed by the Transit Circle at the preceding and following noons, and let the observed values be $-d_1$ and $+d_2$ (*i.e.*, d_1 S., and d_2 N.). Let t_1, t_2 be the corresponding times of transit of the Sun's centre, as observed by the astronomical clock, and let T, the time of transit of any star, be also observed. Then,

$T-t_1 =$ difference of R.A. of star and Sun at first noon,

$T-t_2 = \quad$,, $\qquad$,, $\qquad$,, $\qquad$ at second noon.

Let $T-t_1 = a_1$ and $T-t_2 = a_2$. We have

Increase in Sun's decl. in the day $= d_2-(-d_1) = d_2+d_1$,

,, $\quad$,, $\quad$ R.A. $\quad$,, $\quad = t_2-t_1 = a_1-a_2$,

and both coordinates increase at an approximately uniform rate during the day.

Therefore the $\odot$'s decl. will have increased from $-d_1$ to 0 in a time $d_1/(d_1+d_2)$ of a day, and the corresponding increase in R.A. will be

$$(a_1-a_2) \times d_1/(d_1 + d_2).$$

The Sun is now at Υ, $\therefore \odot$'s R.A. is now $= 0$. Hence,

The star's R.A. $= a_1 - \dfrac{(a_1-a_2)d_1}{d_1+d_2} = \dfrac{a_1 d_2 + a_2 d_1}{d_1+d_2}$.

*139. Flamsteed's Method for finding the First Point of Aries.—The principle of the method now to be described is as follows :—Let S_1, S be two positions of the Sun shortly after the vernal and before the autumnal equinox respectively, and such that the declinations S_1M_1 and SM are equal. Then the right-angled triangles ΥM_1S_1 and $\triangle MS$ will be equal in all respects, and we shall therefore have $\Upsilon M_1 = \triangle M$.

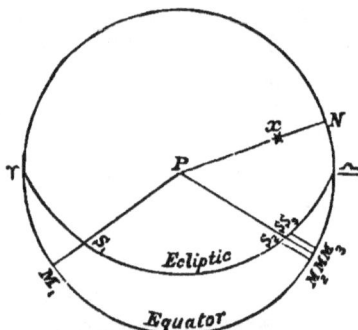

FIG. 52.

At noon, some day shortly after March 21, the Sun is observed with the Transit Circle, say when at S_1. We thus determine its meridian zenith distance z_1, and also the difference between the times of transit of the Sun and some fixed star x, whose R.A. is required. This difference, which is the difference of R.A. of the Sun and star, we shall call a_1. If d_1 be the Sun's declination, and l the observer's latitude, we shall have

$$S_1M_1 = d_1 = l - z_1, \quad M_1N = a_1.$$

We now have to determine MN, the difference of R.A. of the Sun and star shortly before September 23, when the Sun's declination SM is again equal to d_1. But the Sun can only be observed with the Transit Circle at noon, and it is highly improbable that the Sun's declination will again be exactly equal to d_1 at noon on any day. We shall, however, find two consecutive days in September on which the declinations at noon, S_2M_2 and S_3M_3, are respectively greater and less than d_1.

Let z_2 and z_3 be the observed meridian zenith distances at S_2 and S_3; d_2 and d_3 the corresponding declinations $S_2 M_2$, $S_3 M_3$; a_2 and a_3 the observed arcs $M_2 N$ and $M_3 N$, being the differences of R.A. of the Sun and star on the two days.

During the day which elapses between the observations at S_2, S_3, we may assume that the Sun's decl. and R.A. both vary at a uniform rate, so that the change in the decl. is always proportional to the corresponding change in R.A.*

Therefore, $\dfrac{M_2 M}{M_2 M_3} = \dfrac{S_2 M_2 - S M}{S_2 M_2 - S_3 M_3} = \dfrac{d_2 - d_1}{d_2 - d_3}$

$\therefore\ M_2 M = \dfrac{d_2 - d_1}{d_2 - d_3}\, M_2 M_3 = \dfrac{d_2 - d_1}{d_2 - d_3}\, (a_2 - a_3),$

and $\quad MN = M_2 N - M_2 M = a_2 - \dfrac{d_2 - d_1}{d_2 - d_3}(a_2 - a_3).$

Now we have shown that

$$\Upsilon M_1 = M \triangleq :$$

i.e. $\quad \Upsilon N - M_1 N = MN - \triangleq N;$

$\therefore\quad MN + M_1 N = \Upsilon N + \triangleq N = 2\Upsilon N - 180°$

$$= 2\Upsilon N - 12 \text{ hours};$$

$\therefore\quad\quad \Upsilon N = 6\text{h.} + \tfrac{1}{2}\,(M_1 N + MN)$

$$= 6\text{h.} + \tfrac{1}{2}\left\{ a_1 + a_2 - \frac{d_2 - d_1}{d_2 - d_3}(a_2 - a_3) \right\}.$$

This determines ΥN, the star's R.A., in terms of a_1, a_2, a_3, the observed differences between the times of transit of the Sun and star, and d_1, d_2, d_3, the Sun's declinations at the three observations. But we need not even find the declinations, for

$$d_1 = l - z_1, \quad d_2 = l - z_2, \quad d_3 = l - z_3;$$

therefore, substituting, we have

The star's R.A., $\quad \Upsilon N = 6\text{h.} + \tfrac{1}{2}\left\{ a_1 + a_2 - \frac{z_1 - z_2}{z_3 - z_2}(a_2 - a_3) \right\}.$

In applying either of the above methods to the numerical calculation of the right ascension of any star, it is advisable to follow the various steps as we have described them, instead of merely substituting the numerical values of the data in the final formulæ.

* In other words, we assume, as in Trigonometry, that the "principle of proportional parts" holds for the small variations in decl. and R.A. during the day.

*140. The Advantages of Flamsteed's Method.—Among these the following may be mentioned.

1st. The method does not require a knowledge of the latitude, for we do not require to find the Sun's declination. Hence, errors arising from inaccurate determination of the latitude are avoided.

2nd. One great source of error in determining Z.D.'s is the refraction of the Earth's atmosphere. Since the Sun is observed each time in the same part of the sky, z_1, z_2, z_3 will be nearly equally affected by refraction. Hence, the "principle of proportional parts" will hold, so that the small differences in the true Z.D.'s are proportional to the differences in the observed Z.D.'s. Hence we may use the observed Z.D.'s *uncorrected* for refraction.

EXAMPLE.

To find the Right Ascension of *Sirius* and the clock errors in March and Sept., 1891, from the following data, the rate of the clock being supposed correct. (Decl. of *Sirius* = 16° 34′ 2″ S.)

	Mar. 25, 1891.	Sept. 18.	Sept. 19
Decl. of Sun at noon...	1° 48′ 56″	1° 53′ 0″	1° 29′ 43″
Time of transit of Sun	0h. 15m. 36s.	11h. 42m. 42s.	11h. 46m. 17s.
Time of transit of *Sirius*	6h. 39m. 10s.	6h. 40m. 25s.	6h. 40m. 25s.

On Mar. 25, (R. A. of *Sirius*) − (Sun's R.A.) = 6h 39m. 10s. − 0h. 15m. 36s.
$$= 6h. 23m. 34s.$$

Hence, in angular measure, the difference of R.A. is about 96°. Draw the diagram as in Fig. 52, but make the angle S_1PN = 96°; N will therefore lie *between* M_1 and M_2, instead of where represented. Also, since *Sirius* is south of the equator, it should be represented at a point x on PN produced through N. In this figure we shall have

$S_1M_1 = 1° 48′ 56″$; $M_1N = 6h.39m.10s. − 0h.15m.36s. = 6h.23m.34s.$
$S_2M_2 = 1° 53′ 0″$; $NM_2 = 11h.42m.42s. − 6h.40m.25s. = 5h. 2m.17s.$
$S_3M_3 = 1° 29′ 43″$; $NM_3 = 11h.46m.17s. − 6h.40m.25s. = 5h. 5m.52s.$

Also, SM is by construction equal to S_1M_1.

Hence, applying the principle of proportional parts, we have

$$\frac{M_2M}{M_2M_3} = \frac{S_2M_2 - S_1M_1}{S_2M_2 - S_3M_3} = \frac{4′ 4″}{23′ 17″} = \frac{244}{1397},$$

and $M_2M_3 = 3m. 35s. = 215s.$;

∴ $M_2M = 215 \times 244/1397 = 37·5$ seconds;

∴ $NM = 5h. 2m. 17s. + 37s. = 5h. 2m. 54s.$

Now, $NM_1 − NM = N\Upsilon − N\text{≏} = 2N\Upsilon − 12h.$

hence, $\Upsilon N = 6h. + \frac{1}{2}(NM_1 − NM) = 6h. + \frac{1}{2}(6h.23m.34s. − 5h.2m.54s.)$
$= 6h. + \frac{1}{2}(1h. 20m. 40s.) = 6h. 40m. 20s.$

Thus the right ascension of *Sirius* = **6h. 40m. 20s.**

Also, clock error in March = 6h.40m.20s. − 6h.39m.10s. = **+ 1m. 10s.**

,, ,, ,, ,, Sept. = 6h.40m.20s. − 6h.40m.25s. = − **5s.**

141. Precession of the Equinoxes.—Thus far we have treated the first point of Aries as being fixed, and this will evidently be the case if the equator and ecliptic are fixed in direction. But if the right ascensions of various stars are observed over an interval of several years, it will be found that the position of the first point of Aries is slowly changing, and that it moves along the ecliptic in the retrograde direction at the rate of about 50·2″ in a year. This motion is called Precession of the Equinoxes, or, briefly, **Precession.**

Precession is found to be due almost entirely to gradual changes in the direction of the plane of the equator, the ecliptic remaining almost fixed among the stars. Its effect is to produce a yearly increase of 50·2″ in the celestial longitudes of all stars, their latitudes being constant.

In a large number of years the effect of precession will be considerable. Thus, ♈ will perform a complete revolution in the period

$$\frac{360 \times 60 \times 60}{50 \cdot 2} \text{ years, } i.e., \text{ about } 25,800 \text{ years.}$$

At the present time the vernal equinoctial point has moved right out of the constellation Aries into the adjoining constellation Pisces. It still, however, retains the old name of "First Point of Aries." Similarly, the autumnal equinoctial point is in the constellation Virgo, but it is still called the "First Point of Libra."

The rate of precession can be found very accurately by observations of the first point of Aries separated by a considerable number of years. The larger the interval, the larger is the change to be observed, and the less is the result affected by instrumental errors.

*142. Correction for Precession in using Flamsteed's Method.—
During the interval that elapses between the two observations in Flamsteed's method, the right ascension of the observed star will have increased slightly, owing to precession, and the R.A. given by the formula will be the arithmetic mean of the R.A.'s at the times of the two observations.† As the change in R.A. is very approximately uniform, this mean will be the star's R.A. at a time exactly half way between the two observations, i.e., at the summer solstice.

---

† This may be most readily seen by imagining the equator and ecliptic to be at rest, and the change in R.A. to be due to motion of the star.

143. Determination of Obliquity of Ecliptic.—The method now used for finding the obliquity of the ecliptic is similar in principle to that of § 38, but the Sun's meridian zenith distance is observed by means of the transit circle instead of the gnomon.

The obliquity is equal to the Sun's greatest declination at one of the solstices. Since observations with the Transit Circle can only be performed at noon, while the maximum declination will probably occur at some intermediate hour of the day, it will be necessary, in exact determinations, to make observations of the Sun's decl. for several days before and after the solstice. From these it is possible to determine the maximum decl.; the method is, however, too complicated to be described here. For rough purposes the Sun's greatest noon decl. may be taken as the measure of the obliquity.

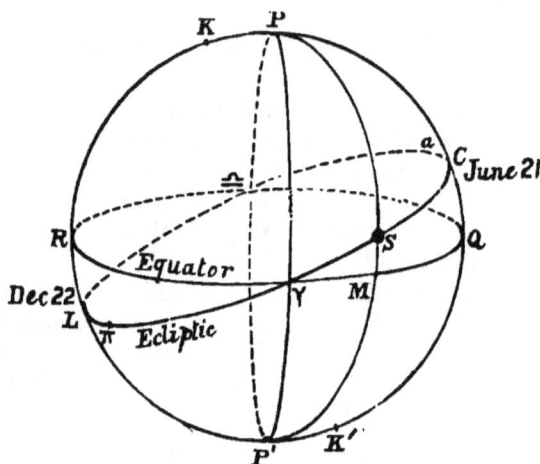

FIG. 53.

144. When the position of ♈ has been determined, the obliquity can also be found by a single observation of the Sun's R.A. and decl. For we thus find the two sides ♈M, MS of the spherical triangle ♈MS, and these data are sufficient to determine both the obliquity M♈S, and the Sun's longitude ♈S.

SECTION III.—*The Earth's Orbit about the Sun.*

145. **Observations of the Sun's Relative Orbit.**—By daily observations with the Transit Circle, the decl. and R.A. of the Sun's centre at noon are found for every day of the year. From these data the Sun's long. is calculated, as in § 144, by solving the spherical triangle ΥSM (Fig. 53). If the obliquity of the ecliptic is also known, we have three data, any two of which suffice to determine the long., ΥS. Thus the accuracy of the observations can be tested, and the Sun's motion at various times of the year can be accurately determined.

Although the determination of the Sun's actual distance from the Earth in miles is an operation of great difficulty, it is easy to *compare* the Sun's distance from the Earth at different times of the year, for this distance is always inversely proportional to the Sun's angular diameter. This property is proved in § 4, but numerous simple illustrations may also be used to show that the angular diameter of any object varies inversely with its distance (see § 4).

The Sun's angular diameter may be readily observed by means of the Heliometer; or, if preferred, any other form of micrometer may be used. The Sun's distances at two different observations will be in the reciprocal ratio of the corresponding angular diameters. Thus, by daily observation, the changes in the Sun's distance during the year may be investigated.

If the circular measure of the Sun's angular diameter is $2r$, then πr^2 is called the Sun's **apparent area**. In fact, this is the area of a disc which would look the same size as the Sun if placed at unit distance from the eye.

EXAMPLE.

The Sun's angular diameter is 31′ 32″ at midsummer, and 32′ 36″ at midwinter. To find the ratio of its distances from the Earth at these times.

The distances being inversely proportional to the angular diameters, we have

$$\frac{\text{Dist. at midsummer}}{\text{Dist. at midwinter}} = \frac{32' \ 36''}{31' \ 32''} = \frac{1956}{1892} = \frac{489}{473} = 1\tfrac{1}{30} \text{ nearly.}$$

Hence the Sun is further at midsummer than at midwinter, in the proportion of very nearly **31 to 30**.

146. Kepler's First and Second Laws.—We may now construct a diagram of the Sun's relative orbit. Let E represent the position of the Earth, $E\Upsilon$ the direction of the first point of Aries. Then, by making the angle ΥES equal to the Sun's longitude at noon, and ES proportional to the Sun's distance, we obtain a series of points S, S'..., S_1..., representing the Sun's position in the plane of the ecliptic, as seen from the Earth at noon on different days of the year. Draw the curve passing through the points S, S'..., S_1,...; this curve will represent the Sun's orbit relative to the Earth, and it will be found that

    I. **The Sun's annual path is an ellipse, of which the Earth is one focus.**

    II. **The rate of motion is everywhere such that the radius vector (*i.e.*, the line joining the Earth to the Sun) sweeps out equal areas in equal intervals of time.**

These laws were discovered by Kepler for the motion of Mars about the Sun, and he subsequently generalized them by showing that the orbits of all the other planets, including the Earth, obeyed the same laws. In their general form they are known as Kepler's First and Second Laws. [See p. 253.]

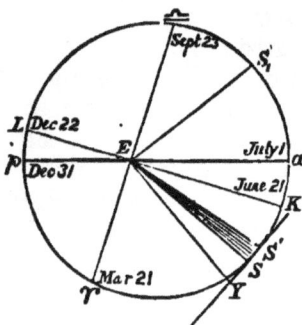

Fig. 54.

147. Perigee and Apogee.—When the Sun's distance from the Earth is least, the Sun is said to be in **perigee.** When the distance is greatest, the Sun is said to be in **apogee**

The positions of perigee and apogee are called the two **Apses** of the orbit; they are indicated at p, a in Fig. 54. The line pEa joining them is the major axis of the ellipse (Ellipse, 4), and is sometimes also called the **apse line.**

148. Verification of Kepler's First Law.—The Sun's angular diameter is observed to be greatest on Dec. 31, and least on July 1; we therefore conclude that these are the days on which the Sun passes through perigee and apogee respectively. The positions of perigee and apogee being thus found, the angle ΥEp is known, which is the long. of perigee.

From the winter solstice to perigee is about 10 days. Hence, during this interval the Sun will have moved through an angle of about 10°;

$\therefore$ longitude of perigee $= 270° + 10° = 280°$ roughly.

To verify that the orbit is an ellipse, it is now only necessary to show that the relation connecting ES and the angle pES is the same as that which holds in the case of the ellipse. If the orbit is an ellipse of eccentricity e, we must have

$ES \times (1 + e \cos pES) = l$ (a constant).   (Ellipse, 3.)

Therefore the Sun's angular diameter must be always proportional to $1 + e \cos pES$.

As the result of numerous observations, it is found that this is actually the case, and the truth of Kepler's First Law for the Sun's orbit relative to the Earth is confirmed.

149. To find e, the eccentricity of the ellipse, the best plan is to compare the greatest and least angular diameters of the Sun, *i.e.*, the diameters at perigee and apogee. Since at these positions pES becomes 0° and 180° respectively, we have, from above,

ang. diam. at p : ang. diam. at $a = 1/Ep : 1/Ea$
$= 1 + e \cos 0° : 1 + e \cos 180° = 1 + e : 1 - e$.

from which proportion e can be found.

Taking the angular diameters at perigee and apogee to be 32′ 36″ and 31′ 32″ (as in the Ex. of § 145), the Sun's distances at those times are in the ratio of 1956″ : 1892″, or 489 : 473;

$\therefore \dfrac{1+e}{1-e} = \dfrac{489}{473}$ $\therefore$ $e = \dfrac{489-473}{489+473} = \dfrac{16}{962} = \dfrac{8}{481}$.

Hence e is very nearly equal to **1/60.**

The Nautical Almanack contains a table giving the Sun's angular diameter daily throughout the year. The average angular diameter may be taken as 32′ approximately.

Owing to the smallness of e, the orbit is very nearly circular, being, really, much more nearly so than is shown in Fig. 54

150. Verification of Kepler's Second Law.—It is found, as the result of observation, that the Sun's increase in longitude in a day, at different times of year, is always proportional to the square of the angular diameter, and is, therefore, inversely proportional to the square of the Sun's distance. From this it may be deduced (as follows) that the area described by the radius vector in one day is always constant.

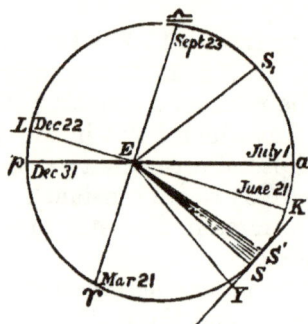

FIG. 55.

Let SS' represent the small arc described by the Sun in a day in any part of the orbit. Then the sector ESS' is the area swept out by the radius vector. This sector does not differ perceptibly from the triangle ESS'; therefore, by trigonometry,

$$\text{area } ESS' = \tfrac{1}{2}ES \cdot ES' \cdot \sin SES'.$$

Since the change in the Sun's distance in one day is imperceptible, we may write ES for ES' in the above formula without materially affecting the result; also, since the angle SES' is small, the sine of SES' is equal to the circular measure of the angle SES'.

Therefore, area $ESS' = \tfrac{1}{2}ES^2 \times \angle SES'$.

But, by hypothesis, the change of longitude SES' varies inversely as ES^2, so that $ES^2 \times \angle SES'$ is constant;

∴ area ESS' is constant,

that is, the area described by the radius vector in a day is constant. Thus, the area described in any number of days is proportional to the number of days, and generally the areas described in equal intervals of time are equal.

151. Deductions from Kepler's Second Law.

(i.) If the circular measure of the Sun's angular diameter is $2r$, then πr^2 is the Sun's **apparent area** (§ 145). Hence *the Sun's daily rate of change of longitude is proportional to the apparent area of its disc.*

(ii.) If Υ, K, $\underline{\frown}$, L represent the Sun's positions at the equinoxes and solstices, we have

$$\angle\, \Upsilon EK = \angle\, KE\underline{\frown} = \angle\, \underline{\frown} EL = \angle\, LE\Upsilon = 90°,$$

and it is readily seen from the figure that

$$\text{area } LE\Upsilon < \text{ area } \underline{\frown}EL < \text{ area } \Upsilon EK < \text{ area } KE\underline{\frown},$$

and the lengths of the seasons, being proportional to these areas, are unequal, their ascending order of magnitude being

Winter,   Autumn,   Spring,   Summer.

Their lengths are, at the present time (1891), about

89d. 0$\frac{1}{2}$h.,   89d. 18$\frac{1}{4}$h.,   92d. 20$\frac{3}{4}$h.,   93d. 14$\frac{1}{2}$h.

(iii.) Since the intensity of the Sun's heat (§ 131) and its rate of motion in longitude both vary as the inverse square of its distance, they are proportional to one another. Hence *the Earth, as a whole, receives equal amounts of heat while the Sun describes equal angles.* In particular, *the total quantities of heat received in the four seasons are equal.*

(iv.) The Sun's longitude changes most rapidly on December 31, and least rapidly on July 1.

(v.) Since the apse line, or major axis, pSa, bisects the ellipse, *the time from perigee to apogee is equal to the time from apogee to perigee.*

*152. **To find the Position of the Apse Line.—** The Sun's distance remains very nearly constant for a short time before and after perigee and apogee, hence it is difficult to tell the exact instant when this distance is greatest or least. For this reason, the following method is generally used:—

The Sun's long. is observed at two points, S, S_1, before and after the apse, when its angular diameters, or its rates of motion in long., are found to be equal. Then $ES = ES_1$, and the symmetry of the ellipse shows that $\angle pES = \angle pES_1$ and $\angle aES = \angle aES_1$. Hence the long. of the apse is the arithmetic mean of the Sun's longitudes at the two observations.

153. **Progressive Motion of Apse Line.—**From such observations, extending over a long period of years, it is found that the apse line is not fixed, but has a forward or direct motion in the ecliptic plane of $11 \cdot 25''$ in a year.

154. The Sun's apparent annual motion may be accounted for by supposing the Earth to revolve round the Sun.

The annexed diagram will show how the Sun's annual motion in the ecliptic, as well as the changes in the seasons, may be accounted for on the theory that the Sun remains at rest while the Earth describes an ellipse round it in the course of the year in a plane inclined at an angle 23° 27½' to the plane of the Earth's equator.

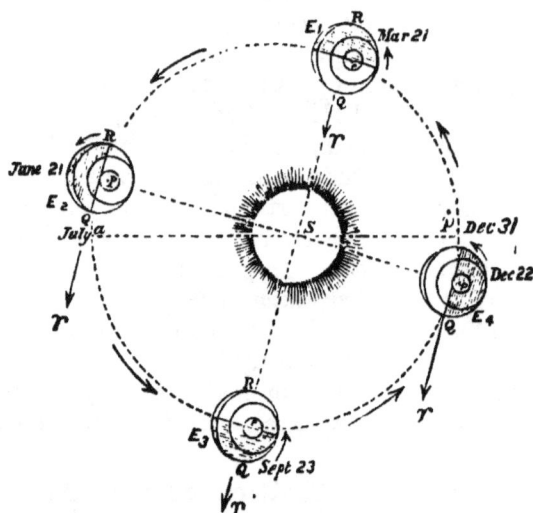

FIG. 56.

The distance of the nearest of the fixed stars is known to be over 200,000 times as great as the Earth's distance from the Sun. Hence, § 5 shows that the directions of the fixed stars will not change to any considerable extent, as the Earth's position varies. We shall, therefore, in the present description, consider the directions of the stars to be fixed. The directions of the various points and circles of the celestial sphere, such as the first point of Aries, will also be fixed.

On March 21, the Earth is at E_1, and the Sun's direction E_1S determines the direction of Υ, the First Point of Aries.

The Sun is vertical at a point Q on the equator, and as the Earth revolves about its axis through P, all points on the equator will come vertically under the Sun. There is night all over the shaded portion of the Earth, day over the rest. The great circle bounding the illuminated part passes through the pole P, and, therefore, bisects the small circle traced out by the daily rotation of any point on the Earth; thus, the day and night are everywhere equal. At the pole P the Sun is just on the horizon.

On June 21, the Earth is at E_2, and the Sun's longitude $\Upsilon E_2 S = 90°$. The Sun is vertical at a point on the tropic of Cancer. Since the arctic circle is entirely in the illuminated part there is perpetual day over the whole arctic zone.

On September 23, the Earth is at E_3, and the Sun's longitude $\Upsilon E_3 S$ is $180°$. The Sun is once more vertical at a point R on the equator, and the day and night are everywhere 12 hours long, as they are at E_1.

On December 22, the Earth is at E_4, and the Sun's longitude $\Upsilon E_4 S$ (measured in the direction of the arrow) is $270°$. The Sun is now at its greatest angular distance south of the equator, and overhead at a point on the tropic of Capricorn; this tropic is not represented, being on the under side of the sphere. Since the arctic circle is entirely within the shaded part there is perpetual night over the whole arctic zone.

155. New Definitions and Facts.—According to the theory of the Earth's orbital motion, Kepler's First and Second Laws must be re-stated thus for the Earth.

I. **The Earth describes an ellipse, having the Sun in one focus.**

II. **The radius vector joining the Earth and Sun traces out equal areas in equal times about the Sun.**

The **ecliptic** is now defined as the great circle of the celestial sphere, whose plane is parallel to that of the Earth's orbit.

The Earth is nearest the Sun on December 31, and is then said to be in **perihelion**. The Earth is furthest from the Sun on July 1, and is then said to be in **aphelion**. Thus, when the Sun is in perigee the Earth is in perihelion, when the Sun is in apogee the Earth is in aphelion. The positions of perihelion and aphelion are indicated by the letters p, a in Fig. 56. The line joining them is the **apse line**.

156. Geocentric and Heliocentric Latitude and Longitude.—Hitherto we have been dealing only with the directions of the celestial bodies as seen from the Earth.

In dealing with the motion of the planets, it is more convenient, as a rule, to define their positions by the directions in which they would be seen by an observer situated at the centre of the Sun.

In every case, the direction of a celestial body may be specified by the two coordinates, celestial latitude and longitude, which measure respectively the arc of a secondary from the body to the ecliptic and the arc of the ecliptic between this secondary and the first point of Aries (§ 17).

These coordinates are called the **Geocentric Latitude** and **Longitude** when employed to define the body's **geocentric** position, or position relative to the centre of the Earth. The names **Heliocentric Latitude** and **Longitude** are given to the corresponding coordinates when employed to define the body's **heliocentric** position, or position relative to the Sun's centre.

When the distance of a fixed star is immeasurably great compared with the radius of the Earth's orbit, its geocentric and heliocentric directions coincide, and there is no difference between the two sets of coordinates. There is a slight difference between the geocentric and heliocentric positions of a few of the nearest fixed stars. But, in the case of the planets, and of comets, the heliocentric latitude and longitude differ entirely from the geocentric, and laborious calculations are required to transform from one system of coordinates to the other.

One fact may, however, be noted. The direction of the Earth as seen from the Sun is always opposite to the direction of the Sun as seen from the Earth. Hence,

The Earth's heliocentric longitude differs from the Sun's geocentric longitude by 180°.

This may be illustrated by referring to Fig. 56. We see that $\Upsilon SE_3 = 0°$, $\Upsilon SE_4 = 90°$, $\Upsilon SE_1 = 180°$, $\Upsilon SE_2 = 270°$; thus, the Earth's longitude is 0° on September 23, 90° on December 22, 180° on March 21, and 270° on June 21.

EXAMPLES.—IV.

1. Describe the phenomena of day and night at a pole of the Earth.

2. Show how to find how long the midwinter Moon when full is above the horizon at a place within the arctic circle of given latitude.

3. Show that the ecliptic can never be perpendicular to the horizon except at places between the tropics.

4. Show that for a place *on* the arctic circle the Sun always rises at 18h. sidereal time from December 21 to June 20, and sets at the same sidereal time from June 20 to December 21.

5. Find the angle between the ecliptic and the equator in order that there should be no temperate zone, the torrid zone and the frigid zone being contiguous.

6. Show how, by observations on the Sun, taken at an interval of nearly six months, the astronomical clock may be set to indicate 0h. 0m. 0s. when ♈ is on the meridian.

7. On March 24, 1878, at noon, the Sun's declination was 1° 29′ 5·1″, and the difference of right ascension of the Sun and a star 6h. 1m. 34·45s. On September 18, 1878, at noon, the Sun's declination was 1° 49′ 30·2″, and it was distant from the star 5h. 27m. 32·97s. in right ascension. On September 19, 1878, at noon, the Sun's declination was 1° 26′ 12·8″, and it was distant from the star 5h. 31m. 8·3s. in right ascension. Find the right ascension of the star and that of the Sun at the first observation.

8. Describe the appearance presented to an observer in the Sun of the parallels of latitude and the meridians of the Earth, any day (i.) between the vernal equinox and the summer solstice, (ii.) between the autumnal equinox and the winter solstice.

9. If a sunspot be situated near the edge of the Sun's disc, describe how its position, relative to the horizon, will change between sunrise and sunset.

10. Describe how the Sun's apparent velocity in the ecliptic varies throughout the year; and give the dates of apogee and perigee. Compare the daily motion in longitude at these dates, having given that the eccentricity of the Earth's orbit is $\frac{1}{60}$.

EXAMINATION PAPER.—IV.

1. What is the astronomical reason for the Earth being divided into torrid, temperate, and frigid zones?

2. Assuming your latitude to be 52°, show by a figure the daily path of the Sun as seen by you on June 21, December 22, and March 21 respectively.

3. Explain the causes of variation in the length of the day on the Earth. Give the dates at which each season begins, and calculate their lengths in days.

4. Discuss the variations in the length of the day at points within the arctic circle; and show how to find, by the Nautical Almanack, the length of the perpetual day.

5. Prove that, in the course of the year, the Sun is as long above the horizon at any place as below it.

6. Explain how it is that winter is colder than summer, although the Sun is nearer.

7. Investigate Flamsteed's method of determining the first point of Aries.

8. From the following observations calculate the Sun's R.A. on March 30, 1872:—

	Sun's declination.	Sun crossed meridian.	α *Serpentis* crossed meridian.
March 30, 1872...	4° 0′ 8·1″	0h. 1m. 4·47s.	15h. 1m. 54·76s.
Sept. 11, 1872 ...	4° 20′ 58·8″	0h. 1m. 4·09s.	4h. 19m. 11·38s.
Sept. 12, 1872 ...	3° 58′ 3·0″	0h. 1m. 4·07s.	4h. 15m. 49·33s.

9. State Kepler's First Law for the orbit of the Earth relative to the Sun, and explain how the eccentricity of the orbit can be found by observations of the Sun's angular diameter.

10. State Kepler's Second Law, and find the relation between the Sun's angular velocity and its apparent area.

CHAPTER V.

ON TIME.

Section I.—*The Mean Sun and Equation of Time.*

157. Disadvantages of Sidereal and Apparent Solar Time.—In Chapter I., Sections II., III., we explained two different ways of reckoning time. One of these, called Sidereal Time, was defined by the diurnal motion of the first point of Aries; the other, called Apparent Solar Time, was defined by the Sun's diurnal motion. We shall now show that neither of these measures of time is suitable for every-day use.

If we were to adopt sidereal time, the time of apparent noon on any day of the year would be measured by the Sun's R.A. on that day, and therefore would get later and later by 24h. during the course of the year.

Thus (*e.g.*), the time of noon would be 0h. on March 21, 6h. on June 21, 12h. on September 23, and 18h. on December 22, and the phenomena of day and night would bear no constant relation to the time.

Apparent solar time is free from these disadvantages, but it cannot be measured by a clock whose rate is uniform, because the length of the solar day is not quite invariable. In § 36 we showed that the difference between a solar and a sidereal day is equal to the Sun's daily increase in R.A., and in § 31 we showed that this increase takes place at a rate which is not quite the same at different times of the year. Hence, the difference between a solar and a sidereal day is not quite constant. But the length of a sidereal day is constant (§ 22). Hence the solar day is not quite constant, and a clock cannot be regulated so as to always mark exactly 0h. 0m. 0s. when the Sun crosses the meridian.

158. The Mean Sun.—Definitions.—To obviate these disadvantages, another kind of time, called **Mean Time,** has been introduced, and this is the time indicated by clocks, and used for all ordinary purposes. Mean Time is defined by means of what is called the **Mean Sun.** This is not really a Sun at all, but simply a point, which is imagined to move round the equator on the celestial sphere.* The hour angle of this moving point measures mean time, just as the hour angle of ♈ measures sidereal time; and the mean Sun has to satisfy the following requirements :—

1st. It must never be very far from the Sun.

2nd. Its R.A. must increase uniformly during the year.

Now the inequalities in the motion in R.A., which render the true Sun unsuitable as a timekeeper, are due to two causes.

1st. The Sun does not move uniformly in the ecliptic, its longitude increasing less rapidly in summer than in winter (§ 151).

2nd. Since the Sun moves in the ecliptic, and not in the equator, its celestial longitude is in general different from its R.A. (§ 31). Hence, even if the Sun were to revolve uniformly, its R.A. would not increase uniformly.

In defining the mean Sun, or moving point which measures mean time, these two causes of irregularity are obviated separately as follows :—

The **Dynamical Mean Sun** is defined to be a point which coincides with the true Sun at perigee, and which moves round the *ecliptic* in the same period (a year) as the true Sun, but at a uniform rate.

Thus, in the dynamical mean Sun, irregularities due to the Sun's unequal motion in longitude are removed, but those due to the obliquity of the ecliptic still remain.

The **Astronomical Mean Sun** is defined to be a point which moves round the *equator* in such a way that its R.A. is always equal to the longitude of the dynamical mean Sun.

---

* The conception of the mean Sun as a *moving point* is important. It would be physically impossible for a *body* to move in this manner.

Since the longitude of the dynamical mean Sun increases uniformly, the R.A. of the astronomical mean Sun increases uniformly. Hence the motion of the latter point *does* give us a uniform measure of time.

The astronomical mean Sun is, therefore, the moving point chosen in defining mean time. It is usually called simply the **Mean Sun.**

159. **Mean Noon and Mean Solar Time.—Equation of Time.**

Mean Noon is defined as the time of transit of the mean Sun.

A **Mean Solar Day** is the interval between two successive mean noons. Like the apparent and sidereal days, it is divided into 24 mean solar hours. During this interval, the hour angle of the mean Sun increases from 0° to 360°. Hence the **mean solar time** at any instant is measured by the mean Sun's hour angle, converted into time at the rate of 1h. per.15°, or 4m. per 1°.

The Sun itself is frequently spoken of as the **True Sun**, or **Apparent Sun**, to distinguish it from the mean Sun. As explained in § 36 the hour angle of the true Sun measures the apparent solar time, and its time of transit is called apparent noon.

The **Equation of Time*** is the name given to the amount which must be added to the apparent time to obtain the mean time.

Thus, the time indicated by a sun-dial (§ 167) is determined by the position of the shadow thrown by the true Sun, and is the apparent solar time; while a clock, which should go at a uniform rate, is regulated to keep mean time. The equation of time will then be defined by the relation,

(Time by clock)=(Time by dial)+(Equation of time).

At apparent noon the sun-dial will indicate 12h., or, as it is more conveniently reckoned, 0h. Hence,

Equation of time = Mean time of apparent noon.

---

* Thus, "equation of time" is not an *equation* at all in the generally accepted sense of the word, but an interval of time (positive or negative).

The equation of time is positive if the Sun is "after the clock," or the true Sun transits after the mean Sun. If the Sun is "before the clock," or the true Sun transits first, the equation of time is negative. The value of the equation of time for every day in the year is given in most almanacks, under the heading "Sun before clock," or "after clock."

160. The equation of time is divided into two parts. The first, which is called the **equation of time due to the eccentricity,** or to the **unequal motion,** is measured by the difference between the hour angles of the true and dynamical mean Suns. The second, or the **equation due to the obliquity,** is measured by the difference of hour angle between the dynamical and astronomical mean Suns.

161. **Equation of Time due to Unequal Motion.—** We shall now trace the variations during the year of that portion of the equation of time which is due to the Sun's unequal motion in the ecliptic. We shall denote this portion by E_1.

Let the true Sun be denoted by S, and the dynamical mean Sun (which moves in the ecliptic) by S_1. If angles are measured in time, then

$$E_1 = \text{(hour angle of } S_1) - \text{(hour angle of } S) = \angle SPS_1 ;$$
$$\therefore E_1 = \text{(R.A. of } S) - \text{(R.A. of } S_1) ;$$

since R.A. and hour angle are measured in opposite directions.

When the Sun is in perigee (p) (on December 31), S_1 coincides with S by definition ; $\therefore E_1 = 0$.

From perigee (p) to apogee (a), the Sun, has described 180°, and the time taken is (§ 151, v.) half that of a complete revolution. Hence, S_1 will also have described 180° ;

$$\therefore \text{at apogee (July 1), } E_1 \text{ is again 0.}$$

Now (§ 151, iv.) S is moving most rapidly at perigee, and most slowly at apogee. Hence, after perigee, S will have got ahead of S_1, and after apogee, S will have got behind S_1. Thus : **From perigee to apogee, E_1 is positive,**

From apogee to perigee, E_1 is negative.

and E_1 **vanishes twice a year,** viz., at perigee and apogee.

162. Equation of Time due to Obliquity.—Let the portion of the equation of time due to the obliquity be denoted by E_2.

Take S_2 on the equator so that $\Upsilon S_2 = \Upsilon S_1$. Then S_2 will be the astronomical mean Sun. Draw PS_1M, the secondary to the equator through S_1. Then

$E_2 =$ hour angle of S_2 —hour angle of S_1

$= \angle S_1PS_2$ (taken positive if S_2 is west of S_1)

$= \angle \Upsilon PS_1 - \angle \Upsilon PS_2 = \Upsilon M - \Upsilon S_2 = \Upsilon M - \Upsilon S_1$,

all angles being supposed converted into time at the rate of 15° to the hour.

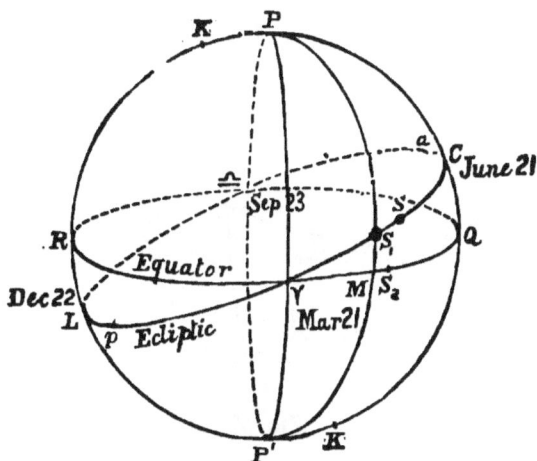

FIG. 57.

At the vernal equinox,* when S_1 is at Υ, S_2 will also be at Υ ; $\therefore E_2 = 0$.

Between the vernal equinox and summer solstice, the angle ΥS_1M will be $< 90°$, and, therefore, $< \Upsilon MS_1$; hence, $\Upsilon M < \Upsilon S_1$;

$\therefore \Upsilon M < \Upsilon S_2$; $\therefore E_2$ **is negative.**

* The vernal and autumnal equinoxes are, strictly, the times when S, and not S_1, coincides with the equinoctial points, but, as S_1 is always near S, the distinction need not be considered here. The same remarks apply to the solstices.

At the summer solstice, S_1 is at C, and S_2 at Q, where $\Upsilon Q = \Upsilon C = 90°$. Hence (Sph. Geom., 21), $\Upsilon QC = 90°$; and M is also at Q; $\therefore E_2 = 0.$

Between the summer solstice and autumnal equinox we shall have $M \triangleq\, < S_1 \triangleq$. But $\Upsilon M \triangleq\, = \Upsilon S_1 \triangleq\, = 180°$; $\therefore \Upsilon M > \Upsilon S_1$; $\therefore \Upsilon M > \Upsilon S_2$; $\therefore E_2$ is positive.

At the autumnal equinox, since $\Upsilon C \triangleq\, = \Upsilon Q \triangleq\, = 180°$, S_1, S_2 will both coincide with $\triangleq$; $\therefore E_2 = 0.$

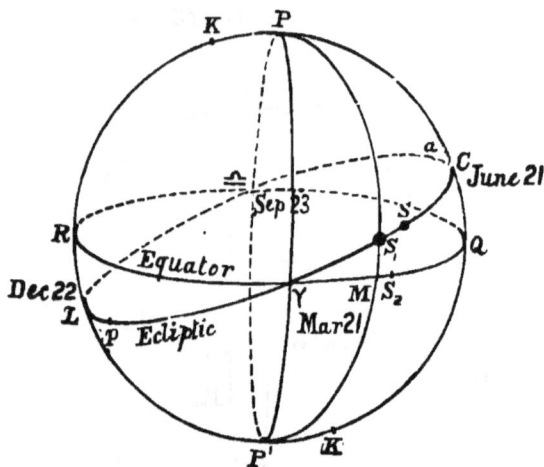

FIG. 58.

In a similar manner we may show that:

From the autumnal equinox to the winter solstice, E_2 is **negative.**

At the winter solstice, $E_2 = 0.$

From the winter solstice to the vernal equinox, E_2 is **positive.**

Collecting these results, we see that

(i.) **From equinox to solstice E_2 is negative.**

(ii.) **From solstice to equinox E_2 is positive.**

(iii.) E_2 **vanishes four times a year, viz., at the equinoxes and solstices.**

163. Graphic Representation of Equation of Time.
—The values of the equation of time at different seasons may now be represented graphically by means of a curved line, in which the abscissa of any point represents the time of year, and the ordinate represents the corresponding value of the equation of time.

In the accompanying figure (Fig. 59) the horizontal line or axis from E_1 to E_1 represents a year, the twelve divisions representing the different months as indicated. The thin curve represents the values of E_1, the portion of the equation of time due to the unequal motion ; this curve is obtained by drawing ordinates perpendicular to the horizontal axis and proportional to E_1. Where the curve is below the horizontal line E_1 is negative.

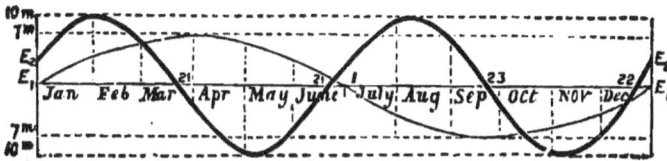

FIG. 59.

The thick curved line is drawn in a similar manner, and represents, on the same scale, the values of E_2, the equation of time due to the obliquity.

In drawing the diagrams to scale, it is necessary to know the maximum values of E_1, E_2. These can be calculated, but the calculations do not depend on elementary methods alone. We shall therefore have to assume the following facts :

The greatest value of E_1 is about   7 minutes,

„   „   „   „ E_2 „   **10**   „

Hence the greatest distances of the thin and thick curves from the horizontal axis should be taken to be about 7 and 10 units of length respectively.

We may now draw the diagram representing E, the total equation of time. We have

$$E = E_1 + E_2.$$

Hence, at every point of the horizontal line we must erect an ordinate whose length is equal to the algebraic sum of the ordinates (taken with their proper sign) of the two curves which represent E_1 and E_2. The extremities of these ordinates will determine a new curve which represents E.

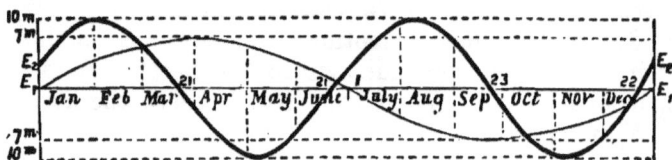

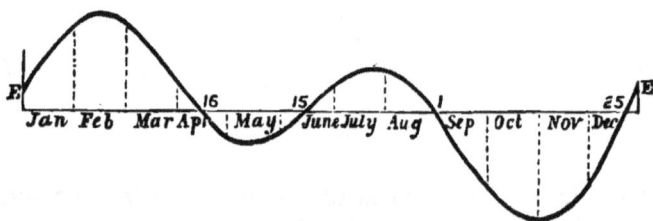

FIG. 60.

This curve is drawn separately in the annexed diagram (Fig. 60). It cuts the horizontal axis in four points. At these points the ordinate vanishes, and E is zero. Hence,

The Equation of Time vanishes four times a year.

164. **Alternative Proof.**—But without representing the values of the equation of time graphically, it can be readily proved that E vanishes four times a year. The proof depends on the fact stated in the last paragraph, that

The greatest equation of time due to the obliquity is greater than the greatest equation due to the eccentricity.

From § 162 it is evident that E_2 must attain its greatest positive value some time between a solstice and the following equinox, and its greatest negative value between an equinox and the following solstice. These maxima occur, in fact, in the months:

February,   May,   August,   November.

Their values, with the proper signs, are respectively about

+10m.,   —10m.,   +10m.,   —10m.

Now, E_1 is never greater than the maximum value of 7m.; hence, whether E_1 is positive or negative, the *total* equation, $E_1 + E_2$, corresponding to either of these maxima, must have the same sign as E_2. Hence, in the year beginning and ending with the date of the maximum value of E_2 in February, E will have the following signs alternately :

+   —   +   —   +

Thus, E changes sign, and therefore vanishes, four times in the year.

165. **Miscellaneous Remarks.**—From Fig. 59 it will be seen that the largest fluctuations in the equation of time occur in the autumn and winter months; during spring and summer they are much smaller.

The days on which the equation of time vanishes are about April 16, June 15, September 1, and December 25.

Between these days E increases numerically, and then decreases, attaining a positive or negative value at some intermediate time. These maxima are :

+14m. 28s. on February 11 ;   —3m. 49s. on May 14 ;
+6m. 17s. on July 26 ;   —16m. 21s. on November 3.

166. **Inequality in the Lengths of Morning and Afternoon.**—If we neglect the small changes in the Sun's declination during the day, the interval from sunrise to apparent noon is equal to the interval from apparent noon to sunset (§ 37). But by morning and afternoon are meant the intervals between sunrise and *mean* noon, and between *mean* noon and sunset respectively. Hence, unless mean and apparent noon coincide, *i.e.*, unless the equation of time vanishes, the morning and afternoon will not be equal in length.

ASTRON.   K

Let r, s be the mean times of sunrise and sunset, E the equation of time. Then

$$12h. - r = \text{interval from sunrise to mean noon.}$$

But apparent noon occurs later than mean noon by E;

$$\therefore 12h. - r + E = \text{interval from sunrise to apparent noon.}$$

Similarly, $s - E = $ interval from apparent noon to sunset;

$$\therefore 12h. - r + E = s - E,$$

or $$r + s = 12h. + 2E,$$

so that **the sum of the times of sunrise and sunset exceeds 12 hours by twice the equation of time.**

The length of the morning is $12h. - r$, and that of the afternoon is s. Now the last relation gives

$$2E = s - (12 - r);$$

$\therefore$ **2 (equation of time)**

$\quad =$ **(length of afternoon) — (length of morning).**

About the shortest day (December 22) the curve representing the equation of time is going upwards, hence E is increasing. But the length of day is changing very slowly (because it is a minimum), hence, for a few days, the half length, $s - E$, may be regarded as constant. Hence, s must increase, and, therefore, the mean time of sunset is later each day. Similarly, it may be shown that sunrise is also later. The afternoons, therefore, begin to lengthen, while the mornings continue to shorten.

Similarly, about June 21, the afternoons continue to lengthen after the longest day, although the mornings are already shortening.

EXAMPLE.—On Nov. 1, the sun-dial is 16m. 20s. before the clock. Given that the Sun rose at 6h. 54m., find the time of sunset.

Time from sunrise to mean noon = 12h. − 6h. 54m. = 5h. 6m.

,,	,,	apparent noon to mean noon	= 0h. 16m. 20s.
∴ ,,	,,	sunrise to apparent noon	= 4h. 49m. 40s.
∴ ,,	,,	apparent noon to sunset	= 4h. 49m. 40s.
∴ ,,	,,	mean noon to sunset	

$\quad\quad = $ 4h. 49m. 40s. − 16m. 20s. = 4h. 33m. 20s.

Hence, the time of sunset was **4h. 33m.**, correct to the nearest minute.

SECTION II.—*The Sun-dial.*

167. The Sun-dial consists essentially of a rod or flat blade, called a **gnomon** or **style** (*OA*, Fig. 61), which is fixed with its edge parallel to the Earth's axis, and therefore pointing in the direction of the celestial pole. The shadow from *OA* is thrown on the dial-plate, which is usually either horizontal or on a wall facing south. The direction of the edge of the shadow determines the hour angle of the Sun, and therefore the apparent time.

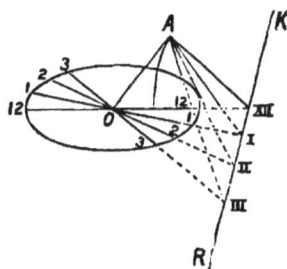

FIG. 61.

The plane through *OA*, the edge of the style, and through the edge of the shadow, evidently passes through the Sun; also it passes through the celestial pole, therefore it will meet the celestial sphere in the Sun's hour or declination circle. Let *OA*XII. be the meridian plane, which is the plane of the shadow at apparent noon, and whose position is supposed known. Then, in order to graduate the plate for the times 1, 2, 3... o'clock, it is only necessary to determine the positions of the planes *OA*I., *OA*II., *OA*III., &c., which make angles of 15°, 30°, 45°, &c., with the meridian plane. Since the Sun's hour angle increases 15° per hour, these planes will be the planes bounding the shadow at 1, 2, 3... o'clock respectively. If we join the points *O*I., *O*II., *O*III., &c., these will be the corresponding lines of shadow in the plane of the gnomon, and will meet the circumference of the dial-plate (which is usually circular) at the required points of graduation 1, 2, 3, &c.

168. Geometrical Method of Graduating the Dial-plate.—To find the planes OA1., OA11., &c., suppose a plane AKR drawn through A perpendicular to OA, meeting the plane of the dial-plate in KR and the meridian plane in Axii. If, in this plane, we take the angles xii.A1., 1.A11., 11.Aiii., &c., each = 15°, the points i., ii., iii...., &c., will evidently determine the directions of the shadow at 1, 2, 3,... o'clock respectively.

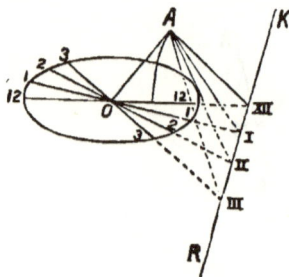

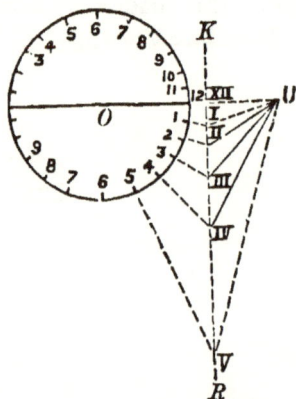

Fig. 62   Fig. 63.

But in practice it is much more convenient to perform the construction **in the plane of the dial itself.** Imagine the plane AKR of Fig. 62 turned about the line KR till it is brought into the plane of the dial, the point A of the plane being brought to U (Fig. 62). Then, by making the angles xii.U1., 1.U11., 11.Uiii., &c., each = 15°, we shall obtain the same series of points i., ii., iii. as before.

If the dial-plate is *horizontal*, and l is the latitude of the place (xii.OA), we have evidently therefore the following construction :—

On the meridian line, measure O xii. = OA sec l, and xii. U = xii. A = Oxii. sin l. Draw Kxii. R perpendicular to OU. Make the angles xii.U1., 1.U11., 11.Uiii., &c., each = 15°, taking i., ii., iii., &c., on KR. Join O1., O11., Oiii., &c., and let the joining lines meet the circumference of the dial in 1, 2, 3, &c. These will be the required points of graduation for 1, 2, 3,... o'clock respectively.

SECTION III.—*Units of Time—The Calendar.*

169. Tropical, Sidereal, and Anomalistic Years.—
Hitherto we have defined a year as the period of a complete
revolution of the Sun in the ecliptic. In order to give a
more accurate definition, however, it is necessary to specify
the starting point from which the revolution is measured.
We are thus led to three different kinds of years.

A **Tropical Year** is the period between two successive
vernal equinoxes, or the time taken by the Sun to perform a
complete revolution relative to the first point of Aries.
The length of the tropical year in mean solar time is very
approximately 365d. 5h. 48m. 45·51s. at the present time.
For many purposes it may be taken as 365¼ days.

A **Sidereal Year** is the period of a complete revolution
of the Sun, starting from and returning to the secondary to
the ecliptic through some fixed star. Thus, after a sidereal
year the Sun will have returned to exactly the same position
among the constellations.

If ♈ were a fixed point among the stars, the sidereal and
tropical year would be exactly of the same length. But ♈
has an annual retrograde motion of 50·22″ among the stars
(§ 141). Consequently, the tropical year is rather shorter
than the sidereal.

An **Anomalistic Year** is the period of the Sun's revolution relative to the apse line—in other words, the interval
between successive passages through perigee.

Owing to the progressive motion of the apse line, the positions
of perigee and apogee move forward in the ecliptic at the rate
of 11·25″ per annum (§ 153). Hence the anomalistic year is
rather longer than the sidereal.

It is easy to compare the lengths of the sidereal, tropical,
and anomalistic years. For, relative to the stars,

In the sidereal year the Sun describes 360°,

In the tropical year it describes $360° - 50·22″$,

In the anomalistic year it describes $360° + 11·25″$;

∴ (**Sidereal year**) : (**tropical year**) : (**anomalistic year**)

= **360°** : **360°−50·22″** : **360°+11·25″**.

From this proportion it will be found that the sidereal year
is about 20 m. longer than the tropical, and 4½m. shorter than
the anomalistic.

128   ASTRONOMY.

170. The Civil Year.—For ordinary purposes, it is important that the year shall possess the following qualifications: 1st. It must contain an exact (not a fractional) number of days. 2nd. It must mark the recurrence of the seasons.

Now the tropical year marks the recurrence of the seasons, but its length is not an exact number of days, being, as we have seen, about 365d. 5h. 48m. 45·51s. To obviate this disadvantage, the **civil year** has been introduced. Its length is sometimes 365, and sometimes 366 days, but its *average* length is almost exactly equal to that of the tropical year.

Taking an ordinary civil year as 365d., four such years will be less than four tropical years by 23h. 15m. 2·04s., or nearly a day. To compensate for this difference, every fourth civil year is made to contain 366 days, instead of 365, and is called a **leap year.** For convenience, *the leap years are chosen to be those years the number of which is divisible by 4, such as* 1892, 1896.

The introduction of a leap year once in every four years is due to Julius Cæsar, and the calendar constructed on this principle is called the **Julian Calendar.**

Now three ordinary years and one leap year exceed four tropical years by 24h.—23h. 15m. 2·04s., *i.e.* 44m. 57·96s. Thus, 400 years of the Julian Calendar will exceed 400 tropical years by (44m. 57·96s.) × 100, *i.e.*, by 3d. 2h. 56m. 36s.

To compensate for this difference, Pope Gregory XIII. arranged that three days should be omitted in every 400 years. This correction is called the **Gregorian correction** and is made as follows: *Every year whose number is a multiple of* 100 *is taken to be an ordinary year of* 365 *days, instead of being a leap year of* 366, *unless the number of the century is divisible by* 4; *in that case the year is a leap year.*

EXAMPLES.—(i.) 1892 is divisible by 4, ∴ the year **1892 is a leap year.** (ii.) 1900 is a multiple of 100, and 19 is not divisible by 4, ∴ **1900 is not a leap year.** (iii.) 2000: the number of the century is 20, and is divisible by 4, ∴ **2000 is a leap year.**

The Gregorian correction still leaves a small difference between the tropical year and the average length of the civil year, amounting to only 1d. 5h. 26m. in 4,000 years.

171. A Synodic Year is a period of 12 lunar months, being nearly 355 days. The name is, however, rarely used.

SECTION IV.—*Comparison of Mean and Sidereal Times.*

172. Relation between Units.—One of the most important problems in practical astronomy is to find the sidereal time at any given instant of mean solar time, and conversely, to find the mean time at any given instant of sidereal time. Before doing this it is necessary to compare the lengths of the mean and sidereal days.

We have seen (§ 169) that a tropical year contains about $365\frac{1}{4}$ mean solar days. In this period both the true and mean Sun describe one complete revolution, or 360° from west to east relative to Υ; or, what is the same thing, Υ describes one revolution from east to west relative to the mean Sun. But the mean Sun performs $365\frac{1}{4}$ revolutions from east to west relative to the meridian at any place. Therefore Υ performs one more revolution, *i.e.*, $366\frac{1}{4}$ revolutions, relative to the meridian.

Now, a sidereal day and a mean solar day have been defined (§§ 22, 159) as the periods of revolution of the mean Sun and of Υ relative to the meridian;

$\therefore$ **$365\frac{1}{4}$ mean solar days = $366\frac{1}{4}$ sidereal days.**

From this relation we have,

$$\text{One mean solar day} = \left(1 + \frac{1}{365\frac{1}{4}}\right) \text{ sidereal days}$$
$$= (1 + \cdot002738) \text{ sidereal days}$$
$$= 24\text{h. } 3\text{m. } 56\cdot5\text{s. sidereal time}$$
$$= 1 \text{ sidereal day} + 4\text{m.} - 4\text{s. nearly;}$$

$\therefore$ one mean solar hour $= 1\text{h.} + 10\text{s.} - \frac{1}{6}\text{s. sidereal time,}$

and 6m. of mean solar time $= 6\text{m.} + 1\text{s. sidereal time nearly.}$

In like manner we have

$$\text{One sidereal day} = \left(1 - \frac{1}{366\frac{1}{4}}\right) \text{mean solar days}$$
$$= (1 - \cdot002730) \text{ mean days}$$
$$= 23\text{h. } 56\text{m. } 4\cdot1\text{s. mean time}$$
$$= 1 \text{ mean day} - 4\text{m.} + 4\text{s. nearly};$$

$\therefore$ one sidereal hour $= 1\text{h.} - 10\text{s.} + \frac{1}{6}\text{s. of mean time,}$

and 6m. sidereal time $= 6\text{m.} - 1\text{s. mean solar time nearly.}$

173. From the results of the last paragraph we have the following approximate rules:—

(i.) **To reduce a given interval of mean time to sidereal time, add** 10*s. for every hour, and* 1*s. for every* 6*m. in the given interval. For every minute so added,* **subtract** 1*s.*

(ii.) **To reduce a given interval of sidereal time to mean time, subtract** 10*s. for every hour, and* 1*s. for every* 6*m. in the given interval. Then* **add** 1*s. for every minute so subtracted.*

EXAMPLE 1.—Express in sidereal time an interval of 13h. 23m. 25s. mean time.

The calculation stands as follows :—

	H.	M.	S.
Mean solar interval ... ... =	13	23	25
Add 10s. per hour on 13h....		2	10
„ 1s. per 6m. on 23m.			4
	13	25	39
Subtract 1s. per 1m. on 2m. 13·8s. ...			2
∴ Required sidereal interval ... ... =	13	25	37

EXAMPLE 2.—Find the mean solar interval corresponding to 14h. 45m. 53s. of sidereal time.

The calculation stands as follows :—

	H.	M.	S.
Given sidereal interval ... ... =	14	45	53
Subtract 10s. per hour on 14h. = 2m. 20s. ⎫		2	28
„ 1s. per 6m. on 46m.(nearly) = 8s. ⎭			
	14	43	25
Add 1s. per 1m. on 2m. 28s. ...⸱ ...			3
∴ Required interval of mean time ... ... =	14	43	28

If accuracy to within a few seconds is not required, the second correction of 1s. per 1m. may be omitted. On the other hand, if the interval consists of a considerable number of *days*, or if accuracy to the decimal of a second is needed, the results found by the rules will no longer be correct. We must, instead, add 1/365¼ of the given mean solar interval to get the sidereal interval, or subtract 1/366¼ of the given sidereal to get the mean solar interval.

In order to still further simplify the calculations, tables have been constructed ; in most cases, these give the quantity to be added or subtracted according as we are changing from mean to sidereal, or from sidereal to mean time.

174. To find the sidereal time at a given instant of mean solar time on a given date at Greenwich.

The Nautical Almanack* gives the sidereal time of mean noon at Greenwich on every day of the year.

Now the given mean time represents the number of hours, minutes, and seconds which have elapsed since mean noon, expressed in mean time. Convert this interval into sidereal time; we then have the sidereal interval which has elapsed since mean noon. Add this to the sidereal time of mean noon; the result is the sidereal time required.

Thus, let m be the mean time at the given instant, measured from the preceding mean noon,

s_0 the sidereal time of mean noon from the Nautical Almanack, and let $k = 1/365\frac{1}{4}$; so that $1+k$ is the ratio of a mean solar unit to the corresponding sidereal unit.

Then, from mean noon to given instant,

Interval in mean time $= m$;
∴ interval in sidereal time $= m + km$
But, at mean noon, sidereal time $= s_0$

∴ at given instant,

required sidereal time, $s = s_0 + m + km$.

If the result be greater than 24h., we must subtract 24h., for times are always measured from 0h. up to 24h.

EXAMPLE.—Find the sidereal time corresponding to 8h.15m. 40s. P.M. on Dec. 20, given that the sidereal time of mean noon was 17h. 55m. 8s.

From mean noon to the given instant, the interval in mean time is 8h. 15m. 40s.

Converting this interval to sidereal time, by the method of § 173, we have

Mean solar interval	= 8h. 15m. 40s.
Add 10s. per hour on 8h.	1m. 20s.
Add 1s. per 6m. on 15m. 40s.	3s.
	8h. 17m. 3s.
Subtract 1s. per 1m. on 1m. 23s.	1s.
∴ Sidereal interval since mean noon	= 8h. 17m. 2s.
But sidereal time of mean noon	= 17h. 55m. 8s.
∴ Sidereal time at instant required	= 26h. 12m. 10s.
Or, deducting 24h., sidereal time is	= 2h. 12m. 10s.

* Or *Whitaker's Almanack*, which may be used if the Nautical is not at hand.

175. To find the mean solar time corresponding to a given instant of sidereal time at Greenwich.

Subtract the sidereal time of mean noon from the given sidereal time; this gives the interval which has elapsed since mean noon, expressed in sidereal time. Convert this interval into mean time; the result is the mean time required.

Let $k' = 1/366\frac{1}{4}$; so that $1 - k'$ is the ratio of a sidereal to a mean solar unit.

Let the given sidereal time $= s,$
and let the sidereal time of the preceding mean noon $= s_0$;
Then, from mean noon to given instant,

     Interval in sidereal time $= s - s_0$;
     $\therefore$ interval in mean time $= (s - s_0) - k'(s - s_0).$
     $\therefore$ **required mean time** $m = (s - s_0) - k'(s - s_0).$

If s be less than s_0, we must add 24h. to s in order that the times s, s_0 may be reckoned from the *same* transit of Υ.

EXAMPLE.—Find the solar time corresponding to 16h. 3m. 42s. sidereal time on May 5, 1891, sidereal time at mean noon being 2h. 52m. 17s.

Sidereal interval since mean noon
           $= $ 16h. 3m. 42s. $-$ 2h. 52m. 17s. $=$ 13h. 11m. 25s.
$\therefore$ Mean solar interval (§ 173)
           $= $ 13h. 11m. 25s. $-$ 2m. 10s. $-$ 2s. $+$ 2s. $=$ 13h. 9m. 15s.

Hence, 13h. 9m. 15s. is the mean time; which, in our usual reckoning, would be called 1h. 9m. 15s., on the morning of May 6 (§ 36). The sidereal time was also 16h. 3m. 42s. a sidereal day or 23h. 56m. 4s. previously, *i.e.*, 1h. 13m. 11s. a.m. on the morning of May 5.

176. To find the mean time corresponding to a given instant of sidereal time at Greenwich (alternative method).—The Nautical Almanack also contains the mean time of "Sidereal Noon," *i.e.*, the mean time when Υ is on the meridian, and when the sidereal clock marks 0h. 0m. 0s. Let this be m_0, and let s be the given sidereal time, k' the factor $1/366\frac{1}{4}$ as before. Then
From sidereal noon to given instant, sidereal interval $= s$;
$\therefore$ „   „   „   „   mean solar „ $= s - k's.$
But, at sidereal noon,   mean time $= m_0$;
$\therefore$ at given instant,

       The required mean time $= m_0 + s - k's.$

177. To find the sidereal time from the mean solar, or the mean time from the sidereal, in any given longitude.—If the longitude is not that of Greenwich, the above methods will require a slight modification, because the sidereal time of mean noon and mean time of sidereal noon are tabulated for Greenwich.

In such cases, the safest plan is as follows:—Find the Greenwich time corresponding to the given local time (§ 96). Convert this Greenwich time from mean to sidereal, or sidereal to mean, as the case may be, and then find the corresponding local time again.

Let the longitude be $L°$ west of Greenwich (L being negative if the longitude is east),

let m_1 be the mean and s_1 the sidereal local time,

m, s the corresponding times at Greenwich,

and let k, k', m_0, s_0 have the same meanings as in §§ 172–4.

By § 96 we have, whether the times be local or sidereal, (Greenwich time)−(local time in long. $L°$ W.) $= \frac{1}{15}L$ h. $= 4L$ m. Therefore, $s - s_1 = \frac{1}{15}L = m - m_1$.

(i.) If m_1 is given and s_1 is required, we have (in hours),
$$m = m_1 + \tfrac{1}{15}L.$$
By § 174, $s = s_0 + m + km = s_0 + m_1 + km_1 + \tfrac{1}{15}L + \tfrac{1}{15}kL$;
∴ $s_1 = s - \tfrac{1}{15}L = s_0 + m_1 + km_1 + \tfrac{1}{15}kL.$

(ii.) If s_1 is given and m_1 is required, we have
$$s = s_1 + \tfrac{1}{15}L.$$
By §§ 175, 176, $m = (s - s_0) - k'(s - s_0)$ or $= m_0 + s - k's$,
i.e., $m = (s_1 - s_0) - k'(s_1 - s_0) + \tfrac{1}{15}L - \tfrac{1}{15}k'L$
$= m_0 + s_1 - k's_1 + \tfrac{1}{15}L - \tfrac{1}{15}k'L$;
∴ $m_1 = m - \tfrac{1}{15}L = (s_1 - s_0) - k'(s_1 - s_0) - \tfrac{1}{15}k'L$
$= m_0 + s_1 - k's_1 - \tfrac{1}{15}k'L.$

EXAMPLE.—Find the solar time when the local sidereal time is 5h. 17m. 32s. on March 21, the place of observation being Moscow (long. 37° 34′ 15″ E.); given that sidereal time of mean noon was 23h. 54m. 52s. at Greenwich.

Reduced to time (§ 23), 37° 34′ 15″ is 2h. 30m. 17s.

∴ Greenwich sidereal time at instant required
$= $ 5h. 17m. 32s. $-$ 2h. 30m. 17s. $= $ 2h. 47m. 15s.

Sidereal interval since Greenwich noon
$= $ 2h. 47m. 15s. $+ $ 24h $- $ 23h. 54m. 52s. $= $ 2h. 52m. 23s.

∴ Greenwich mean time $= $ 2h. 52m. 23s. $- $ 20s. $- $ 9s. $= $ 2h. 51m. 54s.

∴ Moscow mean time $= $ 2h. 51m. 54s. $+ $ 2h. 30m. 17s. $= $ 5h. 22m. 11s

178. Equinoctial Time.—For the purpose of comparing the times of observations made at different places on the Earth, another kind of time has been introduced.

The **Equinoctial Time** at any instant is the interval of time that has elapsed since the preceding vernal equinox, measured in mean solar units.

The advantage of equinoctial time is that it is independent of the observer's position on the Earth, since the instant when the Sun passes through ♈ is a perfectly definite instant of time, and is independent of the place of observation. On the other hand, mean time and sidereal time, being measured from the transits of the mean Sun and of ♈ across the meridian, depend on the position of the meridian—that is, on the longitude of the observer.

The chief disadvantage of equinoctial time is that since the tropical year contains 365d. 5h. 48m. 46s., and not exactly 365 days, the vernal equinox will occur 5h. 48m. 46s. later in the day every year, so that at the end of each tropical year the equinoctial clock will have to be put back 5h. 48m. 46s. Hence also the same equinoctial time will represent a different time of day on the same date in different years.

The disadvantages of using local time are obviated in Great Britain by the universal use of " Greenwich Mean Time."

179. Practical Applications.—In § 41 we showed how to determine roughly the time of night at which a given star would transit on a given day of the year. With the introduction of mean time, in the present chapter, we are in a position to obtain a more accurate solution of the problem.

For the R.A. of any star (expressed in time) is its sidereal time of transit. If this be given, we only have to find the corresponding mean time; this will be the required time of transit, as indicated by an ordinary clock.

In the calculations required in converting the time from one measure to the other, it is advisable *not* to quote the formulæ of §§ 174–177, but to go through the various steps one by one.

If neither the sidereal time of mean noon nor the mean time of sidereal noon is given, we must fall back on the rough method of § 35.

EXAMPLES.

1. Find the solar time at 5h. 29m. 28s. sidereal time on July 1, 1891 ;
mean time of sidereal noon being 17h. 20m. 8s.

Sidereal interval from sidereal noon to the given instant = 5h.29m.28s.
∴ Mean solar interval = 5h. 29m. 28s. −50s. −5s. + 1s. = 5h.28m.34s.
i.e., Mean solar time = 5h. 28m. 34s. + 17h. 20m. 8s. = 22h. 48m. 42s. ;
or, 10h. 48m. 42s. A.M., July 2.

It was also 5h. 29m. 28s., a sidereal day or 23h. 56m. 4s. previously, i.e., 10h. 52m. 38s. a.m. July 1.

2. To find the mean time of transit of *Aldebaran* at Greenwich on
December 12, 1891. Given

	H.	M.	S.
R.A. of *Aldebaran*	4	29	40;
Sidereal time of noon, December 12, 1891	17	23	56.

Since the star's R.A. is less than the sidereal time of noon, we
must increase the former by 24h., in order that both may be measured from the same "sidereal noon."

	H.	M.	S.
∴ Sidereal time of transit + 24h.	28	29	40
Subtract „ „ noon	17	23	56
∴ Sidereal interval from noon to transit	11	5	44
To convert into mean solar units, subtract	0	1	49
∴ Mean Solar interval from noon to transit	11	3	55

∴ *Aldebaran* transits at 11h. 3m. 55s. mean time.

3. To find the (local) sidereal time at New York at 9h. 25m. 31s.
(local mean time) on the morning of September 1, 1891.

Longitude of New York = 74° W.
Sidereal time of mean noon at Greenwich, Sept. 1 = 10h. 42m. 24s.

The given local mean time is measured from midnight, therefore
we must take the time measured from noon as

	H.	M.	S.
August 31, 1891.	21	25	31
Add for 74° west longitude reduced to time	4	56	0
∴ Greenwich mean time is, August 31,	26	21	31
or, September 1,	2	21	31
To convert this interval to sidereal units, add	0	0	24
∴ Sidereal time elapsed since Greenwich noon =	2	21	55
But at Greenwich noon, sidereal time (by data) =	10	42	24
∴ Sidereal time at Greenwich is	13	4	19
Subtract for 74° west longitude,	4	56	0
∴ **Sidereal Time at New York**	8	18	9

4. To find the Paris mean time of transit of *Regulus* at Nice on December 26, 1891.

	H.	M.	S.
Longitude of Paris = 2° 21' E.R.A. of *Regulus* =	10	2	34
„ Nice = 7° 18' E.			
Sidereal time at Greenwich noon =	18	18	48
Here local sidereal time of transit at Nice =	10	2	34
Subtract *east* longitude of Nice, 7° 18', in time	0	29	12
∴ Greenwich sid. time of transit at Nice + 24h. =	33	33	22
Subtract Greenwich sidereal time at noon,	18	18	48
∴ Sidereal interval since Greenwich noon =	15	14	34
To convert to mean solar units, subtract	0	2	30
∴ Greenwich mean time =	15	12	4
Add *east* longitude of Paris, expressed in time =	0	9	24
∴ **Paris mean time of transit** =	15	21	28

That is, 3h. 21m. 28s. in the morning on December 27.

5. Find the R.A. of the Sun at true noon on October 8, 1891, given that the equation of time for that day is −12m. 24s., and that the sidereal time of mean noon on March 21 was 23h. 54m. 52s.

Mean solar interval from mean noon March 21 to mean noon Oct. 8
= 201 days.

Mean solar interval from mean noon to apparent noon on Oct. 8
= −12m. 24s.

∴ interval from mean noon on March 21 to apparent noon on Oct. 8
= 201d. −12m. 24s.

Now, in 365¼ days the mean Sun's R.A. increases 24h., and the increase takes place *quite uniformly*.

	H.	M.	S.
∴ increase in mean Sun's R.A. in 201 days			
= 24h. × 201÷365¼ =	13	12	27
Add mean Sun's R.A. on March 21			
(= sidereal time of mean noon) =	23	54	52
∴ mean Sun's R.A. at mean noon Oct. 8 =	37	7	19
or, subtracting 24h., =	13	7	19
Subtract change of R.A. in 12m. 24s. =			2
∴ mean Sun's R.A. at apparent noon Oct. 8 =	13	7	17
But true Sun's R.A.−mean Sun's R.A			
= equation of time =		−12	24
∴ True Sun's R.A. at apparent noon Oct. 8	= 12h.	54m.	53s,

EXAMPLES.—V.

1. To what angles do Sidereal Time, Solar Time, and Mean Time correspond on the celestial sphere? Are these angles measured direct or retrograde?

2. Draw a diagram of the Equation of Time, on the supposition that perihelion coincides with the vernal equinox.

3. On May 14 the morning is 7·8 minutes longer than the afternoon: find the equation of time on that day.

4. On a sun-dial placed on a vertical wall facing south, the position of the end of the shadow of a gnomon at mean noon is marked on every day of the year. Show that the curve passing through these points is something like an inverted figure of eight.

5. Why are not the graduations of a level dial uniform? Show that they will be so if the dial be fixed perpendicular to the index.

6. Show that if every 5th year were to contain 366 days, every 25th year 367 days, and every 450th year 368 days, the average length of the civil year would be almost exactly equal to that of the tropical year. How many centuries would have to elapse before the difference would amount to a day?

7. Give explicit directions for pointing an equatorial telescope to a star of R.A. 22h., declination $37°$ N., in latitude $50°$ N., longitude $25°$ E., at 10h. Greenwich mean time, when the true Sun's R.A. is 14h. 47m. 17s., and the equation of time is −16m. 14s.

8. If the mean time of transit of the first point of Aries be 9h. 41m. 24·4s., find the time of the year, and the sidereal time of an observation on the same day at 1h. 22m. 13·5s.

9. At Greenwich, the equation of time at apparent noon to-day is −3m. 39·42s., and at apparent noon to-morrow it will be −3m. 35·39s. Prove that the mean solar time at New York corresponding to apparent time 9 A.M. there this morning is 8h. 56m. 20·9s., having given that the longitude of New York is $74°$ 1′ W.

10. Find the sidereal time at apparent noon on Sept. 30, 1878, at Louisville (long. $85°$ 30′ W.) having given the following from the Nautical Almanack:—

At mean noon.

Sun's apparent right ascension.	Equation of time to be added to mean time.
Sept. 30. 12h. 26m. 23·16s.	10m. 0·77s.
Oct. 1. 12h. 30m. 0·51s.	10m. 19·98s.

MISCELLANEOUS QUESTIONS.

1. Explain how to determine the position of the ecliptic relatively to an observer in S. latitude at a given time on a given day.

2. Indicate the position of the ecliptic relatively to an observer at Cape Town (lat. 33° 56′ 3·5″ S.) at noon on August 3.

3. Explain why a day seems to be gained or lost by sailing round the world. State which way round a day seems to be lost, and give the reason why.

4. If the inclination of the ecliptic to the equator were 60°, instead of 23° 27½′, describe what would be the variations in the seasons to an observer in latitude 45°, illustrating your description with a diagram.

5. Describe the changes of position in the point of the Sun's rising at different times of the year, and at different points on the Earth's surface.

6. If the equator and ecliptic were coincident, what kind of curve would be described in space by a point on the Earth's surface, say at the equator, during the course of the year ?

7. Examine when that part of the equation of time due to the eccentricity of the Earth's orbit is positive.

8. On September 22, 1861, the times of transit of α *Lyræ* and of the Sun's centre over the meridian of Greenwich were observed to be 18h. 32m. 51·3s. and 12h. 0m. 23·3s. by a sidereal clock whose rate was correct. Given that the R.A. of α *Lyræ* was 18h. 31m. 43·9s., find the Sun's R.A. and the error of the clock.

9. Define mean time and sidereal time, and compare the lengths of the mean second and the sidereal second.

10. If a, a' are the hour angles in degrees of the Sun at Greenwich, at t and t' hours mean time, show that the equations of time at the preceding and following mean noons, expressed in fractions of an hour, are respectively

$$\frac{a't - at'}{15(t'-t)}, \qquad 24 - \frac{a'(24-t) - a(24-t')}{15(t'-t)}.$$

EXAMINATION PAPER.—V.

1. Define the *dynamical mean Sun* and the *mean Sun*, stating at what points they have the same R.A., and when the former coin-cides with the true Sun. Show that the mean Sun has a uniform diurnal motion, and state how it measures mean time.

2. Define the *equation of time*. Of what two parts is it generally taken to consist? State when each of these parts vanishes, is positive, or negative. Give roughly their maximum values, and sketch curves showing their variations graphically.

3. Show that the equation of time vanishes four times a year.

4. If, on a certain day, the sun-dial be 10 minutes before the clock, what is the value of the equation of time on that day? Will the forenoon of that day or the afternoon be longer, and by how much?

5. Define the terms *solar day, mean solar day, sidereal day*. What is the approximate difference and the exact ratio of the second and third?

6. Define the terms *civil year, anomalistic year, equinoctial time*. Why was this last introduced?

7. Show how to express mean solar time in terms of sidereal time, and *vice versâ*.

8. If the mean Sun's R.A. at mean noon at Greenwich on June 1 be 4h. 36m. 54s., find the sidereal time corresponding to 2h. 35m. 45s. mean time (1) at Greenwich, (2) at a place in longitude 25° E.

9. On what day of the year will a sidereal clock indicate 10h. 20m. at 4 P.M.?

10. In what years during the present century have there been five Sundays in February? When will it next happen?

ASTRON.   L

CHAPTER VI.

ATMOSPHERICAL REFRACTION AND TWILIGHT.

180. Laws of Refraction.—It is a fundamental principle of Optics that a ray of light travels in a straight line, so long as its course lies in the same homogeneous medium; but when a ray passes from one medium into another, or from one stratum of a medium into another stratum of different density, it, in general, undergoes a change of direction at their surface of separation. This change of direction is called **Refraction.***

Let a ray of light SO (Fig. 64) pass at O from one medium into another, the two media being separated by the plane surface AB, and let OT be the direction of the ray after refraction in the second medium. Draw

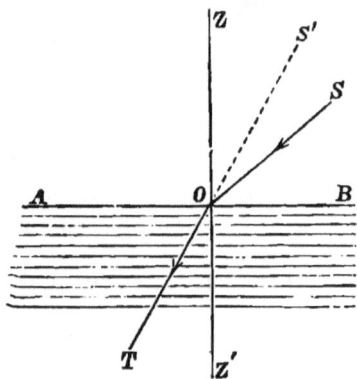

FIG. 64.

ZOZ' the normal or perpendicular to the plane AB at O. Then the three laws of refraction may be stated as follows :—

I. *The incident and refracted rays* SO, OT *and the normal* ZOZ' *all lie in one plane.*

II. *The ratio* $\dfrac{\sin ZOS}{\sin Z'OT}$

is a constant quantity, being the same for all directions of the rays, so long as the two media are the same.†

This constant ratio is called the **relative index of refraction** of the two media, and is usually denoted by the Greek letter μ.

---

* For a fuller description, see Stewart's *Light*, Chap. VI.

† The value of the ratio varies slightly for rays of different colours, but with this we are not concerned in the present chapter.

Thus, if TO be produced backwards to S',

$$\sin ZOS = \mu \sin Z'OT \equiv \mu \sin ZOS',$$

The angles ZOS and $Z'OT$ are usually called the *angle of incidence* and the *angle of refraction* respectively.

III. *When light passes from a rarer to a denser medium, the angle of incidence is greater than the angle of refraction.*

Since $\angle ZOS > \angle Z'OT$, $\sin ZOS > \sin Z'OT$ and $\therefore \mu > 1$.

181. General Description of Atmospherical Refraction.—If the Earth had no atmosphere, the rays of light proceeding from a celestial body would travel in straight lines right up to the observer's eye or telescope, and we should see the body in its actual direction.

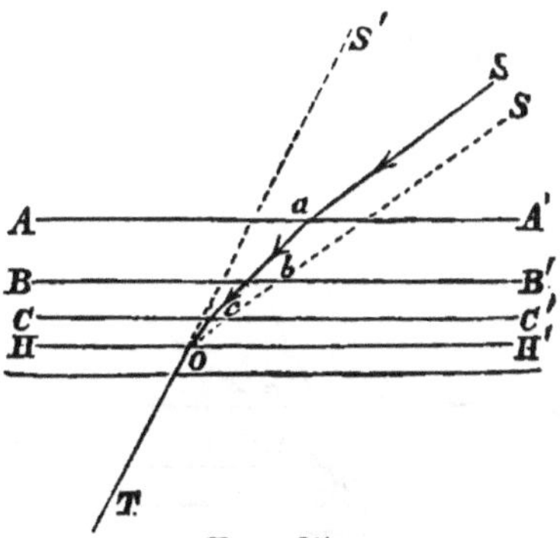

But when a ray Sa (Fig. 65) meets the uppermost layer AA' of the Earth's atmosphere, it is refracted or bent out of its course, and its direction changed to ab.   On passing into a denser stratum of air at BB', it is further bent into the direction bc, and so on; thus, on reaching the observer, the ray is travelling in a direction OT, different from its original direction, but (by Law I.) in the same vertical plane.

The body is, therefore, seen in the direction OS', although its real direction is aS or OS. Also, since the successive horizontal layers of air AA', BB', CC', ... are of increasing density, the effect of refraction is to bend the ray *towards* the perpendicular to the surfaces of separation, that is, towards the vertical.

Hence : **The apparent altitudes of the stars are increased by refraction.**

In reality, the density of the atmosphere increases *gradually* as we approach the Earth, instead of changing abruptly at the planes AA', BB', ... Consequently, the ray, instead of describing the polygonal path $SabcO$, describes a curved path, but the general effect is the same.

FIG. 65.

182. Law of Successive Refractions.—Let there be any number of different media, separated by *parallel planes* AA', BB', CC', HH' (Fig. 66), and let $SabcOT$ represent the path of a ray as refracted at the various surfaces. Then it is a result of experiment that the final direction $S'T$ of the ray is parallel to what it would have been if the ray had been refracted directly from the first into the last medium without traversing the intervening media.

Thus, if a ray SO, drawn parallel to Sa, were to pass directly from the first medium to the last by a single refraction at O, its refracted direction would be the same as that actually taken by the ray Sa, and would coincide with OT.

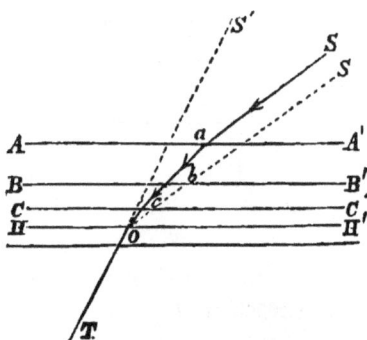 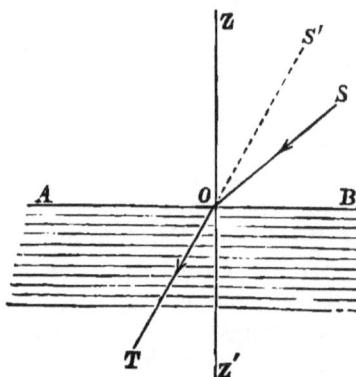

FIG. 66.   FIG. 67.

183. The Formula for Astronomical Refraction.— We shall now apply the above laws to determine the change in the apparent direction of a star produced by refraction.

Since the height of the atmosphere is only a small fraction of the Earth's radius, it is sufficient for most purposes of approximation to regard the Earth as flat, and the surfaces of equal density in the atmosphere as parallel planes. With this assumption, the effect of refraction is exactly the same (§ 182) as if the rays were refracted directly into the lowest stratum of the atmosphere, without traversing the intervening strata.

Let OS (Fig. 67) be the true direction of a star or other celestial body. Then, before reaching the atmosphere, the rays from the star travel in the direction SO. Let their direction after refraction be $S'OT$, then OS' is the *apparent* direction in which the star will be seen, and the angle SOS' is the apparent change in direction due to refraction. The normal OZ points towards the zenith. Hence ZOS is the star's true zenith distance, and ZOS' or $Z'OT$ is its apparent zenith distance, and the first and third laws of refraction show that the star's apparent direction is displaced towards the zenith.

Let $\angle ZOS' = z$, $\angle S'OS = u$, and $\therefore \angle ZOS = z+u$; and let μ be the index of refraction.

By the second law of refraction,
$$\sin(z+u) = \mu \sin z.$$
$$\therefore \quad \sin z \cos u + \cos z \sin u = \mu \sin z.$$

Now the refraction u is in general very small. Hence, if u be measured in circular measure, we know by Trigonometry that $\sin u = u$, and $\cos u = 1$ very approximately. Therefore we have
$$\sin z + u \cos z = \mu \sin z;$$
$$\therefore \quad u = (\mu - 1) \tan z.$$

Let U be the amount of refraction in circular measure when the zenith distance is $45°$. Putting $z = 45°$, we have
$$U = \mu - 1.$$
$$\therefore \quad u = U \tan z.$$

Thus the amount of refraction is proportional to the tangent of the apparent zenith distance.

The last result does not depend on the fact that the refraction is measured in circular measure. Hence, if u'', U'' be the numbers of seconds in u, U, we have
$$u'' = U'' \tan z.$$

The quantity U'' is called the **coefficient of refraction.** Since U is the circular measure of U'', we have
$$U'' = \frac{180 \times 60 \times 60}{\pi} \cdot U = 206265 \, (\mu - 1),$$

whence, if U'' is known, μ can be found, and conversely.

184. Observations on the preceding Formula.—In the last formula u'' represents the correction which must be added to the apparent or observed zenith distance in order to obtain the true zenith distance. By the first law, the azimuth of a celestial body is unaltered by refraction.

Thus the time of transit of a star across the meridian, or across any other vertical circle, is unaltered by refraction. In using the transit circle, there will, therefore, be no correction for observations of right ascension, but in finding the declination the observed meridian Z.D. will require to be increased by U'' tan z.

A star in the zenith is unaffected by refraction, and the correction increases as the zenith distance increases. When a star is near the horizon, the formula $u'' = U''$ tan z fails, since it makes $u'' = \infty$, when z $= 90°$. In this case u is no longer a small angle, so that we are not justified in putting sin $u = u$ and cos $u = 1$. But there is a more important reason why the formula fails at low altitudes, namely, that the rays of light have to traverse such a length of the Earth's atmosphere that we can no longer regard the strata of equal density as bounded by parallel planes. In this case, it is necessary to take into account the roundness of the Earth in order to obtain any approach to accurate results.

For zenith distances less than 75°, the formula is found to give fairly satisfactory results; for greater zenith distances it makes the correction too large.

The coefficient of refraction U'' is found to be about 57", when the height of the barometer is 29·6 inches and the temperature is 50°. But the index of refraction depends on the density of the air, and this again depends on the pressure and temperature. Hence, where accurate corrections for refraction are required, the height of the barometer and thermometer must be read. Any want of uniformity in the strata of equal density, or any uncertainty in determining the temperature, will introduce a source of error; hence it is desirable that the corrections shall be as small as possible. For this reason observations made near the zenith are always the most reliable,

*185. Cassini's Formula.—The law of refraction was also investigated by Dominique Cassini on the hypothesis that the atmosphere is spherical but homogeneous throughout; in this way he obtained the approximate formula

$$u = (\mu - 1) \tan z \, (1 - n \sec^2 z),$$

where n is the ratio of the height of the homogeneous atmosphere to the radius of the Earth.

Cassini's formula may be proved as follows:—Let $SO'O$ be the path of a ray of light from a star S. By hypothesis this ray undergoes a single refraction on entering the homogeneous atmosphere at O'. Let O be the position of the observer, C the centre of the Earth. Produce OO' to S', CO to Z, and CO' to Z'. Let $u = \angle SOS'$ (in circular measure), $z = \angle ZOS'$, $z' = \angle Z'OS'$.

Then, by § 183, if u is small, we have

$$u = (\mu - 1) \tan z';$$

but here z' is not the apparent zenith distance, so that we must express $\tan z'$ in terms of $\tan z$.

Draw CT perpendicular to $O'O$ produced, and $O'N$ perpendicular to COZ.

Then $O'T \tan z' = TC = OT \tan z$;

$$\therefore \quad \frac{\tan z}{\tan z'} = \frac{O'T}{OT} = 1 + \frac{O'O}{OT}$$

$$= 1 + \frac{ON \sec z}{OC \cos z} = 1 + \frac{ON}{OC} \sec^2 z.$$

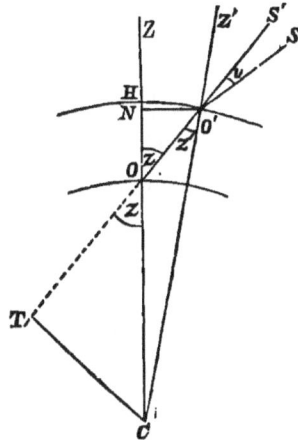

FIG. 68.

But ON is very approximately the height of the homogeneous atmosphere OH, and is therefore $= n \cdot OC$;

$$\therefore \quad \frac{\tan z}{\tan z'} = 1 + n \sec^2 z; \quad \therefore \quad \tan z' = \frac{\tan z}{1 + n \sec^2 z};$$

whence, by substituting in the formula, we have

$$u = (\mu - 1) \frac{\tan z}{1 + n \sec^2 z}$$

$$= (\mu - 1) \tan z \left\{ 1 - n \sec^2 z + n^2 \sec^4 z - n^3 \sec^6 z, \&c. \right\}$$

Now n is very small; we may therefore neglect its square and higher powers; hence we obtain approximately

$$u = (\mu - 1) \tan z \, (1 - n \sec^2 z),$$

which is Cassini's formula.

If the value of n be properly chosen, Cassini's formula is found to give very good results for all zenith distances up to $80°$.

186. To determine the Coefficient of Refraction from Meridian Observations.—Assuming the "tangent law," $u = U \tan z$, the coefficient of refraction U may be found from observations of circumpolar stars as follows.

Let z_1, z_2, the apparent zenith distances of a circumpolar star, be observed at upper and lower culminations respectively. Then the true zenith distances will be

$$z_1 + U \tan z_1 \quad \text{and} \quad z_2 + U \tan z_2.$$

Now, the observer's latitude is half the sum of the meridian altitudes at the two culminations (§ 28), hence if l be the latitude, we have

$$l = \tfrac{1}{2} \left\{ (90° - z_1 - U \tan z_1) + (90° - z_2 - U \tan z_2) \right\},$$

or $\quad 90° - l = \tfrac{1}{2}(z_1 + z_2) + \tfrac{1}{2} U (\tan z_1 + \tan z_2)$ ......(i.).

Now let a second circumpolar star be observed. Let its apparent zenith distances at upper and lower culminations be z' and z''. Then we obtain in like manner

$$90° - l = \tfrac{1}{2}(z' + z'') + \tfrac{1}{2} U (\tan z' + \tan z'') \ \ldots\ldots\text{(ii.)}.$$

Eliminating l from (i.) and (ii.) by subtraction, we have

$$U = \frac{(z_1 + z_2) - (z' + z'')}{(\tan z_1 + \tan z_2) - (\tan z' + \tan z'')}.$$

If the two stars have the same declination, we shall have $z_1 = z'$ and $z_2 = z''$, and the above formula will fail. Hence it is important that the two observed stars should differ considerably in declination; the best results are obtained by selecting one star very near the pole (*e.g.*, the Pole Star) and the other about 30° from the pole.

187. Alternative Method (Bradley's). — Instead of using a second circumpolar star, Bradley observed the Sun's apparent Z.D.'s at noon at the two solstices. Let these be Z_1, Z_2. By § 38, since the true Z.D.'s are

$$Z_1 + U \tan Z_1 \quad \text{and} \quad Z_2 + U \tan Z_2,$$

$$Z_1 + U \tan Z_1 = l - i, \quad Z_2 + U \tan Z_2 = l + i; \quad (i = \text{obliquity.})$$

$$\therefore \quad 2l = Z_1 + Z_2 + U (\tan Z_1 + \tan Z_2) \ldots\ldots\ldots\text{(iii.)}.$$

Eliminating l from (i.), (iii.), we have

$$U(\tan z_1 + \tan z_2 + \tan Z_1 + \tan Z_2) = 180° - (z_1 + z_2 + Z_1 + Z_2),$$

whence U is found.

188. Other Methods of finding the Refraction.— Suppose that at a station on the Earth's equator, either a star on the celestial equator, or the Sun at an equinox, is observed during the day. Its diurnal path from east to west passes through the zenith, and during the course of the day its true zenith distance will change uniformly at the rate of 15° per hour. Thus the true Z.D. at any time is known. Let the apparent Z.D. be observed with an altazimuth. The difference between the observed and the calculated Z.D. is the displacement of the body due to refraction.

By this method we find the corrections for refraction at different zenith distances without making any assumptions regarding the law of refraction.

Except at stations on the Earth's equator, it is not possible to observe the refraction at different zenith distances in such a simple manner. Nevertheless, methods more or less similar can be employed. For this purpose the zenith distances of a known star are observed at different times. The true zenith distance at the time of each observation can be calculated from the known R.A. and declination (§ 26). Hence the refraction for different zenith distances of the star can be determined. This method is very useful for verifying the law of refraction after the star's declination and the observer's latitude have been found with tolerable accuracy. Moreover, it can be employed to find the corrections for refraction at low altitudes when the "tangent law" ceases to give approximate results.

189. Tables of Mean Refraction.—From the results of such observations tables of mean refraction have been constructed by Bessel,* and are now used universally. These are calculated for temperature 50° and height of barometer 29·6 inches; they give the refraction for every 5′ of altitude up to 10°, for larger intervals at altitudes between 10° and 54°, and for every 1° at altitudes varying from 54° to 90°. Other tables give the "Correction for Mean Refraction," which must be added to or subtracted from the mean refraction given in the first table in allowing for differences in the temperature and barometric pressure. The corrections for temperature and pressure are applied separately.

---

* See any book of Mathematical Tables, such as Chambers's.

190. Effects of Refraction on Rising and Setting.

At the horizon the mean refraction is about 33'; consequently a celestial body appears to rise or set when it it is 33' below the horizon. Thus, the effect of refraction is to accelerate the time of rising, and to retard, by an equal amount, the time of setting of a celestial body. In particular, the Sun, whose angular diameter is 32', appears to be just above the horizon when it is really just below.

The acceleration in the time of rising due to refraction can be investigated in exactly the same way as the acceleration due to dip (§ 104). If u'' denotes the refraction at the horizon in seconds, d the declination, x the inclination to the vertical of the direction in which the body rises, the acceleration in the time of rising in seconds

$$= \frac{1}{15}\, u'' \sec x \sec d.$$

Taking the horizontal refraction as 33', or 1980", and putting $x = 0$, $d = 0$, we see that at the Earth's equator at an equinox, the time of sunrise is accelerated by about 2m. 12s. owing to refraction.

When the Sun or Moon is near the horizon, it appears distorted into a somewhat oval shape. This effect is due to refraction. The whole disc is raised by refraction, but the refraction increases as the altitude diminishes; so that the lower limb is raised more than the upper limb, and the vertical diameter appears contracted. The horizontal diameter is unaffected by refraction, since its two extremities are simply raised. Hence, the disc appears somewhat flattened or elliptical, instead of truly circular.

According to the tables of mean refraction, the refraction on the horizon is 33', while at an altitude 30', the refraction is only 28' 23", and at 35' it is 27' 41". Hence, taking the Sun's or Moon's diameter as 32', the lower limb when on the horizon is raised about 5' more than the upper. The contraction of the vertical diameter, therefore, amounts to 5', i.e., about one-sixth of the diameter itself, so that the apparent vertical and horizontal angular diameters are approximately in the ratio of 5 to 6.

191. Illusory Variations in Size of Sun and Moon.
The Sun and Moon generally seem to look larger when
low down than when high up in the sky. This is, however,
merely a false impression formed by the observer, and is not
in accordance with measurements of the angular diameter
made with a micrometer. When near the horizon, the
eye is apt to estimate the size and distance of the Sun and
Moon by comparing them with the neighbouring terrestrial objects (trees, hills, &c.). When the bodies are at
a considerable altitude no such comparison is possible, and a
different estimate of their size is instinctively formed.

**192. Effect of Refraction on Dip, and Distance of
the Horizon.**—Since refraction increases as we approach
the Earth, its effect is always to bend the path of a ray of
light into a curve which is concave downwards (Fig. 69).

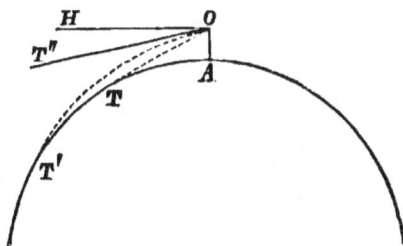

FIG. 69.

Let O be any point above the Earth's surface, and let $T''O$
be the curved path of the ray of light which touches the Earth
at T' and passes through O. Then OT' is the distance of
the visible horizon. Draw the straight tangent OT, then
OT would be the distance of the visible horizon if there
were no refraction; hence, it is evident from the figure that
**The Distance of the horizon is increased by
refraction.**
Draw OT'', the tangent at O to the curved path OT', then
OT'' is the apparent direction of the horizon. Hence, from
the figure we see that
The Dip of the horizon is diminished by refraction.
Both dip and distance are still approximately proportional
to the square root of the height of the observer.

193. Effect of Refraction on Lunar Eclipses and on Lunar Occultations.—In a total eclipse the Moon's disc is never perfectly dark, but appears of a dull red colour. This effect is due to refraction. The Earth coming between the Sun and Moon prevents the Sun's *direct* rays from reaching the Moon, but those rays which nearly graze the Earth's surface are bent round by the refraction of the Earth's atmosphere, and thus reach the Moon's disc.

From observing the "occultations" of stars when the unilluminated portion of the Moon passes in front of them, we are enabled to infer that the Moon does not possess an atmosphere similar to that of our Earth. For the directions of stars would be displaced by the refraction of such an atmosphere just before disappearing behind the disc, and just after the occultation; and no such effect has been observed.

194. Twilight.—The phenomenon of *twilight* is also due to the Earth's atmosphere, and is explained as follows :— After the Sun has set, its rays still continue to fall on the atmosphere above the Earth, and of the light thus received a considerable portion is reflected or scattered in various directions. This scattered light is what we call **twilight,** and it illuminates the Earth for a considerable time after sunset. Moreover, some of the scattered light is transmitted to other particles of the atmosphere further away from the Sun, and these reflect the rays a second time; the result of these second reflections is to further increase the duration of twilight. Twilight is said to end when this scattered light has entirely disappeared, or has, at least, become imperceptible. From numerous observations, twilight is found to end when the Sun is at a depth of about 18° below the horizon.

If the Sun does not descend more than 18° below the horizon, there will be **twilight all night.**

Let l = latitude, d = Sun's declination, then it is easily seen by a figure that the Sun's depth below the horizon at midnight = $90° - d - l$.

This depth is less than 18°, if $l > 72° - d$.

But the greatest value of d is i, or nearly $23\frac{1}{2}°$ (midsummer). Hence, there is twilight all the night about midsummer, at any place whose latitude l is not less than $72° - 23\frac{1}{2}°$, or $48\frac{1}{2}°$.

EXAMPLES.—VI.

1. What would be the effect of refraction on terrestrial objects as seen by a fish under water?

2. For stars near tho zenith show that the refraction is approximately proportional to the zenith distance, and that the number of seconds in the refraction is equal to the number of degrees in the zenith distance. (Take coefficient of refraction = $57''$.)

3. From the summit of a mountain 2400 feet above the level of the sea, it is just possible to see the summit of another, of height 3450 feet, at a distance of 143 miles. Find approximately the radius of the Earth, assuming that the effect of refraction is to alter the distance of the visible horizon in the ratio 12 : 13.

4. Trace the changes in the apparent declination of a star due to refraction in the course of a day, at a place in latitude $45°$ N., the actual declination being $50°$ N.

5. At Greenwich (latitude $51° 28' 31''$ N.) the star α *Cygni* was observed to transit $6° 34' 57''$ south of the zenith. Find the star's declination, employing the results of Question 2.

6. Prove that if the declination of a star observed off the meridian is unaffected by refraction, the star culminates between the pole and the zenith, and that the azimuth of the star from the north is a maximum at the instant considered.

7. Show how the duration of twilight gives a measure of the height of the atmosphere.

8. What is the lowest latitude in the arctic circle at which there is no twilight at midwinter, and what is the corresponding distance from the North Pole in miles?

EXAMINATION PAPER.—VI.

1. What effect has refraction on the apparent position of a star? Show that the greater the altitude of the star the less it is displaced by refraction, and that a star in the zenith is not displaced at all.

2. Prove (stating what optical laws are assumed) that, if the Earth and the layers of the atmosphere be supposed flat, the amount of refraction depends solely on the temperature and pressure at the Earth's surface.

3. Prove the formula for refraction, $r = (\mu - 1) \tan z$. Is this formula universally applicable? Give the reason for your answer.

4. Given that the optical coefficient of refraction of air (μ) = 1·0003, find the astronomical coefficient of refraction (U) in seconds.

5. What is the *refraction error*? How may we approximately determine the correction for refraction from observations made on the transits of circumpolar stars?

6. Show how the constant of refraction (on the usual assumption that the refraction is proportional to the tangent of the zenith distance) might be determined by observing the two meridian altitudes of a circumpolar star whose declination is known.

7. Assuming the tangent formulæ for refraction, find the latitude of a place at which the upper and lower meridian altitudes of a circumpolar star were 30° and 60° ($\sqrt{3} = 1·732$), the coefficient of refraction being 57″.

8. Why is the Moon seen throughout a total eclipse?

9. In the *Scientific American*, June 18, 1887, it was stated by the editor that "The atmosphere by its refraction acts as a lens, producing an apparent increase in the diameter (of the Sun and Moon) near the horizon. When we consider that the atmosphere, as seen from the surface of the globe, is a section of a vast lens whose radius is the semi-diameter of the Earth, it is reasonable to assume a small increase in the size of the objects seen through it, and a still greater increase when seen in the obliquity of the horizon." Why is the above statement altogether incorrect?

10. Find the duration of twilight at the equator at an equinox.

CHAPTER VII.

THE DETERMINATION OF POSITION ON THE EARTH.

SECTION I.—*Instruments used in Navigation.*

195. Among the different uses to which Astronomy has been put, perhaps the most important of all is its application to finding the geographical latitude and longitude of any place on the Earth from observations of celestial bodies. Such observations may be made for either of the following purposes :

1. The determination of the exact latitude and longitude of an observatory. These must be known accurately before the coordinates of a star can be found or observations taken at different observatories can be compared.

2. The construction of maps. The geographical latitude and longitude of a place form a system of coordinates which enable us to represent its exact position on a map.

3. The determination of the exact position of a ship in mid-ocean. This is the most useful application of all ; on a long sea voyage it is necessary to calculate daily the ship's latitude and longitude correct to within a mile or so.

Now, owing to the motion and rocking of a ship, all the astronomical instruments hitherto described are useless at sea. The mariner is therefore obliged to have recourse to others which are unaffected by the unsteadiness of the vessel. The two instruments best fulfilling this condition are the Sextant and the Chronometer, which we shall now describe.

196. The Sextant.—The use of the Sextant is to measure
the angular distance between two objects by observing them
both simultaneously. It consists of a brass framework forming a sector CDE graduated along the circular arc or **limb**
DE; the angle DCE is usually about 60° or rather more.
To the centre C of the arc is fixed an arm BI, capable of
turning about C, and which carries the small mirror B, called
the **index glass.**   Another small mirror A, called the
horizon-glass, is fixed to the arm CD, making an angle of
about 60° with BD.   Of this mirror half the back is usually
silvered, the other half being transparent.   Finally, at T is
fixed a telescope, pointed towards A in such a manner as to
receive the rays of light from the mirror B after reflection at A (Figs. 70, 71).

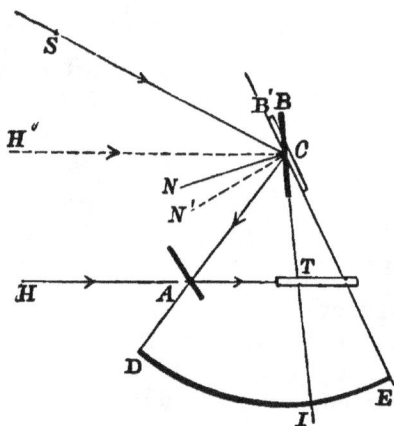

FIG. 70.

On looking through the telescope T we shall see two sets
of images, for objects at H will be seen directly through the
unsilvered part of the mirror A, while objects at S will be
seen after two reflections at the mirrors B and A.   The
mirror is so near the object glass of the telescope as to be
quite out of focus; hence these two sets of images will not
appear separate, but will overlap one another.

The arm BI carries at I an index mark or pointer by which its position can be read off on the graduated scale $D\ddot{E}$. The pointer should read zero when the mirrors A, B are parallel (as in the position $B'E$, Fig. 70). When this is the case, the two images of any very distant object H will coincide. For when a ray of light is reflected in succession at two parallel mirrors, its final direction is parallel to its initial direction.* Hence if $H'CAT$ represents the path of a ray of light from the object H, as reflected in succession at B' and A, the portion AT is parallel to $H'C$, and therefore coincides with the ray HAT, by which the object is seen directly.

Now let it be required to find the angular distance between the two objects H and S. To do this, the mirror B is rotated by means of the arm BI until the image of S (formed by the two reflections) is seen to coincide with H. The angle ECI, through which the mirror B has been turned from its original position, is then half the required angular distance between H, S.

For draw CN', CN perpendicular to the two positions B', B of the mirror respectively. Since in reflection at a plane mirror the angles of incidence and reflection are equal,

$\therefore$ $\angle N'CH = ACN'$ and $\therefore$ $\angle ACH' = 2 \angle ACN'$;

also $\angle NCS = ACN$ and $\therefore$ $\angle ACS = 2 \angle ACN$.

Hence $\angle ACS - \angle ACH' = 2 (\angle ACN - \angle ACN')$,

i.e., $\angle H'CS = 2 . \angle N'CN$

$$= 2 . \angle ECI;$$

or the angular distance between the objects is double the angle ECI.

On the scale ED, every half-degree is marked as 1°. The reading of the pointer I will therefore give *double* the angle ECI, and this is the angular distance required.

The coincidence of the two images in the field of view of the sextant will not be affected by any small displacement of the instrument in its own plane. This peculiarity renders the sextant particularly useful on board ship, where it is impossible to hold the instrument perfectly steady.

---

* See Stewart's *Text-Book of Light*, Chap. IV.

197. Shades, Clamp and Tangent Screw, Reading Glass, Vernier.

For viewing the Sun, the sextant is provided with **shades.** These consist simply of plates of glass blackened for the purpose of reducing the great intensity of the Sun's rays. There are two sets of shades, G, G, hinged to the frame CE in such positions that one set can be inserted between A and C, to deaden the rays from S, while the other set can be turned behind A to deaden the rays from H. They are called respectively the "index shades" and "horizon shades."

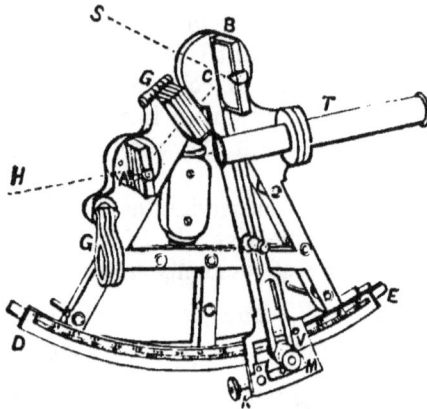

FIG. 71.

The arm or index bar BC is furnished with a **clamp,** by means of which it can be clamped at any desired part of the graduated limb DE. When this has been done the arm can be moved slowly by means of a **tangent screw** K, and in this way can be adjusted with great precision.

The arc DE is usually graduated to divisions of 10',* and is used by means of the lens M, called the "**reading glass.**" But the index bar also carries a scale V called a **Vernier** (§ 198) which, sliding beside the scale on the limb, enables us to read off observations to within 10''.

---

* Of course these divisions are only 5' apart, but in what follows we shall speak of half-minutes as minutes.

◆198. The Vernier is a scale the distance between whose graduations is $10' - 10''$, *i.e.*, $9' 50''$, or $10''$ less than the distance between the graduations on the limb. These graduations are marked $0''$, $10''$, $20''$, &c., being measured in the same direction as on the limb. For example, let us suppose the zero point on the vernier is between the marks $26° 20'$ and $26° 30'$ on the limb. We take the reading by the limb as $26° 20'$. We then look along the vernier scale until we find that *one of the marks on it exactly coincides with one of the marks on the limb*. Suppose that this is the 25th graduation from the zero point of the vernier, *i.e.*, the point marked $4' 10''$. We add this $4' 10''$ to the $26° 20'$ read on the limb, and the sum gives the correct reading, namely, $26° 24' 10''$.

The principle is as follows. Let us denote by P the mark which coincides on the two scales.

Then from zero of vernier scale to P is 25 divisions of *vernier*, *i.e.*, an arc of $25 \times (10' - 10'')$.

Also from $26° 20'$ of scale on limb to P is 25 divisions of *limb*, *i.e.*, an arc of $25 \times 10'$.

∴ from $26° 20'$ on limb to 0 of vernier, represents an arc of

$$25 \times 10' - 25 \times (10' - 10'') \; ; \; i.e., \; 25 \times 10'', \text{ or } 4' 10''.$$

Hence the zero mark of the vernier scale is at a distance $26° 20'$ $+ 4' 10''$ from the zero on the limb, and the reading is $26° 24' 10''$.†

199. The Errors of the Sextant need not be described in detail. If the sextant does not read zero when the two mirrors are parallel, it is said to have an **Index Error,** and a constant correction for index error must be added to all readings made with the instrument. There are also errors due to eccentricity or want of coincidence between the centre about which the index bar turns, and the centre of the limb, errors of graduation, &c.

200. To determine the Index Error of the Sextant.—In all good sextants the graduated limb is continued backwards for about 5° behind the zero point. This portion of the limb is called the "**arc of excess,**" and is used for finding the index error, as follows. The Sun or full Moon is observed; the two images of its disc are brought into contact. Let e be the index-error, r the sextant reading, D the angular diameter of the disc, then we have evidently $D = r + e$. Now let the index bar be moved along the arc of excess until the images again touch, the image which was before uppermost being undermost. If the reading on the arc of excess be $-r'$, we have now $-D = -r' + e$, or $D = r' - e$.

Hence, $\qquad\qquad\qquad 2e = r' - r.$

---

† The simpler forms of mercurial barometer are provided with a vernier by means of which the height of the mercury is read off to the nearest hundredth of an inch. The student will find it of great assistance to carefully examine the vernier in such an instrument

201. To take altitudes at Sea by the Sextant.—

The principal use of the sextant is for finding altitudes.
Now the altitude of a star is its distance from the *nearest*
point of the celestial horizon. To find this, the sextant is so
adjusted that the reflected image of the star appears to lie on
the offing or visible horizon; when the plane of the sextant
is slightly turned, the image of the star should just graze the
horizon without going below it. The sextant reading then
gives the star's angular distance from the nearest point of the
"offing." Subtract the dip of the horizon and the correction for refraction, both of which are given in books of
mathematical tables. The star's true altitude is thus
obtained.

202. To take the Altitude of the Sun or Moon.—

In observing the Sun's altitude, the "index" shades must be
turned into position between the two mirrors, and the instrument adjusted so that the Sun's lower limb appears just to
graze the horizon. The reading of the sextant, when
corrected for dip and refraction, gives the altitude of the
Sun's lower limb. Add the Sun's angular semi-diameter,
which is given in the Nautical Almanack; the altitude of the
Sun's centre is then obtained.

Both the Sun's altitude and its angular diameter may be
obtained by observing the altitudes of the upper and lower
limbs. The difference of the two corrected readings gives the
Sun's angular diameter, and half the sum of the readings
gives the altitude of the Sun's centre.

If this method is used, allowance must be made for the
change in the Sun's altitude between the observations. For
this purpose, three observations must be made. First take
the altitude of the Sun's lower limb, then of the upper limb,
and lastly, again of the lower limb. Also note the time
of each observation. The difference between the first and
third readings determines the Sun's motion in altitude; from
this, by a simple proportion, the change in altitude between
the first and second observations is found, and thus the altitude of the lower limb at the second observation is known.
We can now find the Sun's angular diameter, and the altitude
of its centre at the second observation.

Let $t_1 =$ time of 1st observation, when $a =$ alt. of *lower* limb;
$t_2 =$ time of 2nd observation, when $b =$ alt. of *upper* limb;
$t_3 =$ time of 3rd observation, when $a' =$ alt. of *lower* limb;
Then in time $t_3 - t_1$, the alt. of lower limb increases $a' - a$.

$\therefore$ in time $t_2 - t_1$ it increases $(a' - a) \times \dfrac{t_2 - t_1}{t_3 - t_1}$.

Hence if a_2 denote the alt. of lower limb at second observation,

$$a_2 = a + (a' - a)\frac{t_2 - t_1}{t_3 - t_1} = \frac{(t_3 - t_2)a + (t_2 - t_1)a'}{t_3 - t_1}.$$

This finds a_2, and we then have

Sun's angular diameter $= b - a_2$.

Alt. of Sun's centre at second observation $= \dfrac{1}{2}(b + a_2)$.

In taking the altitude of the **Moon**, the altitude of the **illuminated** limb must be observed, and the angular semi-diameter, as given in the "Nautical Almanac," must be added or subtracted, according as the lower or upper limb is illuminated.

203. Artificial Horizon for Land Observations.—
Owing to the absence of a well-defined offing on land, an **artificial horizon** must be used. This is simply a shallow dish of mercury, protected in some manner from the disturbing effect of the wind. The sextant is used to observe the angular distance between a star and its image as reflected in the mercury. Half this angular distance is the star's apparent altitude; correcting this for refraction, the true altitude is obtained (*cf.* § 65).

As the limb of the sextant is generally an arc of not more than 70°, the instrument will not measure angular distances of more than 140°, and it can, therefore, only be used with an artificial horizon for altitudes of under 70°. For greater altitudes the zenith sector must be used.

At sea, where altitudes are measured from the offing, this objection does not apply. On account of the motion of the vessel an artificial horizon is useless; hence, no observations can be taken when the offing is ill-defined, which frequently happens, especially at night. The mariner is, for this reason, chiefly dependent upon observations of the Sun and Moon, and such stars of the first magnitude, or planets, as are visible about dusk.

204. The Chronometer is the form of timepiece used on
board ship, and in all observations in which clocks are un-
available, owing to their want of portability. In principle,
the chronometer is simply a large and very accurately constructed watch; its rate of motion being controlled, not by a
pendulum, but by a balance-wheel, which oscillates to and
fro under the influence of a steel hair-spring. In order that
the chronometer may go at a uniform rate, the balance-wheel
is constructed in such a manner that its time of oscillation is
unaffected by changes of temperature. If the wheel were
made of one continuous piece of metal, any increase of temperature would cause the whole to expand, and the couple
exerted by the spring would not reverse its motion so readily,
so that the time of oscillation would be increased. To

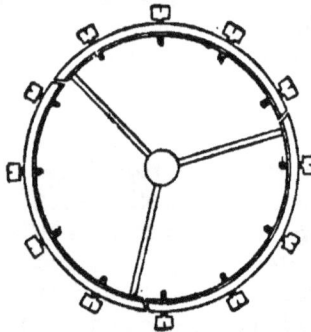

FIG. 72.

obviate this, the rim of the wheel is made in several (generally
three) disconnected arcs, each being formed of steel within
and of brass without. When the temperature rises, the sup-
porting arms or spokes expand, pushing the arcs outward;
but in each arc the outer half of brass expands more than the
inner half of steel, and this causes it to curl inwards,
bringing the extremity actually nearer the centre than it was
before. The arcs carry small screw weights, and by adjusting
these nearer to or further from the supports, the compensation can be arranged with great accuracy.*

* The student who has read a little Rigid Dynamics will notice
that the compensation must be so arranged that the "moment of
inertia," of the balance-wheel is unaffected by the temperature.

Another peculiarity of the chronometer consists in the "detached escapement." The action of the main spring, while keeping up the oscillations, must not affect their periodic time, and to secure this condition the escapement is so arranged that the balance wheel is only acted on during a very small portion of each oscillation.

The chronometer is usually suspended in a framework, in such a manner that when the vessel rolls the instrument always swings into a horizontal position; the framework also serves to protect it from violent shaking.

205. Error and Rate of the Chronometer.—A chronometer is constructed to keep Greenwich mean solar time. As in the case of the astronomical clock, the amount that a chronometer is slow when it indicates noon is called its **error**, and the amount which it loses in 24 hours is called its **rate**. If the chronometer is fast, the error is negative; if it gains, the rate is negative.

The essential qualification of a good chronometer is that its rate must be quite uniform. It is not necessary that the rate shall be *zero*, provided that its amount is known, since a correction can easily be applied to obtain the correct time from the chronometer reading. During sea voyages extending over a large number of days, the correction for rate may become considerable, and there is no very satisfactory method of finding the chronometer error at sea; for this reason the instrument is **rated**, *i.e.*, has its rate determined by comparisons with a standard clock, whenever the ship is in port. Moreover, many ships carry several chronometers, which serve to check each other; if the rate of one should vary slightly, this change would be detected by comparison with the others.

Many of the best chronometers used in the Navy and elsewhere are tested at the Greenwich Observatory. They are there kept in a special room, in which they can be subjected to artificial variations of temperature, with a view of ascertaining whether the compensation for temperature is perfect or not. The chronometers are compared daily with the standard clock. The process of rating is performed by two assistants who have acquired the power of counting the beats of the clock while reading off the errors of one chronometer after another. In this manner, about a hundred chronometers can be rated in half an hour.

SECTION II.—*Finding the Latitude by Observation.*

206. The methods of finding latitude may be conveniently classified as follows:—

A. *Meridian Observations.*

(1) By a single meridian altitude of the Sun or a known star.
(2) By meridian altitudes of two stars, one north and one south of the zenith, taken with the sextant.
(3) By two observations of a circumpolar star.

B. *Observations not made on the Meridian.*
(" *Ex-meridian Observations.*")

(4) By a single observed altitude, the local time being known.
(4A) By " circum-meridian altitudes."
(4B) By observing the altitude of the Pole Star.
(5) By observations of two altitudes.
(6) By the Prime Vertical instrument.

We now proceed to examine the various methods in detail, but it must be premised that the " ex-meridian " methods cannot be thoroughly explained without spherical trigonometry.

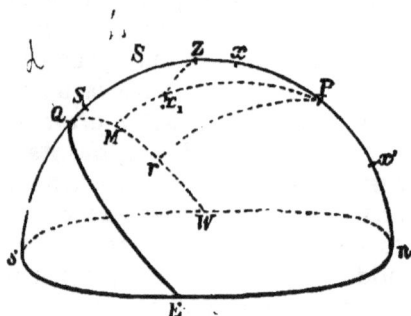

FIG. 73.

207. **Latitude by a Single Meridian Altitude.**—Let S (Fig. 73) represent the position of the Sun or a star of known declination when southing.

Let the meridian altitude sS be observed, and let it be $= a$; also let z be the meridian Z.D. ZS, so that $z = 90° - a$. Let d be the known N. decl. QS, and l the required N. latitude QZ.

EXAMPLE.

On April 11, 1891, in longitude 80° 12′ E. (roughly) with an artificial horizon, the meridian reading of the sextant for the Sun's lower limb was observed to be 107° 59′ 48″. Barometer 30·7 inches, thermometer 72°. Find the latitude, having given the following data :—

		o	′	″	
⊙'s (Sun's) decl. at Greenwich noon, Ap. 11	=	8	19	4	⎫ From
Hourly variation of decl.	=			55·1	⎬ Nautical
⊙'s semi-diameter	=		15	59	⎭ Almanack.

Mean refraction at altitude 54°	=	41	⎫ From
Correction for barometer	=	+1	⎬ Tables.
„ for thermometer	=	−2	⎭

The calculation is best arranged as follows :—

		o	′	″
(i.) Double observed alt. of lower limb	=	107	59	48
∴ observed alt.	=	53	59	54
Corrected refraction at this alt.				
(which is nearly 54°)... ...	=			40 (−)
∴ true alt. of lower limb	=	53	59	14
Ang. semi-diam.	=		15	59 (+)
Merid. alt. ⊙'s centre	=,	54	15	13
Subtract from		90		
Merid. Z.D. of ⊙'s centre	=	35	44	47 S.............(i.)

		M.	S.
(ii.) Long. 8° 12′ E. in time ...	=	32	48

∴ time of observation ...   . = 32 48 *before* Greenwich noon.

		o	′	″
⊙'s decl. at Greenwich noon April 11	=	8	19	4 N. (increasing).
Variation in 30m. before noon ...	=		27	(−)
„ 2m. 48s. (about) ...	=		3	(−)

		o	′	″
∴ ⊙'s decl. at time of observation ...	=	8	18	34 N.
Add ⊙'s merid. Z.D. from (i.)	=	35	44	47 S.

Required north latitude   ...   . = 44° 3′ 21″.

211. To find the latitude by sextant observations of the meridian altitudes of two stars which culminate on opposite sides of the zenith.—This is really only a modification of the first method. Two stars of known declination are selected which culminate, one south and the other north of the zenith, at very nearly the same altitude. The latitude is calculated independently from observations of the meridian altitudes of either star, and the mean of the two results is taken as the correct latitude.

This method possesses the following advantages :—

1st. There is no need to correct the observed altitudes for dip of the horizon ;

2nd. The result is unaffected by any constant instrumental errors (index error, &c.) which affect both altitudes equally;

3rd. The correction for refraction is reduced to a minimum, or even entirely eliminated, if the altitudes are almost equal.

For let d_1, d_2 be the north declinations of the two stars ;
z_1 (south) and z_2 (north) their true meridian Z.D.'s ;
a_1 and a_2 their observed meridian altitudes ;
u_1 and u_2 the corrections for refraction;
D the dip of the horizon ;
e the correction for constant instrumental errors.

For true meridian altitudes of the two stars we have

$$90^\circ - z_1 = a_1 + e - D - u_1, \qquad 90^\circ - z_2 = a_2 + e - D - u_2.$$

The two observations give, therefore, for the latitude (by § 204)

$$l = d_1 + z_1 = d_1 + 90^\circ - a_1 - e + D + u_1,$$
$$l = d_2 - z_2 = d_2 - 90^\circ + a_2 + e - D - u_2.$$

Therefore, taking the mean of the two results,

$$l = \tfrac{1}{2}(d_1 + d_2 + z_1 - z_2) = \tfrac{1}{2}\{d_1 + d_2 + (a_2 - a_1) - (u_2 - u_1)\},$$

a result involving no corrections beyond the difference of refractions, $u_2 - u_1$.

Moreover, if the altitudes a_1 and a_2 are greater than 45°, and their difference $(a_2 - a_1)$ is less than a degree, then $\tfrac{1}{2}(u_2 - u_1)$ is $< 1''$, and therefore the refraction correction may be entirely neglected.

212. Latitude by Circumpolars.—This method has already been mentioned in § 28, but we will here repeat the investigation for convenience.

Let x, x' (Fig. 74) represent the positions of a circumpolar star at its upper and lower transits. Let its meridian altitudes nx and nx' be observed, and let their *corrected* values be a_1 and a_2 respectively. Since

$$Px = \text{star's N.P.D.} = Px',$$
$$\therefore \quad nP = \tfrac{1}{2}(nx + nx'),$$
or
$$l = \tfrac{1}{2}(a_1 + a_2).$$

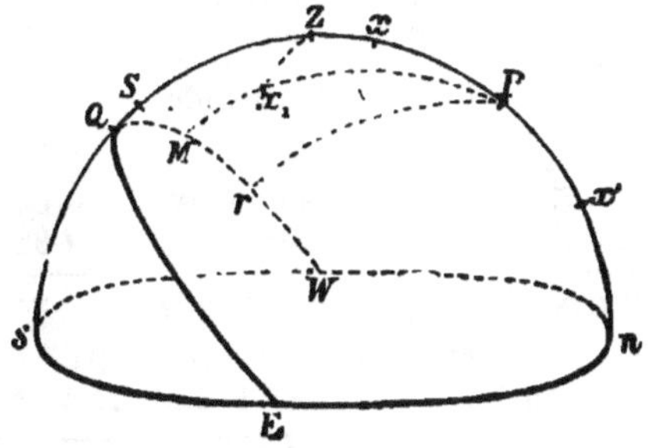

FIG. 74.

In this formula no knowledge of the star's declination is required, but the observed altitudes require to be corrected for refraction, dip, &c.

The circumpolar method is most useful in determining the latitude of a fixed observatory, because this must be done before the declination of any star can be determined. The transit circle is used to determine the meridian altitudes at the two culminations.

By observing two or more circumpolars the correction for refraction may be found, as in § 186, and the observed altitudes may then be corrected for refraction.

As the declinations of a large number of stars are given in astronomical tables, the circumpolar method is never used at sea. It would possess no advantage, and would have the disadvantage of requiring a correction for the change in the ship's place between the two culminations.

EXAMPLES.

1. The observed meridian altitude of β *Ceti* (decl. 18° 36′ 44·5″ S.) is 36° 43′ 12″, and that of a *Ursæ Minoris* (decl. 88° 41′ 53·1″ N.) at its upper culmination is 36° 9′ 57″, both altitudes being measured from the "offing," and the dip being unknown. Find the latitude, given Refraction at alt. 36° = 1′ 20″; at alt. 37° = 1′ 17″.

This is an example of the method of § 211. The calculation stands thus :—

β *Ceti* (south).				a *Ursæ Minoris* (north).		
36°	43′	12″	Observed altitudes	36°	9′	57″
− 0	1	18	Refraction corrections	− 0	1	19·5
36	41	54	Corrected Altitudes	36	8	37·5
90	0	0		90	0	0
+ 53	18	6 S.	Zenith Distances	− 53	51	22·5 N.
− 18	36	44·5 S.	Declinations	+ 88	41	53·1 N.
34	41	21·5 N.	Calculated Latitudes	34	50	30·6 N.

Thus, lat. by star north of zenith = 34° 50′ 30·6″ N.

     „   „   south   „   = 34 41 21·5 N.

2) 69 31 52·1

Mean latitude   = 34° 45′ 56″ N.

Here, owing to dip, one of the calculated latitudes is 4′ 34·6″ too great, and the other is 4′ 34·5″ too small, but the mean of the two results is the correct latitude.

2. The observed altitudes of β *Ursæ Minoris* at lower and upper culmination are 29° 58′ 15″ and 60° 45′ 3″. Find approximately the latitude, assuming the coefficient of refraction to be 57″.

By the "tangent formula," refraction at altitude 30° (approx.)

= 57″ tan 60° = 57″ × √3 = 57″ × 1·732 = 1′ 39″.

Refraction at alt. 60° = 57″ tan 30° = 57″ × √3/3 = 1′ 39″÷3 = 33″.

Hence true alt. at lower culmination = 29° 58′ 15″ − 1′ 39″ = 29° 56′ 36″

   „   „   „ upper   „   = 60 45 3 − 33″ = 60 44 30

2) 90 41 6

∴ Required North Latitude = 45° 20′ 33″

LATITUDE BY EX-MERIDIAN OBSERVATIONS.

213. To find the latitude by a single altitude, the local time being known.—If the local time be known, a single altitude of the Sun or a known star is sufficient to determine the latitude.

For let S be the observed body, Z the zenith, P the pole.†

Then in the spherical triangle PZS, the known local time enables us to find the hour angle ZPS. For, if the Sun be observed, its hour angle ZPS

$$= 15 \times (\text{apparent local time})$$
$$= 15 \times (\text{mean local time} - \text{equation of time});$$

and if a star be observed, its hour angle ZPS

$$= 15 \times (\text{local sidereal time} - \text{star's R.A.}).$$

Also $ZS =$ observed body's Z.D. $= 90° - (\text{observed altitude})$;

$PS =$,,   ,, N.P.D. $= 90° - (\text{known decl.})$.

Hence, ZS, PS, and the angle ZPS are known. These data completely fix the spherical triangle ZPS, and from them ZP can be found by Spherical Trigonometry.

Hence the latitude is found, being $= 90° - ZP$.

*214. By Circum-meridian Altitudes.—This is a particular case of the method last described. In attempting to find the latitude by meridian observations, it may happen that passing clouds prevent the body from being observed at the instant of transit. In this case the latitude can be found from the observed altitude when very near the meridian. The hour angle ZPS is then small, and the difference between the observed and meridian altitudes is also small. This difference is called the "Reduction," and is found by approximate methods.

The best results are obtained by taking a number of altitudes of the body before and after passing the meridian.

*215. By a Single Altitude of the Pole Star.—The N.P.D. of *Polaris* is only about $1° 16\frac{1}{4}'$. Hence, if its altitude is observed at any time, the latitude may be found by adding to, or subtracting from, this altitude, a small correction, never greater than about $1° 16\frac{1}{2}'$.

---

† The student will have no difficulty in illustrating §§ 213-216 with diagrams. For § 213, Fig 75 may be copied.

This correction consists of three parts, which are given by three tables in the Nautical Almanack. The first two corrections depend on the sidereal time, and on the observed altitude; the third is due to variations in the R.A. and N.P.D. of *Polaris*, due to precession (§ 141), etc.

*216. Latitude by observation of Two Altitudes.—By observing the altitudes of two known stars, both the latitude and the local sidereal time can be found.

The same method can be employed to determine the latitude by two observations of the Sun's altitude, separated by a known interval of time.

The necessary calculations are very complicated, involving Spherical Trigonometry, and they cannot be materially simplified even by the use of tables.

A very useful *geometrical* construction, enabling us, from the two observed altitudes, to indicate the exact position of a ship on a globe without calculation, will be detailed in Section VI. of this chapter.

217. Latitude by the Prime Vertical Instrument.—The latitude of a fixed observatory may be found by means of an instrument similar to the Transit Circle, but whose telescope turns in the plane of the prime vertical instead of the meridian. A star will cross the middle wire of such an instrument when its direction is either due east or west; the times of the two transits are observed. Let S, S' be the positions of a known star at its eastern and western transits, Z the zenith, P the pole. The sidereal interval between the two transits determines the angle SPS', and this is evidently twice the angle ZPS. Hence $\angle ZPS$ is known. Also PS, the star's N.P.D., is known, and PZS is a right angle. Therefore, the spherical triangle ZPS is completely determined, and the colatitude ZP can be found.

The times of the transits are unaffected by refraction, and this fact constitutes the principal advantage of the method.

The observations may be performed by an altazimuth, whose horizontal circle is clamped so that the telescope moves in the prime vertical. The instrument must be so adjusted that the interval of time between the first transit and culmination is equal to the interval between culmination and the second transit. The culmination must be observed with a Transit Circle.

SECTION III.—*To find the Local Time by Observation.*

218. In determining the longitude of a place on the Earth, the first step is to find the local time by observations of the hour angle of a known celestial body. If the time indicated by a chronometer or clock at the instant of observation be also noted, we shall find the difference between the true local time and the indicated time. This difference is the error of the clock on local time.

In § 167 we described one instrument for observing local time—the Sun-dial. This cannot, however, be used except for very rough observations, as the boundary of the shadow cast by the style is not sufficiently well defined to admit of accurate measurements. Moreover the Sun-dial is not portable.

For this reason the local time is usually found by one or other of the following methods :—

1st. By meridian observations.
2nd. By equal altitudes.
3rd. By a single altitude, the latitude being known.
4th. By observation of two altitudes.

219. **Local Time by Meridian Observations.**—In a fixed observatory, the local sidereal time is found by means of the Transit Circle, as explained in §§ 24, 54. The transit of a known star is observed; the local sidereal time of transit is equal to the star's R.A., and is therefore known.

Or by observing the transit of the Sun's centre, the time of *apparent* local noon may be found. The equation of time is the mean time of apparent noon, and is given in the "Nautical Almanack"; hence the local mean time is found.

These methods are not available at sea, as the Transit Circle cannot be used. It might be thought that we could use a sextant to ascertain the instant when the body's altitude is greatest, but, for a short interval before and after the transit, the altitude remains very nearly constant; it is therefore impossible to tell with any degree of accuracy when it is a maximum.

On the other hand, a slight error in the time of observation does not affect the altitude perceptibly, so that the meridian altitude may be observed with great accuracy, as in § 208.

ASTRON.   N

220. Method of Equal Altitudes.—When it is required to find the local time from observations taken with a sextant, the simplest method is as follows :—Observe the altitude of any celestial body some time before it culminates. After the body has passed the meridian, observe the instant of time when its altitude is again the same as it was at the first observation. Half the sum of the times of the two observations gives the time of transit.

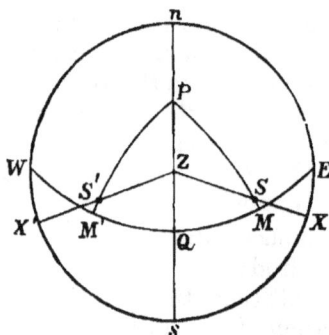

FIG. 75.

For let S, S' be the two observed positions of the body, Z the zenith, and P the pole.

The altitudes of SX, $S'X'$ being equal, the zenith distances are equal;

$$\therefore ZS = ZS'.$$

Also $$PS = PS',$$

and the spherical triangles ZPS, ZPS' have ZP in common.

$$\therefore \angle SPZ = \angle ZPS'.$$

Now let t_1 and t_2 be the times of the two observations, t the time of transit.

Then $t - t_1$ is the time taken to describe the angle SPZ;

$t_2 - t$,,   ,,   ,,   ,, ZPS'.

Since the two angles are equal,

$$\therefore t - t_1 = t_2 - t;$$
$$\therefore t = \tfrac{1}{2}(t_1 + t_2).$$

From the time of transit the local time can be found, as in the last article.

221. In observing the Equal Altitudes with a Sextant, the following method is used :—At the first observation clamp the index bar at an altitude slightly greater than that of the body. Continue to observe the body as it rises, till its image is in contact with the horizon, and note the instant of time (t_1) at which this happens. Keep the index bar clamped until the second observation ; commence observing the body again just before it has reached the same altitude again, and note the instant of time (t_2) when its image is again in contact with the horizon. The two observed times (t_1, t_2) are the times of equal altitude.

If an artificial horizon be used, we must observe the two instants of time (t_1, t_2) when the two images are in contact.

222. Equation of Equal Altitudes.—If the Sun be the observed body, its declination will, in general, change slightly between the two observations ; hence PS will not be exactly equal to PS', and the angles SPZ, ZPS' will not be quite equal. For this reason a small correction must be applied, in order to allow for the effect of the change of declination. This correction is called the **Equation of Equal Altitudes,** and may be found from tables which have been calculated for the purpose.

At Sea allowance must also be made for the change of position of the ship between the two observations, and this correction is also effected by means of tables.

223. The method of Equal Altitudes possesses the following advantages :—

1st. The results are unaffected by errors of graduation of the sextant, for the actual readings are not required.

2nd. The semi-diameter of the observed body need not be known.

3rd. The observed altitudes, being equal, are equally affected by refraction, and no refraction correction need therefore be made.

4th. The dip of the horizon need not be known, provided that it is the same at both observations.

224. With a Gnomon, the time of apparent noon can be roughly found in a very simple manner. A rod is fixed vertically in a horizontal plane, and on the latter are drawn several circles, concentric with the base of the rod. Let the times be observed, before and after noon, when the extremity of the shadow cast by the rod just touches one of these circles. At these two instants the Sun's altitudes are, of course, equal, and therefore the time of apparent noon is the arithmetical mean between the observed times.

EXAMPLE.—The shadow of a vertical stick at Land's End (long. 5° 40′ W.) is observed to have the same length at 9h. 27m. A.M. and 3h. 1m. 40s. P.M., Greenwich time. Find the equation of time on the day of observation.

Greenwich mean time of local apparent noon is

$\frac{1}{2}$ { 9h. 27m. 0s. + 3h. 1m. 40s. − 12h. } =   14m. 20s.

But, by § 96, Greenwich mean time of local mean noon =   22m. 40s.

∴ Eqn. of time = local mean time of apparent noon =   − 8m. 20s.

*225. The Latitude may also be found by the method of equal altitudes, though the calculations require Spherical Trigonometry. For this purpose, the altitude at either observation must be read off on the sextant, and corrected for refraction, dip, &c. The zenith distance SZ is therefore known. The angle SPZ is also known, being half the angle described in the interval $t_2 - t_1$, and PS, being the complement of the declination, is also known. The spherical triangle ZPS is therefore completely determined, and ZP, which is the complement of the latitude, can be found.

226. Local Time by a Single Altitude, the Latitude being known.—This is the converse of the method for finding the latitude described in § 213. If the altitude of a known body, S, be observed in known latitude, we know ZS, SP, PZ, which are the complements of the observed altitude, the declination, and the latitude respectively; hence the hour angle SPZ, and therefore also the local time, may be found.

*227. Local Time by Two Altitudes.—The method of § 216 determines, not only the latitude, but also the hour angles of the bodies at the two observations, and these determine the local time. The method of equal altitudes is in reality only a particular case

SECTION IV.—*Determination of the Meridian Line.*

228. Before setting up a transit circle or equatorial in a fixed observatory, it is necessary to know with considerable accuracy the direction of the meridian line, *i.e.*, the line joining the north and south points of the horizon. At sea, the directions of the cardinal points are determined by a mariner's compass; but here, too, it is of great use, on long voyages, to determine the **variation of the compass,** or the deviation of the magnetic needle from the meridian line. This deviation is different at different parts of the Earth.

There are three ways of finding the meridian line : first, by two observations of a celestial body at equal altitudes; second, by a single observation of the azimuth; third, by one or more observations of the Pole Star.

229. **By Equal Altitudes.**—When a body has equal altitudes before and after culmination, the corresponding azimuths are equal and opposite.

For if S, S' denote the two positions of the body, the triangles ZPS, ZPS' are equal in all respects ;

$\therefore \ \angle PZS = \angle PZS'$ and

$\therefore \ \angle sZS = \angle sZS'$.

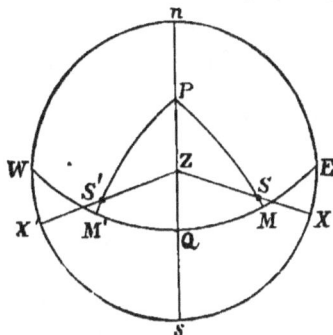

230. **At Sea,** the Sun's azimuth, or compass bearing, may be observed when rising and when setting; the meridian

FIG. 76.

line bisects the angle between the two directions (§ 29).

231. **On Land,** we may observe the directions of the shadow cast by a vertical rod on a horizontal plane when it has equal lengths; for this purpose we mark the points at which the end of the shadow just touches a circle concentric with the base of the rod (*cf.* § 224). Bisecting the angle between the two directions, the north and south points are found.

If greater accuracy is required, an altazimuth may be used. The readings of the horizontal circle are taken when the altitudes of a star are equal; the meridian reading is the arithmetical mean of the two readings. While observing the equal altitudes, the vertical circle must be kept clamped.

*232. By a Single Observation.—If the direction of the vertical plane through a single celestial body S be observed at any instant, the direction of the meridian line may be found by means of Spherical Trigonometry.

For if any three parts of the triangle ZPS are known, the triangle is completely determined, and the angle PZS can be found.

The azimuth $sZS = 180° - PZS$, and is then known hence the meridian line ZS is found.

Now the sides PS, ZS, ZP are the complements of the declination, the altitude, and the latitude; and the hour angle ZPS is known, if the local time be known. Any three of these data are sufficient to determine the angle PZS.

Thus, for example, the Sun's direction, *either* at sunrise or at sunset, determines the meridian line, if either the local time or the latitude is known.

233. By Observations of the Pole Star.—The direction of the meridian may be very accurately determined by observations of the star *Polaris*. If the azimuthal readings of this star be observed at the two instants when it is furthest from the meridian, east and west, respectively the reading for the meridian is half their sum. The observations may be made with an altazimuth. The azimuth at either observation is a maximum, and it remains very nearly constant for a short interval before and after attaining its maximum. Hence, a slight error in the time of observation will not perceptibly affect the azimuth. The same method is applicable to any star which culminates between the pole and the zenith.

The most accurate method is, however, that employed in finding the deviation error of the Transit Circle (§ 59). If the telescope always moves in the plane of the meridian the interval from upper to lower culmination, and the interval from lower to upper culmination, will both be exactly twelve sidereal hours. If not, the small amount by which the vertical plane swept out by the telescope is east or west of the meridian, can be found by observing the amount by which the two intervals are greater and less than 12h.

SECTION V.—*Longitude by Observation.*

234. In Section III. of the present chapter we showed how the local time can be found by observing the celestial bodies. When this has been done, the longitude of the place of observation may be found by comparing the observed local time with the corresponding Greenwich time.

For in § 96 we showed that if the longitude of a place west of Greenwich be $L°$, then

(Greenwich time) — (local time) $= \frac{1}{15}L$ h. $= 4L$ m. ; whence, knowing the difference of the two times, L may be found.

The methods of finding Greenwich mean time, and hence longitude, may be classified as follows :—

A. *Methods available at Sea.*

(1) By the chronometer.
(2) By the method of lunar distances.
(3) By celestial signals.

B. *Methods suitable for Land Observations.*

(4) By repeated transmission of chronometers.
(5) By the chronograph.
(6) By terrestrial signals.
(7) By Moon culminating stars or by the Moon's meridian altitude.

235. **Longitude by the Chronometer.**—By reading the chronometer used on board ship, and making the necessary corrections for error and rate, the Greenwich mean time at any instant may be found. If, then, the local mean time is determined by observing the Sun, or one of the other celestial bodies, and the observations are timed by the chronometer, the difference between the local and Greenwich mean times will be found, and this determines the ship's longitude measured from Greenwich.

EXAMPLE 1. — At apparent noon a chronometer indicates 19h. 33m. 25s., Greenwich mean time, and the equation of time is −2m. 1s. To find the longitude.
Here the local mean time is ... ...   −2m. 1s.
∴ Greenwich mean time−local mean time   ...   = 19h. 35m. 26s.
Mult. by 15, we have long. W. of Greenwich   ...   = 293° 51′ 30″
or sub. from 360°, long. E. of Greenwich   ...   ...   = 66° 8′ 30″

EXAMPLE 2.—Find the longitude, from the following data:—
Sun's computed hour angle = 75°E. Time by chronometer = 23h.7m.31s.
Equation of time = + 3m.55s.   Correction for error and rate,—1m.18s.

(i.) Here ⊙'s hour angle in time   ... ...   = 5h. before noon
     ∴ apparent local time ... ...   = 19h. 0m. 0s.
     Equation of time ... ...   =   3   55
                                                       ――――――――――――
     ∴ mean local time... ... ...   = 19h. 3m. 55s.

(ii.) Observed time   ... ...   = 23h. 7m. 31s.
      Correction   ... ... =   −1   18
                                                        ――――――――――――
      Greenwich time   ... ...   = 23   6   13
                                                          19   3   55
                                                        ――――――――――――
      W. Long. in time ... ...   = 4   2   18
                                                                 15
                                                        ――――――――――――
      ∴ required long. ... ...   = 60° 34′ 30″ W.

EXAMPLE 3.—On June 29, from a ship in the North Atlantic
Ocean, the Sun was observed to have equal altitudes when the
chronometer indicated 11h. 27m. 26s. and 6h. 48m. 32s.   At noon on
June 25, the chronometer was 3s. too fast, and it gains 8s. a day.
The equation of time on June 29 at 3 p.m. was + 2m. 58s.   To
find the ship's longitude.

	H.	M.	S.
The process stands as follows :—			
Chronometer time of first observation ... ... =	11	27	26
„ „ „ second observation + 12h. ... =	18	48	32
2) 30	15	58
		15	7	59

Hence the chronometer time of local apparent noon   =   3   7   59

Correction for chronometer error June 25 = − 3s. ⎫
    „   „   „   rate in 4 days = −32s. ⎬ =   −36
    „   „   „   „   „ 3 hours = − 1s. ⎭

∴ Greenwich time of local apparent noon ... ...   =   3   7   23
Subtract equation of time (since mean noon occurs
    first) ... .   =   −2   58
                                                         ――――――――――――
∴ Greenwich time of local mean noon... ... ...   =   3   4   25
                                                                    15
                                                         ――――――――――――
∴ longitude west of Greenwich   ... ...   = 46°   6′ 15″

236. Method of Lunar Distances.—If from any cause the ship's chronometer should stop, or its indications should become unreliable, the Greenwich time may be found by observations of lunar distances. In this method the Moon, by its rapid motion among the stars, takes the place of a chronometer, its position relative to the neighbouring stars determining the Greenwich time. The Moon moves through 360° in 27½ days; hence it travels at the relative rate of about 33' per hour, or rather over 1″ in every 2s., and this motion is sufficiently rapid to render it available as a timekeeper.

For this purpose, tables of lunar distances are given in the Nautical Almanack. These tables give the angular distances of the Moon's centre from the Sun or from such bright stars or planets as are in its neighbourhood, calculated for every third hour of Greenwich mean time, and for every day of the year.

The angular distance of the Moon's bright limb from one of the given stars may be observed by means of a sextant. By adding or subtracting the Moon's semi-diameter, as given in the Nautical Almanack, and correcting as explained below, the angular distance of its centre may be found. During the interval of three hours between the times given in the Nautical Almanack, the angular distance changes at an approximately uniform rate, and therefore the Greenwich time of the observation may be computed by proportional parts.

237. Clearing the Distance.—One of the great draw-backs of the lunar method consists in the laborious calculations necessary for what is called **"clearing the distance."** The angular distance between the Moon and the star will be affected by refraction, and this alone requires a correction to be applied to the observed lunar distance ; but there is another correction, for what is called **parallax,** which is equally important. This latter correction depends on the fact that the Moon's distance from the Earth is only about 60 times the Earth's radius, and at this comparatively small distance the direction of the Moon *cannot* be considered as independent of the observer's position on the Earth, as has been done with the fixed stars* (§ 5).

---

* Indeed, if a star happens to be behind the Moon's disc, it may sometimes appear on opposite sides of the Moon to two observers at nearly opposite points on the Earth.

180   ASTRONOMY.

For this reason, the lunar distances of a star, as tabulated
in the Nautical Almanack, are the angles which the Moon
and star subtend **at the centre of the Earth.** They are,
therefore, sometimes called the **geocentric** lunar distances.
Hence it is necessary to calculate the Moon's geocentric
position from that observed, before the Greenwich time of the
observation can be determined.

The correction for **parallax,** will be dealt with more fully
in the next chapter. Suffice it to mention here that the
parallax, like the refraction correction, depends only on
the Moon's zenith distance, and therefore, the only data
needed for clearing the distance are the altitudes of the two
bodies at the time of observation. The calculations are then
greatly simplified by the use of tables.

238. **Advantages and Disadvantages of the Lunar
Method.**—The method of lunar distances was introduced at
a time when chronometers were very imperfectly constructed,
and could not be relied on during a moderate voyage. At the
present time, owing to the high degree of accuracy attained
in the construction of chronometers, combined with the
reduction in the length of sea voyages since the introduction
of steam, the lunar method has been almost entirely superseded by the use of chronometers. It is still used, however,
for the occasional correction of a chronometer if the voyage
be extremely long ; and explorers rely upon it mainly.

The principal disadvantages of using lunar distances are :

1st. The calculations necessary for clearing the distance
are very tedious, and not such as could be performed readily
by a seaman possessing little or no knowledge of mathematics.
Moreover, the corrections are often considerable.

2nd. A slight error in the observed lunar distance would
introduce a considerable error in the estimated longitude.
The best sextants are only divided to every 10″, and an error
of 10″ in the observed lunar distance would introduce an error
of 20s. in the computed Greenwich time. This would give,
in the longitude, an error of 5′, or of 5 geographical miles at
the equator. Even this degree of accuracy would be difficult
to attain in practice, while the rate of a well-constructed
chronometer can be depended upon to within 1s. per day.

EXAMPLE. — On Nov. 14, the cleared angular distance of the Moon's centre from *Aldebaran* was found to be 32° 44′ 52″. Find the Greenwich time, having given the following data :—

ANGULAR DISTANCE OF THE MOON FROM *Aldebaran*.

Date.	Position of Star.	6 P.M.	9 P.M.	Midnight.
Nov. 14.	East.	33° 32′ 57′	31° 44′ 14″	29° 55′ 32″

The calculation stands as follows :—

Ang. dist. at 6 P.M.   = 33° 32′ 57″
„ „ at observation = 32 44 52
──────────────
Decrease since 6 P.M. = 0 48 5

Ang. dist. at 6 P.M. = 33° 32′ 57″
„ „ at 9 P.M. = 31 44 14
──────────────
Decrease in 3 h. = 1 48 43

∴ In 3h. the Moon's angular distance from *Aldebaran* decreases 1° 48′ 43″, or 6523″;

∴ the time in which it decreases 48′ 5″, or 2885″, is

$$= 3h. \times \frac{2885}{6523} = 1h.\ 19m.\ 37s.$$

∴ Greenwich time of observation = 6h. + 1h. 19m. 37s.

= 7h. 19m. 37s.

239. Longitude by Celestial Signals.—The eclipses of Jupiter's satellites begin and terminate at times which can be calculated beforehand; it would, therefore, appear possible to ascertain the Greenwich time by observing the instants at which a satellite disappears into, or emerges from, the shadow cast by the planet. But, as the disappearance and emergence take place gradually, it is impossible to employ this method with accuracy to the determination of longitude. The same objection applies still more forcibly in the case of eclipses of the Moon.

By observing the occultations of stars behind the disc of the Moon, we have another way of determining the Greenwich time and finding the longitude. This is merely a particular case of the method of lunar distances, since at the instant of disappearance, the star's apparent (uncorrected) distance from the Moon's centre is equal to the Moon's semi-diameter.

METHODS OF FINDING LONGITUDE ON LAND.

240. Longitude by repeated transmission of Chronometers. The chronometer method of comparing longitudes can be employed with far greater accuracy on land, on account of the possibility of taking repeated journeys to and fro in order to effect the comparison of the local times. The rate of the chronometer is determined by observing its error at the first station, both before and after taking it to the second.

Suppose, for example, that it is required to find the difference of longitude between two stations, A and B. A chronometer is compared with the standard clock at A, and its error is noted. It is then carried to B, and its indications are compared with those of a clock regulated to keep local time. It is then again brought back to A, and compared a second time with the standard clock. The increase in the chronometer error during the whole interval serves to determine the rate of the chronometer. We can now correct for error and rate the time indicated by the chronometer at A, and thus determine the difference between the local times at A and B. By converting this difference into angular measure at the rate of $15°$ to the hour, the required difference of longitude of the two stations is determined.

It is probable that the rate of the chronometer may not be the same while it is being shaken about on its journey as while it is at rest. This difference of rate may be allowed for by comparing the chronometer with the local clock soon after arrival, and again before departing. The total loss while at rest is thus found, and by subtracting we have the total loss during the two journeys. The only assumption which it is necessary to make is that the rate is the same on the outward journey as on the return journey.

In order to obtain a result as free from error as possible, a number of journeys to and fro are performed, and several chronometers are used on each journey. The most accurate result is found by taking the mean of the calculated values for the difference of longitude.

EXAMPLES.

At 17h. by a chronometer, the Greenwich mean time was found to be 16h. 59m. 57·2s.   It was taken to a place A, and indicated 4h., when the local mean time was 3h. 47m. 46·9s.; and when it indicated 11h., the Greenwich time was 11h. 0m. 9·7s.   To find the longitude of A in time and in angle.

Here, at 17h., the chronometer error by Greenwich time was $-2\cdot8$s.

$$\text{,, ,, } 24+11\text{h.} \qquad\qquad \text{,,} \qquad \text{,,} \qquad \text{,,} \qquad \text{,,} \qquad +9\cdot7\text{s}$$

$\therefore$ in 18h. the chronometer lost 12·5s. ;

$\therefore$ the loss in 11h. $= \dfrac{11}{18} \times 12\cdot5$s. $= 7\cdot64$s. nearly;

$\therefore$ the Greenwich time, when the chronometer indicated 4h., was

$$= 4\text{h.} -2\cdot8\text{s.} +7\cdot64\text{s.} = 4\text{h. 0m. } 4\cdot84\text{s.},$$

and the local time at the same instant was $= 3$h. 47m. 46·9s.

$\therefore$ required longitude $= 12$m. 17·9s. W. $= 3°\ 4'\ 28''$ W.

2. As a ship starts from Liverpool, its chronometer indicates 0h., and is correct by Greenwich mean time.   After 16 days, as it reaches Quebec, the chronometer indicates 7h. 0m. 23s., and Quebec time is 2h. 5m. 42s.   Nearly seven days afterwards, the ship departs at Quebec noon, the chronometer then reading 4h. 54m. 39s.; and when it reaches Liverpool, after a voyage of just over fourteen days, it is found to be 17s. slow by Greenwich mean time.   Find the longitude of Quebec.

By Quebec time, the ship stayed in port 7d.$-$2h. 5m. 42s.

$$= 6\text{d. } 21\text{h. } 54\text{m. } 18\text{s.}$$

By chronometer, the ship stayed in port 7d.4h.54m.39s.$-$7h.0m.23s.

$$= 6\text{d. } 21\text{h. } 54\text{m. } 16\text{s.}$$

$\therefore$ in 7 days in port, chronometer lost   2s.

But in 37 days altogether, ,,   ,,   17s.

$\therefore$ in 30 days at sea,   ,,   ,,   15s.

$\therefore$ in 16 days, from Liverpool to Quebec, it lost   8s.

But chronometer time on arrival was   7h. 0m. 23s.

$\therefore$ Greenwich time was   7h. 0m. 31s.

And local time was   2h. 5m. 42s.

The difference $=$ longitude of Quebec (in time) $= 4$h. 54m. 49s.

$\therefore$ Longitude of Quebec (in angle) $= 73°\ 42'\ 15''$ W.

241. Longitude by the Chronograph.—When two observatories are in telegraphic communication, the local time may be readily signalled from one to the other by means of the electric current, and the difference between the longitudes thus determined.

This method is employed in connection with the chronographic method of recording transits, the chronographs being connected by the telegraph line, so that a transit is recorded nearly simultaneously at both stations.

Let us call the two stations A and B. When the star crosses the meridian at A, the observer presses the button of his chronograph. Let t_1, t_2 be the times of transit at A as thus recorded at A and B respectively. When the same star crosses the meridian at B, the times of transit are again recorded at A and B. Let these recorded times be T_1 and T_2 respectively.

The transmission of the signal from one station to the other is not quite instantaneous, because a small interval of time must always elapse before the current has attained sufficient strength to make the signal at the distant station. Let this interval be x. Then the transit at A will be recorded too late at B by the amount x, and the transit at B will be recorded too late at A by the same amount x.

When this correction is applied, the true times of the two transits, as determined by the chronograph record at A, will be t_1 and $T_1 - x$. Hence, if L denote the difference of longitude in time measured westwards from A to B, the chronograph record at A gives

$$L = T_1 - x - t_1.$$

Again, the true times of the two transits, as determined by the chronograph record at B, will be $t_2 - x$ and T_2. Hence the chronograph record at B gives

$$L = T_2 - (t_2 - x) = T_2 - t_2 + x.$$

By addition, we have

$$2L = T_1 - t_1 + T_2 - t_2; \quad \therefore L = \tfrac{1}{2}(T_1 - t_1 + T_2 - t_2),$$

a result which does not involve x.

Thus we see that, by using both chronograph records, and taking the mean of the separately calculated differences of longitude, the corrections due to the time occupied by the passage of the signals are entirely eliminated.

*242. Elimination of Personal Equation.—In the above investigation we have taken no account of the personal equations of the two observers. But if e is the correction for personal equation of the observer at A, and E is that of the observer at B, the observed times t_1, t_2 must both be increased by e, and T_1, T_2 must both be increased by E. Introducing these corrections, the formula gives

$$L = \tfrac{1}{2}(T_1 - t_1 + T_2 - t_2) + (E - e).$$

To eliminate the corrections, let the two observers change places, and repeat the operations, and let the new recorded times of transit be denoted by accented letters. The correction E must now be applied to the times t_1', t_2', and the correction e must be applied to T_1' and T_2'. Therefore

$$L = \tfrac{1}{2}(T_1' - t_1' + T_2' - t_2') + (e - E).$$

By again taking the mean of the two results we get

$$L = \tfrac{1}{4}\{(T_1 - t_1 + T_2 - t_2) + (T_1' - t_1' + T_2' - t_2')\},$$

a result in which personal equation is eliminated.

243. Longitude by Terrestrial Signals.—Before the introduction of the electric telegraph and the chronometer, other signals had to be used. Among such signals may be mentioned flashes of light and rockets visible simultaneously from two stations at a considerable distance apart. The heliograph, in which signals are transmitted by flashes of reflected sunlight, forms another means of determining differences of longitude between two stations visible one from the other; and this method is still often found very useful in surveying a country. A flash of lightning and the bursting of a meteor have also occasionally been used, but they are far too uncertain in their occurrence to be of much value. The local time of the signal is noted at each place, and the difference of these times gives the difference of longitudes.

The signals must in every case be seen, not heard, as an explosion, even if audible at two distant stations, would not be heard simultaneously at both, owing to the comparatively small velocity of sound. Where the distance between the two stations is great, a chain of intermediate stations must be established, and the local time of each station compared with that of the next; this method was used in most of the earliest determinations of longitude. Now such methods are entirely superseded by the use of the chronometer and the electric telegraph.

244. Longitude by Moon culminating Stars.—Here, as in the method of lunar distances, the Moon's position determines the Greenwich time, but instead of observing the Moon's angular distance from a neighbouring star, we observe the difference of right ascension between the Moon and the star by taking their times of transit with a transit circle.

The method is not available at sea, because transits cannot be taken with a sextant. It can be used to determine, by means of a portable transit circle, the longitude of a temporary observatory set up in a country where there is no means of telegraphic communication with the outer world. Its great advantage over the method of lunar distances is that it does not involve the laborious process of "clearing the distance," because the times of passage across the meridian are unaffected by parallax and refraction.

The necessary data for the calculations are given in the Nautical Almanack. The time of transit of the star determines the local sidereal time at the place, and when the observatory clock is thus corrected, the time of the Moon's transit is its R.A. The tables in the Nautical Almanack give the Moon's R.A. at the time of its transit at Greenwich. The increase of R.A. is proportional to the time which elapsed between the transits at Greenwich and at the place of observation, and hence the Greenwich time of the local transit is known. Hence, the longitude may be found.

*245. Longitude by Meridian Altitude of the Moon.—Another method of finding the longitude is sometimes used, namely to find the Greenwich time by observations of the Moon's declination. For this purpose, the Moon's meridian altitude is observed with a transit circle and its declination deduced (§ 24). The Nautical Almanack contains the Moon's declination for every 3h. of Greenwich time; from this the Greenwich time of observation may be found by proportional parts. But the method is difficult to employ, because the observations are affected by the same sources of error, arising from parallax and refraction, as in the method of lunar distances, and there is also a correction for dip in observations made at sea. Moreover, the Moon's daily motion in declination is so small (the greatest variation being about 5° per day), that a slight error in the computed declination would very considerably affect the calculated value of the longitude.

SECTION VI.— *Captain Sumner's Method.*

246. We shall now show that, by taking two altitudes of the Sun with a sextant, and noting the Greenwich times of observation with a chronometer, we can construct a ship's position on a terrestrial globe geometrically.

The Sub-Solar Point.—We can at once find the position on the terrestrial globe of a place at which the Sun is in the zenith on a given day, at a given instant of Greenwich time. For, evidently, the latitude of the place is equal to the Sun's declination, and is, therefore, known ; while the longitude west of Greenwich is equal to the Greenwich apparent time, which may be found by subtracting the equation of time from the mean time. The place is called the **Sub-Solar Point.**

The Circle of Position.—Assuming the Earth to be spherical, the Sun's Z.D. at any place is equal to the angular distance of the place from the sub-solar point. (For it is evidently the angle between the directions of the zeniths at the given place and at the sub-solar point.) Hence, the places at which the Sun has a given Z.D. all lie on a small circle of the terrestrial globe, whose pole is at the sub-solar point, and whose angular radius is equal to the Sun's Z.D. This circle is the **circle of position.**

Geometrical Construction for the Position of the Ship. — If, then, two altitudes of the Sun be observed, and the Greenwich times noted with a chronometer, we can find the sub-solar points, and thus construct the circles of position, and we know that the ship lies on each circle. The ship must, therefore, be at one of the two points in which the two circles cut. To decide which is the actual position, the Sun's azimuth must be very roughly estimated at the two observations. On the globe it will be easy to see at which of the two places the Sun had the observed azimuths. Thus the ship's exact position on the globe is found. It is easy to allow for the ship's motion between the observations.

If two *stars* are observed, the two **substellar points** (or places at which the stars are in the zenith) can be constructed. For the latitude of either is equal to the corresponding star's decl., and its longitude is equal to the star's hour angle at Greenwich = sidereal time − star's R.A.

The ship's place can now be found by drawing the circles of position as before.

ASTRON.   O

EXAMPLES.—VII.

1. At noon on the longest day a circumpolar star is passing over the observer's meridian, and its zenith distance is the same as that of the Sun's centre; at midnight it just grazes the horizon. Find the latitude.

2. On January 2, 1881, on a ship in the North Atlantic in longitude 48° W., it was observed that the Sun's meridian altitude was 15° 21′ 45″. The Sun's declination at noon at Greenwich on the same day was 22° 54′ 33″, and the hourly variation 13·78″. Find the ship's latitude.

3. Show how to find the latitude by observing the *difference* of the meridian zenith distances of two known stars which cross the meridian on opposite sides of the zenith at nearly equal distances from it. Explain whether the stars chosen should be near to or remote from the zenith. Give also the advantages and disadvantages of this method of finding the latitude, as compared with the method of circumpolars.

4. On a certain day the observed meridian altitude of α *Cassiopeiæ* (declination 55° 49′ 11·1″ N.) was 85° 10′ 18″. The eye of the observer was 18 feet above the horizon, and the error for refraction for the altitude of the star is 5″; determine the latitude.

5. The deck of a ship (stationary) is 25 feet from the sea, and the dip of the horizon at 1 foot is 1′; if the two meridian altitudes of a circumpolar star from the sea horizon be 60° 2′ and 29° 58′, find the latitude.

6. At the winter solstice the meridian altitude of the Sun is 15°. What is the latitude of the place? What will be the meridian height of the Sun at the equinoxes and at the summer solstice?

7. Describe the altazimuth, and show how it can be used to find the time of apparent noon and the azimuth of the meridian by the method of equal altitudes.

8. A vertical rod is fixed exactly in the centre of a circular fountain basin, and it is observed that on the 25th of July the extremity of the shadow exactly reaches the margin of the water at 10h. 7m. A.M., and at 2h. 25m. P.M. The equation of time on that day is + 6m. What is the error, compared with local time, of the watch by which these observations were taken?

9. In the railway station at Ventimiglia is a clock one face of which indicates Paris time, the other Roman time. It is observed that, when the former indicates 12h. 39m. 4s., the latter indicates 1h. 19m. 40s. The longitude of Paris being 2° 21′ E., find the longitude of Rome.

10. In Question 9, what is the corresponding local time at Ventimiglia, the longitude being 7° 35′ E. ?

11. A chronometer is set by the standard clock at Greenwich at 6 A.M. It is then taken to Shepton Mallet, and indicates noon when the local time is 11h. 49m. 50s. The chronometer is then brought back to Greenwich, and indicates 9 P.M., when the correct time is 8h. 59m. 55s. Find the longitude of Shepton, supposing the chronometer rate uniform.

12. In applying the lunar method, find the error in the calculated longitude of the observer due to an error of 1′ in the tables of the Moon's longitude.

13. Amerigo Vespucci is said to have found his longitude in latitude 10° N. in the following manner. At 7.30 P.M. the Moon was 1° E. of Mars, at midnight the Moon was $5\frac{1}{2}$° E. of Mars. The Nuremberg time of conjunction of the Moon and Mars was midnight. Hence he calculated that his longitude was $82\frac{1}{2}$° W. of Nuremberg. Discuss the accuracy of the method, and point out the necessary corrections.

14. A chronometer whose rate is uniform is found at Greenwich to have an error of δ_1 hours when the time which it indicates is t_1. It is then taken to a place A, and when it indicates t_2 it is found that the excess of the observed *local time* of the place A over t_2 is δ_2 hours. It is now again brought back to Greenwich, and the chronometer time and error are observed to be t_3 and δ_3 hours respectively. Prove that the longitude of A east of Greenwich is

$$15 \ (\delta_2 t_3 + \delta_3 t_1 + \delta_1 t_2 - t_2 \delta_3 - t_3 \delta_1 - t_1 \delta_2)/(t_3 - t_1) \text{ degrees.}$$

15. The sidereal times of transit of a certain star across the meridian of an observatory A, as recorded at A, and by a telegraphic signal at B, are t_1, t_2 respectively. The sidereal times of transit of the same star across the meridian of B, recorded by telegraphic signal at A, and at B, are T_1, T_2 respectively. If the signals take the same time to travel in either direction, show that the difference of the longitudes of B and A in angular measure

$$= \tfrac{15}{2}(T_1 - t_1 + T_2 - t_2).$$

16. The altitudes of two known *stars* are observed at a given instant of time. Show how to find on a terrestrial globe the places at which the stars are vertically overhead, and give a geometrical construction for the place of observation.

17. In Question 16, find the condition that there should be two, one, or no possible positions of a ship at which the altitudes of the known stars have certain given values.

18. If longitude is found by lunar distances, and latitude by meridian altitudes, find the latitude in which an error of 1′ in the sextant reading will introduce the same error in both observations if estimated not in angle, but in miles on the Earth's surface.

EXAMINATION PAPER.—VII.

1. Give a description of the Sextant, and explain how to use it for taking altitudes (1) at sea, (2) on land.

2. How does a Chronometer differ from an ordinary watch? What are its *error* and *rate?*

3. Prove that a single meridian altitude of a star, whose declination is known, will determine the latitude. Why is a zenith sector sometimes preferred to a transit circle for this purpose?

4. Show how the latitude is determinable by two meridian observations of a circumpolar star. Why is this method not generally applicable on board ship?

5. Show how to find the latitude of a place (1) by observing the Sun's altitude at a given time; (2) by the Prime Vertical Instrument.

6. Describe the method of equal altitudes for finding the time of transit of a celestial body. If the times be observed by the ship's chronometer, show how to find the longitude.

7. What methods are available for the determination of Greenwich time at sea? Describe the method of taking lunar distances.

8. How is the difference of longitude determined by electric telegraph? Explain how the personal equation and the time of transmission of the signal are eliminated.

9. Contrast the method of Moon-culminating Stars with that of Lunar Distances in respect of the instruments employed, and of the intricacy of the calculations involved. What other celestial signals have been proposed, and what is their disadvantage?

10. Knowing the Greenwich time, show how to construct graphically on a globe the position of the ship without any calculation whatever.

CHAPTER VIII.

THE MOON.

Section I.—Parallax—The Moon's Distance and Dimensions.

247. Definitions.—By the **Parallax** of a celestial body is meant the angle between the straight lines joining it to two different places of observation.

In § 5 we stated that the fixed stars are seen in the same direction from all parts on the Earth; hence such stars have no appreciable parallax. The Moon, Sun, and planets, on the other hand, are at a (comparatively) much smaller distance from the Earth, and their parallax is a measurable quantity. The distance of the Moon from the Earth's centre is about 60 times the radius of the Earth. The effects of parallax in connection with the method of Lunar Distances have already been mentioned (§ 237).

To avoid the necessity of specifying the place of observation, the direction of the Moon or any other celestial body is always referred to the **centre of the Earth.** The direction of a line joining the body to the Earth's centre is called the body's **geocentric** direction. The angle between the geocentric direction and the direction of the body relative to any given observatory is called the body's **Geocentric Parallax**, or more shortly, its **Parallax.** Thus the geocentric parallax is the angle subtended at the body by the radius of the Earth through the point of observation.

The **Horizontal Parallax** is the geocentric parallax of a body when on the horizon of the place of observation.

248. General Effects of Geocentric Parallax.—
Assuming the Earth to be spherical, let C (Fig. 77)
be the Earth's centre, O the place of observation, and M the
centre of the Moon or other observed body. Then the angle
OMC is the geocentric parallax of M.

Produce CO to Z; then OZ is the direction of the zenith
at O, and ZOM is therefore the zenith distance of M as seen
from O (corrected of course for refraction). Now

$$\angle ZOM = \angle ZCM + \angle OMC ;$$

therefore the apparent zenith distance of M is increased by
the amount of the geocentric parallax. Conversely to find
$\angle ZCM$ we must subtract the parallax OMC from the
observed zenith distance ZOM.

The azimuth is unaltered by parallax, because OM, CM
lie in the same plane through OZ.

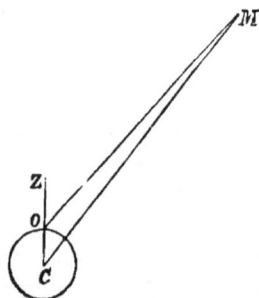

Fig. 77.

249. To find the Correction for Geocentric Parallax.
In Fig. 77, let

$a = CO = $ Earth's radius,
$d = CM = $ Moon's (or other body's) geocentric distance,
$z = ZOM = $ observed zenith distance of M,
$p = OMC = $ parallax of M.

By Trigonometry, since the sides of $\triangle OMC$ are proportional to the sines of the opposite angles,

$$\therefore \quad \frac{\sin CMO}{\sin COM} = \frac{CO}{CM},$$

that is
$$\frac{\sin p}{\sin z} = \frac{a}{d}.$$

Therefore
$$\sin p = \frac{a}{d}\sin z.$$

Let P be the horizontal parallax of M. Then, when $z = 90°$, $p = P$, and therefore the last formula gives

$$\sin P = \frac{a}{d}\sin 90° = \frac{a}{d}.$$

Hence, by substitution,

$$\sin p = \sin P \cdot \sin z.$$

This formula is exact. But the angles p and P are in every case very small, and therefore their sines are very approximately equal to their circular measures. Hence we have the approximate formula

$$p = P \cdot \sin z,$$

or, **The parallax of a celestial body varies as the sine of its apparent zenith distance.**

The last formula holds good no matter what be the unit of angular measurement. Thus if p'', P'' denote the numbers of seconds in p, P respectively, we have, by reducing to seconds,
$$p'' = P'' \sin z.$$

EXAMPLES.

1. Supposing the Sun's horizontal parallax to be 8·8″, to find the correction for parallax when the Sun's altitude is 60°.

Here $z = 90° - 60° = 30°$, $P'' = 8·8''$, and therefore
$$p'' = P'' \sin 30° = 8·8'' \times \tfrac{1}{2} = 4·4''.$$

2. To find the corrections for the Moon's parallax for altitudes of 30° and 45°, the Moon's horizontal parallax being 57′.

In the two cases we have respectively $z = 60°$ and $z = 45°$, and the corresponding corrections are
$$p'' = 57' \sin 60° = 57' \times \tfrac{1}{2}\sqrt{3} = 28'30'' \times \sqrt{3}$$
$$= 1710'' \times 1·7320 = 2961·7'' = 49'21·7'',$$
and
$$p'' = 57' \sin 45° = 57' \times \tfrac{1}{2}\sqrt{2} = 28'30'' \times \sqrt{2}$$
$$= 1710'' \times 1·4142 = 2418·3'' = 40'18·3''.$$

250. Relation between the Horizontal Parallax and Distance of a Celestial Body.—In the last paragraph

we showed that $\qquad \sin P = \dfrac{a}{d}.$

This formula may be proved independently by drawing MA to touch the Earth at A. M is on the horizon at A; the $\angle CMA$ is therefore the horizontal parallax P, and we have immediately

$$\sin P = \sin CMA = CA/CM = \quad /d.$$

Since P is small, we have approximately

Circular measure of $P = a/d$.

and therefore in seconds

$$P'' = \frac{180 \times 60 \times 60}{\pi} \frac{a}{d} = 206265\,\frac{a}{d},$$

which shows that, **The horizontal parallax of a body varies inversely as its distance from the Earth.**

Fɪɢ. 78.

If we know the Earth's radius a and the distance d, the last formula enables us to calculate the horizontal parallax P''. Conversely, if we know the horizontal parallax of a body we can calculate its distance.

EXAMPLE 1.—Given that the Moon's distance is 60 times the Earth's radius, to find the Moon's horizontal parallax.

We have $\qquad \dfrac{a}{d} = \dfrac{1}{60};$

$\therefore \qquad$ circular measure of $P = \dfrac{1}{60}$ approximately.

Now the unit of circular measure $= 57{\cdot}2957°$;

$\therefore \qquad P$ (in angular measure) $= \dfrac{1}{60} \times 57{\cdot}2957° = 57{\cdot}2957'$

$$= 57' \, 17{\cdot}7'',$$

and this is the required horizontal parallax.

EXAMPLE 2.—Given that the Sun's parallax* is 8·8″, to find the Sun's distance, the Earth's radius being 3,960 miles.

The circular measure of 8·8″ is $= \dfrac{8\cdot8 \times \pi}{180 \times 60 \times 60}$,

and, by the formula, we have, for the Sun's distance in miles,

$$d = \frac{a}{\text{circ. meas. of } P} = \frac{3960 \times 180 \times 60 \times 60}{8\cdot8 \times \pi}.$$

Taking $\pi = 3\frac{1}{7}$, and calculating the result correct to the first three significant figures, we find the Sun's distance d

$$= \textbf{92,800,000 miles} \text{ approximately.}$$

It would be useless to carry the calculations beyond the third figure, for, of course, the values of the Earth's radius and Sun's parallax are only approximate; moreover, we should have to use the more accurate value of π, viz., 3·141592......

251. Comparison between Parallax and Refraction.

—It will be noticed that while parallax and refraction both produce displacements of the apparent position of a body along a vertical circle, the displacement due to parallax is directed *away* from the zenith, and is always proportional to the *sine* of the zenith distance, while that due to refraction is directed *towards* the zenith, and is proportional to the *tangent* of the zenith distance, provided the altitude is not small. Also the correction for parallax is inversely proportional to the distance of the body, and is imperceptible, except in the case of members of our solar system; while the correction for refraction is independent of the body's distance, and depends only on the condition of the atmosphere.

The Moon's horizontal parallax is about 57′, while the horizontal refraction is only 33′. Hence, by the combined effects of parallax and refraction, the Moon's apparent altitude is diminished, or its Z.D. increased. The time of rising is, therefore, on the whole retarded, and the time of setting accelerated. The effect of parallax on the times of rising and setting may be investigated by the methods of §§ 104, 190.

For all other bodies, including the nearest planets, the correction for refraction far outweighs that due to parallax.

---

* When astronomers speak of the *parallax* of the Sun, Moon, or a planet, without further specifying the observation, the *horizontal parallax* is always to be understood.

252. To find the Moon's Parallax by Meridian Observations.—The Moon's parallax may be conveniently determined as follows. Let A and B be two observatories situated on the same meridian, one north, the other south of the equator. Let M denote the Moon's centre, and let x be a star having no appreciable parallax, whose R.A. is approximately equal to that of the Moon, their declinations being also nearly equal.

Let the Moon's meridian zenith distances ZAM and $Z'BM$ be observed with the transit circles at A and B, and let xAM and xBM, the differences of the meridian Z.D.'s of the Moon and star at the two stations, be also observed.

Let $z_1 = \angle ZAM,$ $z_2 = \angle Z'BM.$
$a_1 = \angle xAM,$ $a_2 = \angle xBM.$
$P =$ Moon's required horizontal parallax.

By § 249, we have, approximately,
$$\angle AMC = P \sin z_1, \quad \angle BMC = P \sin z_2.$$

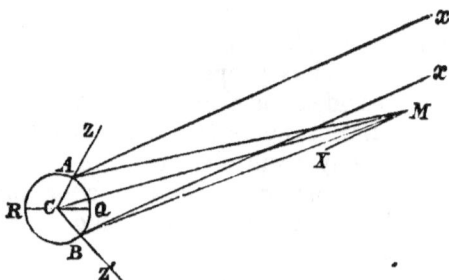

FIG. 79.

$$\therefore \quad \angle AMB = P (\sin z_1 + \sin z_2) \ \ldots\ldots\ldots \ \text{(i.)}.$$
Moreover, if MX be drawn parallel to Ax or Bx,
$$\angle XMA = \angle MAx = a_1 ;$$
$$\angle XMB = \angle MBx = a_2 ;$$
$$\therefore \quad \angle AMB = a_1 - a_2 \ \ldots\ldots\ldots\ldots\ldots \ \text{(ii.)}.$$
From (i.) and (ii.),
$$P (\sin z_1 + \sin z_2) = a_1 - a_2 ;$$
$$\therefore \quad P = \frac{a_1 - a_2}{\sin z_1 + \sin z_2} ;$$
whence the Moon's parallax, P, may be found.

253. If the two observatories are not on the same meridian, allowance must be made for the change in the Moon's declination between the two observations. Let the stations be denoted by A, B, and let B' be the place on the meridian of A, which has the same latitude as B. Then, if the Moon's meridian Z.D. be observed at B, we can, by adding or subtracting the change of declination during the interval, find what would be the meridian Z.D. if observed from B'. Moreover, the star's meridian Z.D. is the same both at B and at B'. Hence it is easy to calculate what would be the angles at B' corresponding to the observed angles at B. From the former, and the observed angles at A, we find the parallax P, as before.

To ensure the greatest accuracy, it is advisable that the difference of longitude of the two stations should be so small that the correction for the Moon's motion in declination is trifling. It is necessary, however, that $a_1 - a_2$ should be large; for this reason the stations should be chosen one as far north and the other as far south of the equator as possible. The observatories at Greenwich and the Cape of Good Hope have been found most suitable.

The principal advantage of the above method is that the probable errors arising from any uncertainty in the corrections for refraction are diminished as far as possible.

For, since the Moon and observed star have nearly the same declination, the corrections for refraction to be applied to a_1, a_2, their small *differences* of Z.D., are very small indeed. The errors are not of so much moment in the denominator $\sin z_1 + \sin z_2$, as the latter is not itself a small quantity.

From such observations, the mean horizontal parallax of the Moon has been found to be **57' 2·707"**.

This value corresponds to a mean distance of 60·27 times the equatorial radius of the Earth, or **238,840** miles. The distance and parallax of the Moon are not, however, quite constant; their greatest and least values are in the ratio of (roughly) 19 : 17. For rough calculations, the Moon's distance may be taken as 60 times the Earth's radius.

Neither this method nor the next (§ 254) gives accurate results for the Sun, for the brilliancy of the rays renders all stars in its neighbourhood invisible

254. To find the Parallax of a Planet from Observations made at a Single Observatory.—The parallax of Mars, when nearest the Earth, has also been determined by the following method, depending on the Earth's rotation.

Since the apparent altitude of a body is always diminished by parallax, it can easily be seen by a figure, that, shortly after a planet has risen, its R.A. and longitude appear greater than their geocentric values (the planet being displaced eastwards), while shortly before setting they appear less than their geocentric values (the displacement being westwards). The planet's position, relative to certain fixed stars, is observed soon after rising and before setting by means of an equatorial furnished with a micrometer or heliometer.

The observed change of position is due partly to parallax and partly to the planet's motion relative to the Earth's centre during the interval between the observations, which produces displacements far greater than those due to parallax. But by repeating the observations on successive days, the planet's rate of motion can be accurately determined, and the displacements due to parallax can thus be separated from those due to relative motion. Refraction need not be allowed for; because it affects those stars with which the planet is compared, as well as the planet itself.

This method can be used for the Moon, but the Moon's motion is so rapid that the calculations are more complicated.

*255. Effect of the Earth's Ellipticity.—The effect of parallax is made rather more complicated by the spheroidal form of the Earth. For, by § 249, the *magnitude* of the horizontal parallax at any place depends on its distance from the Earth's centre, and since this distance is not the same for all places on the Earth, the horizontal parallax is not everywhere the same. Again, the *direction* in which the body is displaced is away from the line (produced) joining the centre of the Earth with the observer (§ 248). But this line does not pass exactly through the zenith (§ 117). Hence the displacement is not in general along a vertical, so that the azimuth as well as altitude is very slightly altered by parallax.

256. The Equatorial Horizontal Parallax is the geocentric parallax of a body seen on the horizon of a place at the Earth's *equator*. It is generally adopted as the measure of the parallax of a celestial body. Its sine is equal to

(Earth's *equatorial* radius)/(body's geocentric distance).

257. Relation between Parallax and Angular Diameter.—In Fig. 80 it will be seen that the angle CMA, which measures the parallax of M, also measures the Earth's angular semi-diameter as it would appear from M. Thus, *the Moon's parallax is the angular semi-diameter of the Earth as it would appear if observed from the Moon.*

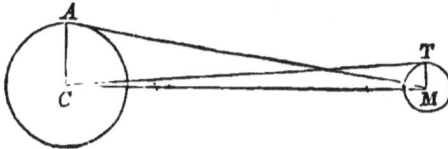

FIG. 80.

258. To Find the Moon's Diameter.—Let a, c be the radii of the Earth and Moon respectively, measured in miles, d the distance between their centres, P the Moon's horizontal parallax, m the Moon's angular semi-diameter as it would appear if seen from the Earth's centre. Then, from Fig. 80,

$$\sin P = \frac{a}{d}, \qquad \sin m = \sin TCM = \frac{TM}{CM} = \frac{c}{d};$$

$\therefore c : a = \sin m : \sin P = m : P$ approximately;

i.e. (rad. of Moon) : (rad. of Earth)
$\qquad = (\, \mathbb{C}\,$'s ang. semi-diam.) : ($\mathbb{C}$'s hor. parx.).

Hence, knowing the Moon's horizontal parallax and its angular diameter, the Moon's radius can be found.

The Moon's mean angular diameter $2m$ is observed to be about $31' 5''$. From this the Moon's actual diameter is readily found to be about 2160 miles, or $\frac{3}{11}$ of the Earth's diameter.

The surfaces of spheres are proportional to the squares, and the volumes to the cubes of their radii. Hence the Moon's superficial area is about $\frac{9}{121}$, or $\frac{3}{40}$, and its volume about $\frac{27}{1331}$, or $\frac{1}{50}$ of that of the Earth.

EXAMPLE.—To find the Moon's diameter in miles, given
$\qquad\qquad \mathbb{C}$'s angular diameter $= 31' 7''$,
$\qquad\quad \mathbb{C}$'s equatorial horizontal parallax $= 57' 2''$,
$\qquad\quad$ Earth's equatorial radius $= 3963$ miles.

$\therefore \mathbb{C}$'s diameter $2c = a \times \dfrac{2m}{P} = 3963 \times \dfrac{31' 7''}{57' 2''} = 3963 \times \dfrac{1867}{3422} = 2162.$

Thus the Moon's diameter is 2162 miles.

SECTION II.—*Synodic and Sidereal Months—Moon's Phases—Mountains on the Moon.*

259. **Definitions.**—In § 40 we defined the *lunation* as the period between consecutive new Moons, and showed that it was rather longer than the period of the Moon's revolution relative to the stars. We shall now require the following additional definitions, most of which apply also to the planets.

The **elongation** of the Moon or planet is the difference between its celestial longitude and that of the Sun. If the body were to move in the ecliptic its elongation would be its angular distance from the Sun.

The Moon or planet is said to be in **conjunction** when it has the same longitude as the Sun, so that its elongation is zero. The Moon is in conjunction at *new Moon* (§ 40). The body is in **opposition** when its elongation is 180°. In both positions it is said to be in **syzygy**. The body is said to be in **quadrature** when its elongation is either 90° or 270°.

The period between consecutive conjunctions is called the **synodic period** of the Moon or planet. The Moon's synodic period is, therefore, the same as a lunation; it is also called a **Synodic Month**. In this period the Moon's elongation increases by 360°, the motion being *direct*.

The period of revolution relative to the stars is called the **sidereal period**; that of the Moon, the **Sidereal Month.**

The average length of the **Calendar Month** in common use is slightly in excess of the synodic month (*cf.* § 171).

260. **Relation between the Sidereal and Synodic Months.**

Let the number of days in a year be Y, in a sidereal month M, and in a synodic month S.

In M days the Moon's longitude increases   360°;
∴ in 1 day the Moon's longitude increases $360°/M$.
Similarly in 1 day the Sun's longitude increases $360°/Y$,
and the Moon's elongation increases $360°/S$.
Now, from the definition,

(Moon's elongation) = (Moon's long.) − (Sun's long.),

and their daily rates of increase must be connected by the same relation ;

$$\therefore \frac{360}{S} = \frac{360}{M} - \frac{360}{Y};$$

$$\therefore \quad \frac{1}{S} = \frac{1}{M} - \frac{1}{Y}, \text{ or } \frac{1}{M} = \frac{1}{S} + \frac{1}{Y};$$

i.e., $\dfrac{1}{\text{sider. month}} = \dfrac{1}{\text{synod. month}} + \dfrac{1}{\text{year}}.$

EXAMPLE.—Find (roughly) the length of the sidereal month, given that the synodic month (S) = 29½d., and the year (Y) = 365¼d.

Here we have $\dfrac{1}{M} = \dfrac{1}{29\frac{1}{2}} + \dfrac{1}{365\frac{1}{4}}.$

To simplify the calculations, we put the relation into the form

$$\frac{29\frac{1}{2} \times 365\frac{1}{4}}{29\frac{1}{2} + 365\frac{1}{4}} = 29\frac{1}{2} \times \frac{365\frac{1}{4}}{394\frac{3}{4}} = 29\frac{1}{2} \times \left(1 - \frac{29\frac{1}{2}}{394\frac{3}{4}}\right)$$

$$= 29\cdot5 - 29\cdot5 \times \frac{118}{1579} = 29\cdot5 - 2\cdot20 = 27\cdot3.$$

Hence the sidereal month is very nearly **27½** days.

261. To determine the Moon's Synodic Period.— An eclipse of the Sun can only happen at conjunction, and an eclipse of the Moon at opposition, and the middle of the eclipse determines the exact instant of conjunction or opposition, as the case may be. Hence, by observing the exact interval of time between the middle of two eclipses, and counting the number of lunations between them, the length of a single lunation, or synodic period, can be found with great accuracy expressed in mean solar units of time.

The records of ancient eclipses enable us to find a still closer approximation to the mean length of the lunation. From modern observations, the length of a lunation has been found with sufficient accuracy to enable us to tell the exact number of lunations between these ancient eclipses and a recent lunar eclipse (this number being, of course, a *whole number*). By dividing the known interval in days by this number, the mean length of the synodic period during the interval can be accurately found. At the present time the length of a lunation is 29·5305887 days, or 29d. 12h. 44m. 2·7s. nearly.

From this the length of the Moon's sidereal period is calculated, as in § 260, and found to be 27d. 7h. 43m. 11·5s. nearly

262. Phases of the Moon.—The acccompanying diagrams will show how tho phases of the Moon are accounted
for on the hypothesis that the Moon is an opaque body
illuminated by the Sun. In the upper figure the central
globe represents the Earth, the others represent the Moon in
different parts of its orbit, while the Sun is supposed to be at
a great distance away to the right of the figure.* The half
of the Moon that is turned towards the Sun is illuminated, the other half being dark. The Moon's appearance
depends on the relative proportions of the illuminated and
darkened portions that are turned towards the Earth.

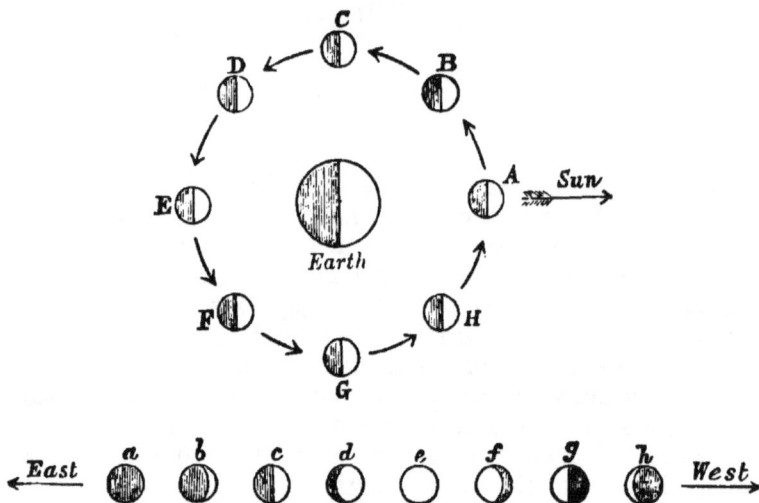

FIG. 81.

The lower figures, *a, b, c, d, e, f, g, h,* represent the appearances of the Moon relative to the ecliptic, as seen from the
Earth when in the positions represented by the corresponding
letters in the upper figure.

---

* The Sun's distance is about 390 times the Moon's. If the
former be represented by an inch, the latter will be represented by
about 11 yards.

At *A*, *a* the Moon is in conjunction, and only the dark part is towards the Earth. This is called **New Moon**.

At *B*, *b* a portion of the bright part is visible as a crescent at the western side of the disc. The Moon's appearance is known as **horned**. The points or extremities of the horns are called the **cusps**.

At *C*, *c* the Moon's elongation is 90°, and the western half of the disc, or visible portion, is illuminated, the eastern half being dark. The Moon is then said to be **dichotomized**. This is called the **First Quarter**. The Moon's age is about 7½ days.

At *D*, *d* more than half the disc is illuminated. The Moon's appearance is then described as **gibbous**.

At *E*, *e* the Moon is in opposition. The whole of the disc is illuminated. This is called **Full Moon**. The Moon's age is about 15 days.

At *F*, *f* a portion of the disc at the western side is dark. The Moon is again gibbous, but the bright part is turned in the opposite direction to that which it has at *D*, *d*.

At *G*, *g* the Moon's elongation is 270°. The eastern half of the disc is illuminated, and the western half is dark. The Moon is again dichotomized. This is called the **Last Quarter**. The Moon's age is about 22 days.

At *H*, *h* only a small crescent in the eastern portion is still illuminated. The Moon is now again horned, but the horns are in the opposite direction to those in *B*, *b*.

Finally, the Moon comes round to conjunction again at *A*, and the whole of the part towards the Earth is dark.

From new to full Moon, the visible illuminated portion increases, and the Moon is said to be **waxing**. From full to new, the illuminated portion decreases, and the Moon is said to be **waning**.

It will be noticed from a comparison of the figures that the illuminated portion of the visible disc is always that nearest the Sun. Moreover, its area is greater the greater the Moon's elongation.*

---

* The phases of the Moon may be readily illustrated experimentally, by taking an opaque ball, or an orange, and holding it in different directions relative to the light from the Sun or a gas-burner.

263. Relation between Phase and Elongation.—Let M (Fig. 82) be the centre of the Moon, MS the direction of the Sun, $E'ME$ that of the Earth. Draw the great circles AMB perpendicular to ME, and CMD perpendicular to MS; the former is the boundary of the part of the Moon turned towards the Earth, and the latter is the boundary of the illuminated portion. Hence the visible bright portion is the lune AMC. The angle of the lune, $\angle AMC$, is equal to $\angle E'MS$ (Sph. Geom. 16). The area of a spherical lune is proportional to its angle. Hence,

$$\frac{\text{area of visible illuminated part}}{\text{area of hemisphere}} = \frac{\angle AMC}{180°} = \frac{\angle E'MS}{180°}$$
$$= \frac{180° - \angle EMS}{180°}.$$

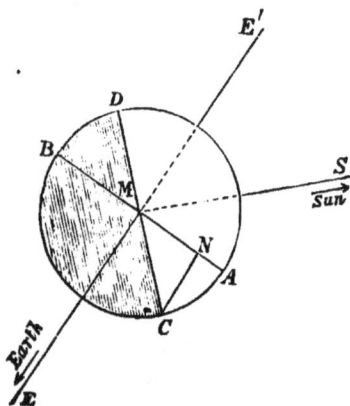

FIG. 82.

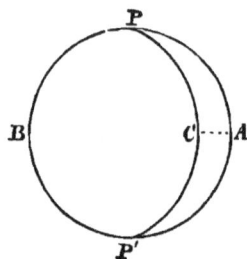

FIG. 83.

But this does not give the "*apparent area*" of the bright part. For, as in § 145, the apparent area of a body is the area of the disc formed by projecting the body on the celestial sphere. If N denote the projection of the point C on the plane AMB (so that CN is perpendicular to BA), the arc AC will be seen in perspective as a line of length AN, and the bright part will be seen as a *plane lune* (Fig. 83), whose boundary PCP' optically forms the half of an ellipse whose major axis is PP', and minor axis $2MN$.

It may be shown that

area of half-ellipse PCP' : area of semicircle PAP'

$$= MN : MA$$

and $\therefore$ area $APCP'$: area $APBP' = AN : AB$

$$= 1 - \cos AMC : 2 = 1 - \cos E'MS : 2.$$

Hence the apparent area of the bright part is proportional to $1 - \cos SME'$.

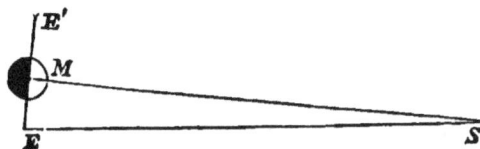

FIG. 84.

The angle SME' differs from the Moon's elongation SEM by the small angle ESM (Fig. 84); *i.e.*, the angle which the Moon's distance subtends at the Sun. This angle is very small, being always less than $10'$. Hence the area of the phase is very approximately proportional to $1 - \cos(\text{Moon's elongation})$.

264. Determination of the Sun's Distance by Aristarchus.—From observing the Moon's elongation when dichotomized, Aristarchus (B.C. 270 *circ.*) made a computation of the Sun's distance in the following manner. When the Moon is dichotomized, $\angle SME = 90°$, the Moon's elongation $\angle SEM = 90° - \angle ESM$, and $\cos SEM = EM/ES$. Hence, by observing the angle SEM, the ratio of the Sun's distance to the Moon's was computed.

But this method is incapable of giving reliable results, owing to the impossibility of finding the exact instant when the Moon is dichotomized. The Moon's surface is rough, and covered with mountains, and the tops of these catch the light before the lower parts, while throwing a shadow on the portions behind them. Hence the boundary of the bright part is always jagged, and is never a straight line, as it would be at the quarters, if the surface of the Moon were perfectly smooth. In fact, Aristarchus estimated the Sun's distance as only about 19 times that of the Moon, whereas they are really in the proportion of nearly 400 to 1.

**265. Earth-Shine on the Moon. — Phases of the
Earth.**—When the Moon is nearly new, the unilluminated
portion of its surface is distinctly visible as a disc of a dull-
red colour. This appearance is due to the light reflected
from the Earth as "Earth-shine," which illuminates the
Moon in just the same way that the moonshine illuminates
the Earth at full Moon. From § 258, the Earth's superficial
area is greater than the Moon's in the proportion of about
40 : 3. Consequently the Earth-shine on the Moon is more
than 13 times as bright as the moonshine on the Earth.

The Earth, as seen from the Moon, would appear to pass
through phases similar to those of the Moon, as seen from
the Earth. The Earth's and Moon's phases are evidently
supplementary. Thus, when the Moon is new the Earth
would appear full, and *vice-versâ*; when the Moon is in the
first quarter, the Earth would appear in the last quarter.

Owing, however, to twilight, the boundary of the Earth's
illuminated portion would not be so well defined as in the
case of the Moon ; there would be a gradual shading off from
light to darkness, extending over a belt of breadth 18° on
beyond the bright part. The entire absence of twilight on
the Moon is one of the strongest evidences against the existence of a lunar atmosphere similar to that of our Earth.

266. Appearance of Moon relative to the Horizon.—
We are now in a position to represent, in a diagram, the
Moon's position and appearance relative to the horizon at a
given time of day and year when the Moon's age is given.

The ecliptic having been found, as explained in § 41,
the age of the Moon determines the Moon's elongation,
as in § 40. Measuring this angle along the ecliptic, we find
the Moon's position roughly ; for the Moon is never very far
from the ecliptic (*cf.* § 40). The elongation also determines
the phase, and enables us to indicate the appearance of
the disc. The bright side or limb is always turned towards
the Sun. The cusps, therefore, point in the reverse direction,
and the line joining them is perpendicular to the ecliptic.

We can also trace the changes in the direction of the
Moon's horns relative to the horizon, between its time of
rising and setting.

Take, for example, the case when the Moon is a few (say three) days old. The Moon is then a little east-of the Sun; therefore the bright limb is at the western side of the disc, and the horns point eastward. Hence, at rising, the horns are pointed downwards, and at setting they are pointed upwards (Fig. 85).

FIG. 85.   FIG. 86.

When the Moon is waning, the reverse will be the case (Fig. 86).

267 Heights of Lunar Mountains.—We stated in § 264 that the Moon's surface is covered with mountains, and that in consequence the bounding line between the illuminated and dark portions of the disc is always jagged and irregular; while the mountains themselves throw their shadows on the portions of the surface behind them. These circumstances have led to the two following different ways of measuring the height of the lunar mountains

First Method.—If a tower is standing in the middle of a perfectly level plain, it is evident from trigonometry that the length of the shadow, multiplied by the tangent of the Sun's altitude, gives the height of the tower. The same will be true in the case of the shadow cast by a mountain, provided we measure the length of the shadow from a point vertically underneath the summit. Now, in the case of the Moon it is possible, from knowing the Moon's age, to calculate exactly what would be the altitude of the Sun as it would be seen from any point of the lunar surface. The apparent length of the shadows of the mountains can be measured, in angular measure, by means of a micrometer; from this their actual length can be calculated, allowance being, of course, made for the fact that we are not looking vertically down on the shadows, and hence they appear foreshortened. In this way, the height of the mountains can be found.

The principal disadvantage of this method is, that if the surface of the Moon surrounding the mountain should be less flat than it has been estimated, there will be a corresponding error in the height of the mountain. In particular, it would be impossible to apply the method to find the heights of mountains closely crowded together.

268. Second Method.—In treating of the Earth in § 104, we showed that one effect of the dip of the horizon is to accelerate the times of rising, and to retard the times of setting of the Sun and stars. We also showed how to calculate the amount of the acceleration if the dip be known. Conversely, if the acceleration in the time of rising be known, the dip of the horizon can be calculated, and from this the height of the observer above the general level of the Earth may be found.

Now precisely the same method may be applied to measure the heights of lunar mountains. When the Moon is waxing the Sun is gradually rising over those parts of the Moon's surface which are turned towards the Earth. The tops of the mountains catch the rays before the lower parts, and, therefore, stand out bright against the dark background of the unilluminated parts below. Similarly, when the Moon is waning, the summits of the mountains remain as bright specks after the lower portions are plunged in shadow. By noticing the exact instant at which the Sun's rays begin or cease to illuminate the summit, this acceleration or retardation, due to dip, may be calculated, and the height of the mountain determined.

If the Moon's surface around the mountain is fairly level, the distance of the mountain from the illuminated portion at the instant of disappearance determines *the distance of the visible horizon* as seen from the mountain. This distance can be calculated from measurements made with a micrometer (proper allowance being made for foreshortening if the mountain is not in the centre of the disc).

Hence the height (h) of the mountain may be calculated by the formula of § 101 (i.), viz., $h = d^2/2a$, where d is the estimated distance of the horizon, and a the *Moon's* radius.

SECTION III.—*The Moon's Orbit and Rotation.*

269. The Moon's Orbit about the Earth can be investigated by a method precisely similar to that employed in the case of the Sun (see § 145). The Moon's R.A. and decl. may be observed daily by the Transit Circle. The observed decl. must be corrected for refraction and parallax (neither of which affect the R.A., since the observations are made on the *meridian*). We thus find the positions of the Moon on the celestial sphere relative to the Earth's *centre* for every day at the instant of its transit across the meridian of the observatory.

Instead of observing the Moon's parallax daily, the Moon's distances from the Earth's centre on different days, may be compared by measuring the Moon's angular diameters, with the heliometer. Here, however, another correction for parallax is required. For the observed angular diameters are inversely proportional to the corresponding distances of the Moon from the *observer*, and not from the *centre* of the Earth.

This correction is by no means inconsiderable. Thus, for example, if the Moon be vertically overhead, its distances from the observer and from the Earth's centre will differ by the Earth's radius, *i.e.*, by about $\frac{1}{60}$ of the latter distance, and its angular diameter will, therefore, be increased in the proportion of about 60 to 59.

Having thus determined the direction and distance of the Moon's centre, relative to the Earth's centre, for every day in the month, the Moon's orbit may be traced out in just the same way as the Sun's orbit was traced out in § 146. It is thus found that the motion obeys approximately the following laws :—

(i.) *The Moon's orbit lies in a plane through the Earth's centre, inclined to the plane of the ecliptic at an angle of about* **5° 8′.**

(ii.) *The orbit is an ellipse, having the Earth's centre in one focus, the eccentricity of the ellipse being about* $\frac{1}{18}$.

(iii.) *The radius vector joining the Earth's and Moon's centres traces out equal areas in equal intervals of time.*

The period of revolution is, of course, the sidereal lunar month, as defined in Section II., namely, about 27¼ days.

The laws which govern the Moon's motion are thus identical with Kepler's laws for the Earth's orbital motion round the Sun (§ 155).

270. The Eccentricity of the Moon's Orbit is found by comparing the Moon's greatest and least distances, which are inversely proportional to its least and greatest (geocentric) angular diameters respectively. The latter are in the ratio of about 17 to 19, and it is inferred that the eccentricity is about $(19-17)/(19+17)$, or $\frac{1}{18}$ (*cf.* § 149).

The terms **perigee, apogee, apse line** are used in the same sense as in § 147. *Perigee* and *apogee* are the points in the orbit at which the Moon is nearest to and furthest from the Earth respectively. Both are called the *apses* or *apsides*, the line joining them being called the *apse line, apsidal line* or *line of apsides*, according to choice. It is the major axis of the orbit.

As in § 151, it follows that the Moon's angular motion in its orbit is swiftest at **perigee**, and slowest at apogee.

271. Nodes.—The points in which the Moon's orbit, or its projection on the celestial sphere, cuts the ecliptic are called the Moon's **Nodes** (*cf.* § 40). The line joining them is called the **Nodal Line.** It is the line of intersection of the planes of the Moon's orbit and ecliptic. That node through which the Moon passes in crossing from south to north of the ecliptic is distinguished as the **ascending node,** the other is distinguished as the **descending node.**

272. Perturbations.—As the result of observations extending over a large number of lunar months, it is found that the Moon does not describe exactly the same ellipse over and over again, and that, therefore, the laws stated in § 269 are only approximate. The actual motion can, however, be represented by supposing the Moon to revolve in an ellipse, the positions and dimensions of which are very slowly varying. This mode of representing the motion may be illustrated by imagining a bead to revolve on a smooth elliptic wire which is very slowly moved about and deformed.

The complete investigation of these small changes or **perturbations,** as they are called, belongs to the domain of Gravitational Astronomy. It will be necessary here to enumerate the chief perturbations, on account of the important part they play in determining the circumstances of eclipses.

273. Retrograde Motion of the Moon's Nodes.—The Moon's nodes are not fixed, but have a retrograde motion along the ecliptic of about 19° in a year. This phenomenon closely resembles the retrograde motion of ♈ (Precession, § 141), but is far more rapid. Its effect is to carry the line of nodes, with the plane of the Moon's orbit, slowly round the ecliptic, performing a complete revolution in 6793·391 days, or rather over 18·6 years.

One result of this nodal motion is that the angle of inclination of the Moon's orbit to the equator is subject to periodic variations. When the Moon's ascending node coincides with the first point of Aries, the angle between the Moon's orbit and the equator will be the *difference* of the angles they make with the ecliptic, *i.e.* about 23° 28' – 5° 8' or 18° 20'. When, on the contrary, the ascending node coincides with the first point of Libra, the angle between the orbit and the equator will be the *sum* of the angles they make with the ecliptic, *i.e.*, 23° 28' + 5° 8' or 28° 36'. The period of fluctuation is the time of revolution of the Moon's nodes relative to the first point of Aries, and is a few days (nearly five) greater than their sidereal period of revolution, on account of precession.

274. Progressive Motion of Apse Line.—The line of apsides is not fixed, but has a direct motion in the plane of the Moon's orbit, performing a complete revolution in 3232·575 days, or about nine years. A similar progressive motion of the apse line of the Earth's orbit about the Sun was mentioned in § 153. The latter motion is, however, much less rapid, its period being about 108,000 years.

275. Other Perturbations.—The inclination of the Moon's orbit to the ecliptic is not quite constant. It is subject to small periodic variations, its greatest and least values being 5° 13' and 5° 3'.

In addition there are variations in the eccentricity of the orbit, in the rates of motion of the nodes, and in the length of the sidereal period. All of these render the *accurate* investigation of the Moon's orbit one of the most complicated problems of Astronomy.

276. The Moon's Rotation.—It is a remarkable fact that the Moon always turns the same side of its surface to the Earth. Whether we examine the markings on its surface with the naked eye, or resolve them into mountains and streaks with a telescope, they always appear very nearly the same, although their illumination, of course, varies with the phase.

From this it is evident that the Moon rotates upon its axis in the same "sidereal" period as it takes to describe its orbit about the Earth, *i.e.*, once in a sidereal month. It might, at a first glance, appear as if the Moon had no rotation, but such is not the case. To explain this, let us consider the phenomena which would be presented to an observer if situated on the Moon in the centre of the portion turned towards the Earth.

The Earth would always appear directly overhead, *i.e.*, in the observer's zenith. But as the Moon describes its orbit about the Earth, the direction of the line joining the Earth and Moon revolves through 360°, relative to the fixed stars, in a sidereal month. Hence the direction of the observer's zenith on the Moon must also revolve through 360° in a sidereal month, and therefore the Moon must rotate on its axis in this period.

The Moon would be said to describe its orbit *without rotation*, if the same points on its surface were to remain always directed towards the same fixed *stars*. Were this the case, *different* parts of the surface would become turned towards the *Earth* as the Earth's direction changed, and *this is not what actually occurs.*

It thus appears that, to an observer on the Moon, the directions of the stars relative to the horizon would appear to revolve through 360° once in a sidereal lunar month. Thus, the sidereal month is the period corresponding to the sidereal day of an observer on the Earth. In a similar way, the Sun's direction would appear to revolve through 360° in a synodic month. This, therefore, is the period corresponding to the solar day on the Earth, as is otherwise evident from the fact that the Moon's phases determine the alternations of light and darkness on the Moon's surface, and that they repeat themselves once in every synodic month.

277. Librations of the Moon.—Libration in Latitude.—If the axis about which the Moon rotates were perpendicular to the plane of the Moon's orbit, we should not be able to see any of the surface beyond the two poles (*i.e.*, extremities of the axis of rotation). In reality, however, the Moon's axis, instead of being exactly perpendicular to its orbit, is inclined at an angle of about 6½° to the perpendicular, just as the Earth's axis of rotation makes an angle of about 23° 28' with a perpendicular to the ecliptic. The consequence is that during the Moon's revolution the Moon's north and south poles are alternately turned a little towards and a little away from the Earth; thus, in one part of the orbit we see the Moon's surface to an angular distance of 6° 44' beyond its north pole, in the opposite part we see 6° 44' beyond the south pole. This phenomenon is called the Moon's **libration in latitude.** It makes the Moon's poles appear to nod, oscillating to and fro once in every revolution relative to the nodes.

Libration in latitude may be conveniently illustrated by the corresponding phenomenon in the case of the Earth's motion round the Sun, as represented in Fig. 56 (§ 154). At the summer solstice the whole of the Arctic circle is illuminated by the Sun's rays, and therefore an observer on the Sun (if such could exist) would see the Earth's surface for a distance of 23° 28' beyond the north pole. Similarly, at the winter solstice an observer on the Sun would see the whole of the Antarctic circle, and a portion of the Earth's surface extending 23° 28' beyond the south pole.

278. Libration in Longitude.—Owing to the elliptical form of the orbit, the Moon's angular velocity about the Earth is not quite uniform, being least at apogee and greatest at perigee. But the Moon rotates about its polar axis with perfectly uniform angular velocity equal to the *average* angular velocity of the orbital motion (so that the periods of rotation and of orbital motion are equal).

Thus, at *apogee* the angular velocity of rotation is slightly *greater* than that of the orbital motion, and is, therefore, greater than that required to keep the same part of the Moon's surface always turned towards the Earth. In consequence, the Moon will appear to gradually turn round, so as to show a little more of the *eastern* side of its surface.

At *perigee*, the angular velocity of rotation is *less* than that of the orbital motion, and is, therefore, not quite sufficient to keep the same part of the Moon's surface always turned towards the Earth. In consequence we shall begin to see a little further round the *western* side of the Moon's disc.

This phenomenon is called **libration in longitude.** Its maximum amount is 7° 45′ ; thus, during each revolution of the Moon relative to the apse line, we alternately see 7° 45′ of arc further round the eastern and western sides of the disc than we should otherwise.

279. Diurnal Libration.—The phenomenon known as **diurnal libration** is really only an effect of parallax. If the Moon were vertically overhead, and if we were to travel eastwards, we should, of course, begin to see a little further round the eastern side of the Moon's surface. If we were to travel westwards we should begin to see a little further round the western side. Now, the rotation of the Earth carries the observer round from west to east. Hence, when the Moon is rising we see a little further round its western side, and when setting we see a little further round its eastern side, than we should from a point vertically underneath the Moon.

Similarly an observer in the northern hemisphere would always see rather more of the Moon's northern portion, and an observer in the southern hemisphere would see rather more of the southern portion than an observer at the equator.

The greatest amount of the diurnal libration is equal to the Moon's horizontal parallax, and is therefore about 57′. We see 57′ round the Moon's western corner when rising, and 57′ round the eastern corner when setting.

An observer at any given instant sees *not quite half* (49·998 per cent.) the Moon's surface. The visible portion is bounded by a cone through the observer's eye enveloping the Moon, and is less than a hemisphere by a belt of breadth equal to the Moon's angular semi-diameter, *i.e.*, about 16′.

280. General Effects of Libration.—In consequence of the three librations, about 59 per cent. of the Moon's surface is visible from the Earth at some time or other, instead of rather under 50 (49·998) per cent., as would be the case if there were no libration. At the same time only about 41 per cent. of the surface is always visible from the Earth. The remainder is sometimes visible, sometimes invisible.

To an observer on the surface of the Moon the result of libration in latitude and longitude would be that the Earth, instead of remaining stationary in the sky, would appear to perform small oscillations about its mean position. It would really appear to describe a series of ellipses. The motion of the different parts of the Earth across its disc in the course of the Earth's diurnal revolution would be the only phenomenon resulting from the cause which produces diurnal libration.

281. Metonic Cycle.—A problem of great historic interest in the study of the lunar motions is the finding of a method of ready prediction of the Moon's phases. From the earliest times there have been religious festivals regulated (as Easter still is) by the Moon's phases; but the direct calculation, from first principles, of the phase for a given day would be long and tedious.

This difficulty was overcome by the discovery of the so-called Metonic Cycle by Meton and Euctemon, B.C. 433. They found that after a cycle of nineteen years the new and full Moons recurred on the same days of the year. To show this it is necessary to prove that nineteen years is nearly an exact multiple of the synodic month. Now, 1 tropical year = $365 \cdot 2422$ days; $\therefore$ 19 years = $6939 \cdot 60$ days, and 1 synodic month = $29 \cdot 5306$ days; $\therefore$ 235 months = $6939 \cdot 69$ days; $\therefore$ 19 years differs from 235 lunations by $\cdot 09$ days, *i.e.*, 2h. 10m. nearly.

If we define the **Golden Number** of a year as the remainder when (1 + the number of the year A.D.) is divided by 19, and the **Epact** as the Moon's age on the 1st of January, we see that two years which have the same Golden Number have corresponding lunar phases on the same days, and in particular have the same epact.

Hence, the Golden Number of the year 1 B.C. (which might be more consistently called 0 A.D.) is evidently 1; and it happens that that year had new Moon on January 1, and, therefore, its epact is zero. But twelve lunar months contain $354 \cdot 37$ days, and fall short of the average year ($365 \cdot 25$ days) by $10 \cdot 88$ days, which is nearly $\frac{11}{30}$ lunations. Hence, the epact is greater by $\frac{11}{30}$ of a lunation each year; and since whole months are not counted in estimating the Moon's age, it is (in months) the fractional part of

$$\tfrac{11}{30}\{\text{Golden Number} - 1\};$$

or, in days, the remainder when $11\{\text{Golden No.} - 1\}$ is divided by 30.

Thus the Golden Number of 1892 is the remainder when 1893 is divided by 19, *i.e.*, 12. Hence, the epact is the remainder when $11\{12 - 1\}$ is divided by 30, *i.e.*, 1; hence, the Moon is one day old on January 1, 1892, and new on December 31, 1891.

In the epact, fractions of a day are never reckoned. Owing to the extra day in leap year, the rule is sometimes a day wrong; but it is near enough for fixing the ecclesiastical calendar.

282. Harvest Moon. — The full Moon which occurs nearest the autumnal equinox is called the **Harvest Moon.** Owing to the Moon's direct motion in its orbit the time of moonrise always occurs later and later every day, but in the case of the harvest Moon the daily retardation is less than in the case of any other full Moon, as we shall now show. To simplify our rough explanations we suppose the Moon to be moving in the ecliptic.

The Moon's R.A. determines the time at which the Moon crosses the meridian (*cf.* § 24). In consequence of the orbital motion the R.A. increases continuously, just as in the case of the Sun (§ 30), only the increase is more rapid (360° per month instead of per year). Therefore the Moon transits later and later every night.

When the Moon is in the first point of Aries it is passing from south to north of the equator, and its declination is increasing most rapidly. Now, the arguments of §§ 123–125 are applicable to the Moon as well as the Sun, and they show that, as the declination increases, there is, in north latitudes, a corresponding increase in the length of time that the Moon is above the horizon. The effect of this increase is to lengthen the interval from the Moon's rising to its transit; this lengthening tends to counterbalance, more or less, the retardation in the time of transit, thus reducing the retardation in the time of *moonrise* to a *minimum.*

Similarly it may be shown that whenever the Moon passes the first point of Libra, the daily retardation of moonrise will be a *maximum,* while that of the time of *setting* will be a *minimum.* These phenomena, therefore, recur once each lunar month.

Now, at harvest time the Sun is near ♎ ; hence, when the Moon is near ♈ it is full; and the minimum retardation of the Moon's rising, therefore, takes place at full Moon. And since the Moon is then opposite the Sun, it rises at sunset. Both these causes make the phenomenon more conspicuous in itself than at other times, and as the continuance of light is useful to the farmers when gathering in their harvest, the name Harvest Moon has been applied.

At the following full Moon the phenomena are similar but less marked. But as it is now the hunting season, the Moon is called the " Hunter's Moon."

EXAMPLES.—VIII.

1. If a, a' be the true and apparent altitudes of a body affected by parallax, prove the equation $a = a' + P \cos a'$.

2. If the Sun's parallax be 8·80'', find the Sun's distance.

3. If in our latitude, on March 21, the Moon is in its first quarter, about what time may it be looked for on the meridian, and how long does it remain above the horizon?

4. Show that from a study of the Moon's phases we can infer the Sun to be much more distant than the Moon. Prove that if the synodic period were 30 days, and the Sun only twice as distant as the Moon, the Moon would be dichotomized after only 5 days instead of 7½.

5. Taking the usual values of the Sun's and the Moon's distances, calculate, roughly, the mean value of the angle ESM when the Moon is dichotomized.

6. Under what conditions is the line of cusps perpendicular to the horizon? Consider specially the appearance to an observer on the Arctic circle.

7. There was an eclipse of the Moon on Jan. 28, 1888, central at 11.10 in the evening. What is the Moon's age on May 21 of that year?

8. Find approximately the position and appearance of the Moon, relatively to the horizon, in latitude 50° N., in the middle of November at 10 P.M., when it is ten days old.

9. At a place in the temperate zone can the Sun or the Moon be longer above the horizon?

10. What would be the effect on the Harvest Moon (i.) if the polar axis of the Earth were perpendicular to the ecliptic, or (ii.) if the Moon were to move in the ecliptic?

EXAMINATION PAPER.—VIII.

1. What is *parallax*, and under what conditions is the parallax of a heavenly body greatest? Show by some simple illustrations that as the distance of an object increases, its parallax lessens.

2. Prove the formula $\sin p = \sin P \sin z$, where P is the Moon's horizontal parallax, and p its parallax when its zenith distance is z.

3. How is the distance of the Moon determined by observations made in the plane of the meridian? Why cannot the Sun's parallax be accurately determined in this way?

4. Show that we can calculate the Moon's sidereal period given its synodic period and the length of the year. Find it, given that these are 29½ and 365¼ days respectively.

5. Describe the *phases of the Moon*, and find an expression for the phase when the Moon is at a given elongation. Show how an observation of the Moon, when at its first quarter, would help us to find the ratio of the distances of the Moon and the Sun.

6. Describe some methods for determining the heights of lunar mountains.

7. Describe the phenomena of the Moon's motion. Given that the Moon moves in a plane inclined at 5° to the ecliptic, find the lowest north latitude of a place where the full Moon can never rise at the summer solstice.

8. Explain (and illustrate by figures) how it is that we see more than half the Moon's surface, and define the terms *node, phase, libration*.

9. Describe the general appearance presented by the solar system to an observer situated at the centre of the Moon's hemisphere turned towards the Earth. When would the Earth be partially eclipsed to such an observer?

10. Explain the phenomenon called the *Harvest Moon*, and show that from a similar cause the daily retardation in the *sidereal* time of *sunrise* is least at the vernal equinox.

CHAPTER IX.

ECLIPSES.

SECTION I.—*General Description of Eclipses.*

283. Eclipses are of two kinds, lunar and solar. If at full Moon the centres of the Sun, Earth, and Moon are very nearly in a straight line, the Earth, acting as a screen, will stop the Sun's rays from reaching the Moon, and the Moon will, therefore, be either wholly or partially darkened. This phenomenon is called a **Lunar Eclipse.**

On the other hand, if the three centres are nearly in a straight line when the Moon is new, the Moon, by coming between the Earth and the Sun, will cut off the whole or a portion of the Sun's rays from certain parts of the Earth's surface. In such parts the Earth will be darkened, and the Sun will appear either wholly or partially hidden. This phenomenon is a **Solar Eclipse.**

If the Moon were to move exactly in the ecliptic we should have an eclipse of the Moon at every opposition, and an eclipse of the Sun at every conjunction, for at either epoch the centres of the Earth, Sun, and Moon would be in an exact straight line. In consequence, however, of the Moon's orbit being inclined to the ecliptic at an angle of about $5\frac{1}{4}°$, the Moon at " syzygy " (conjunction or opposition) is generally so far on the north or south side of the ecliptic that no eclipse takes place. An eclipse only occurs when the Moon at syzygy is very near the ecliptic, and, therefore, not far from the line of nodes (§ 271).

ASTRON.   Q

284. Different Kinds of Lunar Eclipse.—Eclipses of the Moon are of two kinds, total and partial. Let S, E be the centres of the Sun and Earth respectively. Draw the common tangents ABV and $A'B'V$ to the two globes, meeting on SE produced in V, and draw also the other pair of tangents $AB'K'$, $A'BK$ cutting at U, between S and E. If the figure be supposed to revolve about SE, the tangents will generate cones, enveloping the Sun and Earth, and having their vertices at U and V. The space BVB', inside the inner cone, is called the **umbra**; the space between the inner and outer cone is called the **penumbra.** * The character of the lunar eclipse will vary according to the following conditions :—

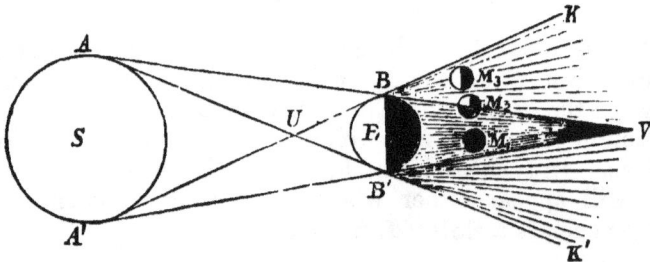

FIG. 87.

(i.) If at opposition, the Moon falls entirely within the umbra or inner cone BVB', as at M_1, no portion of the Moon's surface then receives any direct rays from the Sun, and the Moon is therefore plunged in darkness (except for the light which reaches it after refraction by the Earth's atmosphere, as explained in § 193). The eclipse is then said to be **total.**

(ii.) If the Moon falls partly within and partly without the umbra BVB', as at M_2, the portion within the umbra receives no light from the Sun, and is, therefore, obscured, while the remaining portion receives light from part of the Sun's surface about A, and is, therefore, partially illuminated. The eclipse is then said to be **partial.**

---

* For further description of the formation of the umbra and penumbra, see Wallace Stewart's *Text-Book of Light*, § 5,

(iii.) If the Moon falls entirely within the "penumbra," or outer cone, as at M_3, it receives the Sun's rays from A, but not from A'. There is no true eclipse, but only a diminution of brightness (sometimes called a "*penumbral eclipse*").

A lunar eclipse is visible simultaneously from all places on that hemisphere of the Earth over which the Moon is above the horizon at the time of its occurrence.

Near the boundary of the hemisphere there are two strips in the form of lunes, comprising those places respectively at which the Moon sets and rises during the eclipse ; at such places only its beginning or end is seen.

285. Phenomena of a Total Eclipse of the Moon.— As the Moon gradually moves towards opposition, the first appearance noticeable is the slight darkening of the Moon's surface as it enters the penumbra. This darkening increases very gradually as the Moon approaches the umbra, or true shadow. At **"First Contact"** a portion of the Moon enters the umbra, and the eclipse is then seen as a partial eclipse, the dark portion being bounded by the circular arc formed by the boundary of the umbra. As the Moon advances, the dark portion increases till the whole of the Moon is within the umbra, and the eclipse is total. When the Moon begins to emerge at the other side of the umbra, the eclipse again becomes partial, and continues so until **"Last Contact,"** when the Moon has entirely emerged from the umbra, after which the Moon gradually gets brighter and brighter till it finally leaves the penumbra.

In the case of a partial eclipse, the umbra merely appears to pass over a portion of the Moon's disc, which portion is greatest at the middle of the eclipse.

286. Effects of Refraction on Lunar Eclipses.—In § 193 it was stated that, owing to atmospheric refraction, the Moon's disc appears of a dull-red colour during the totality of the eclipse. A still more curious phenomenon is noticed when an eclipse occurs at sunset or sunrise. The refraction at the horizon increases the apparent altitudes of the Sun and Moon in the heavens, so that both appear above the horizon when they are just below. Hence a total eclipse of the Moon is sometimes seen when the Sun is shining.

287. Different Kinds of Solar Eclipse.—An eclipse of the Sun may be either **total, annular,** or **partial.** To explain the difference between the first two kinds of eclipse, let us suppose that the observer is situated exactly in the line of centres of the Sun and new Moon, so that both bodies appear in the same direction. Then, if the Moon's angular diameter is greater than the Sun's, the whole of the Sun will be concealed by the Moon; the eclipse is then said to be **total.** If, on the other hand, the Sun has the greater angular diameter, the Moon will conceal only the central portion of the Sun's disc, leaving a bright ring visible all round; under such circumstances, the eclipse is said to be **annular.** Lastly, if the observer is not exactly in the line of centres, the Moon may cover up a segment at one side of the Sun's disc; the eclipse is then **partial.**

Now, the Moon's angular diameter varies, according to the distance of the Moon, from 28′ 48″ at apogee to 33′ 22″ at perigee, the corresponding limits for the Sun's diameter being 31′ 32″ at apogee, and 32′ 36″ at perigee. Hence, both total and annular eclipses of the Sun are possible. Thus, when the Sun is in apogee and the Moon in perigee an eclipse must be either total or partial; when the Sun is in perigee and the Moon in apogee, an eclipse must be annular or partial.

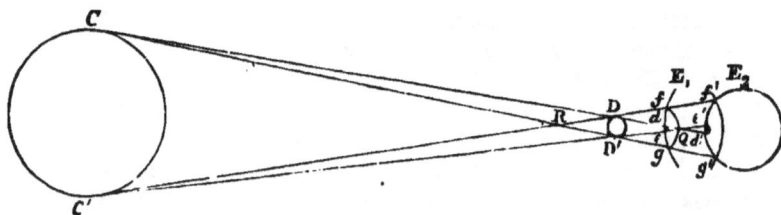

FIG. 88.

288. Circumstances of a Solar Eclipse.—Fig. 88 shows the different circumstances under which a solar eclipse is seen from different parts of the Earth. Draw the common tangents CDQ, $C'D'Q$, CRD', $C'RD$ to the Sun and Moon, forming the enveloping cones DQD' and fRg; these constitute respectively the boundaries of the umbra and penumbra of the Moon's shadow. First let the umbra DQD' meet the Earth's surface (E_1) before coming to a point at Q, the curve

of intersection being *de*. Also let the penumbra *fRg* meet
the Earth's surface in the curve *fg*. Then from any place on
the Earth within the space *de* the Sun appears totally eclipsed.
At a place elsewhere within the penumbra *fg*, the Sun appears
partially eclipsed, a portion only being obscured by the Moon.

Next let the umbra *DQD'* come to a point *Q* before
reaching the Earth *E₂*. Then, if the cone of the umbra be
produced to meet the Earth in *d'e'*, an observer anywhere
within the space *d'e'* sees the eclipse as an annular eclipse.
At any place elsewhere within the penumbra *f'g'*, the eclipse
appears partial, as before. At parts of the Earth which fall
without the penumbra there is no eclipse. Hence a solar
eclipse is only visible over a part of the Earth's surface,
and its circumstances are different at different places.

As the Sun and Moon move forward in their relative orbits,
and the Earth revolves on its axis, the two cones of the
Moon's shadow travel over the Earth, and the eclipse becomes
visible from different places in succession   The inner cone
traces out on the Earth a very narrow belt, over which
the eclipse is seen as a total or annular eclipse, according
to circumstances. The outer cone, or penumbra, sweeps out
a far broader belt, including that part of the Earth's surface
where the eclipse is visible as a partial eclipse.

A total or annular eclipse of the Sun, like a total eclipse
of the Moon, always begins and ends as a partial eclipse, the
totality or annular condition only lasting for a short period
about the middle of the eclipse. The maximum duration
of totality at the Equator is just under eight minutes.

In the case of an annular eclipse, there are two internal,
as well as two external, contacts, and the eclipse remains
annular during the interval between the internal contacts.
This may sometimes be rather more than twelve minutes.

Owing to the limited area of the belt over which a solar
eclipse is visible, the chance that any eclipse may be visible
at any given place is far smaller than in the case of a lunar
eclipse. The chance of an eclipse being *total* at any place is
very small indeed. The last eclipse visible as a total eclipse
in England occurred in 1724 ; the next will take place on
June 29th, 1927. One or more partial eclipses are visible at
Greenwich in nearly every year.

Section II.—*Determination of the Frequency of Eclipses.*

289. **To Find the Limits of the Moon's geocentric position consistent with a Solar or Lunar Eclipse.**

In Fig. 89, let the plane of the paper represent any plane through the Sun's and Moon's centres; and let ABV and $A'B'V$ represent the common tangents bounding the cone of the Earth's true shadow. Let AUB' be the other common tangent, which goes (nearly) through B'; and let the line SE, joining the centres of the Sun and Earth, meet the common tangents in V and U. Let T, t, t' be those points on ABV and AB' whose distance from E is equal to that of the Moon.

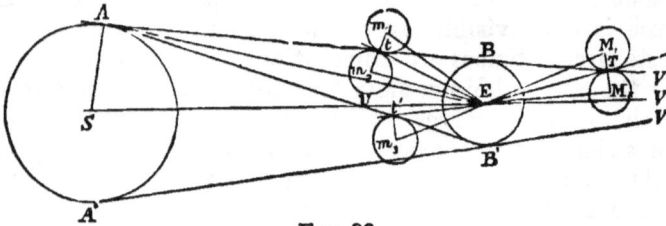

Fig. 89.

Then, if M_1, M_2 denote the positions of the Moon's centre, when touching the cone BV externally and internally at T, it is evident that a **lunar eclipse** occurs whenever the full Moon is nearer the line of centres than M_1. Hence, if m denote the Moon's angular semi-diameter TEM_1, the Moon's angular distance from EV must be less than VEM_1, or $VET + m$.

Similarly, the lunar eclipse is **total** when the Moon is not further from the line of centres than M_2; for this the Moon's (geocentric) angular distance from the line of centres must be not greater than VEM_2, or $VET - m$.

Let m_1, m_2 be the centres of the Moon at internal and external contact with AB near t. There is evidently a **solar eclipse** visible at some point of the Earth's surface (such as B) as a partial eclipse, if the Moon's angular distance from the Sun is less than SEm_1, or $SEt + m$.

Supposing the Moon's distance to be such that its angular radius is less than that of the Sun, there is an **annular** eclipse whenever the Moon lies wholly within the cone AVA', as at m_3. This requires the Moon's geocentric angular distance from the Sun to be less than SEm_3, or $SEt - m$.

If, however, the Moon is so near that its angular radius is greater than that of the Sun, the angle it subtends is greater than ABA', and therefore there is a **total** eclipse at B whenever the edge of the Moon reaches the internal tangent $A'B$. Taking m_3 to represent the corresponding position of the Moon when touching the other tangent AB' at t' (for the sake of clearness in the figure), we see that, in order that there may be a total eclipse somewhere on the Earth's surface, the geocentric angular distance between the Moon's and Sun's centres must be less than SEm_3 or $SEt' + m$.

Now, as the cone AVA' tapers to a point at V, the breadth of its cross section is greater near m_1, m_2, m_3 than near M_1, M_2, and when the Moon is in syzygy, its angular distance from EV or $ES = $ its latitude. Hence the limits of latitude are greater for a solar than for a lunar eclipse, and therefore the probability of the occurrence of a solar eclipse is greater than the probability of a lunar eclipse. This explains why, on the whole, **solar eclipses are more frequent than lunar.**

*290. We shall now calculate the angles VEM_1, VEM_2, SEm_1, SEm_2, SEm_3. Let p, P denote the horizontal parallaxes of the Moon and Sun respectively; m, s their respective angular semi-diameters (Fig. 89). We have $s = \angle SEA$,
$$p = \angle BTE = \angle BtE = \angle B't'E, \quad P = \angle BAE = \angle B'AE,$$
and $m = \angle TEM_1 = \angle TEM_2 = \angle tEm_1 = \angle tEm_2 = \angle t'Em_3$.

For the lunar eclipses we have, from the triangle TEA,
$$\angle ETB + \angle EAB = 180° - \angle TEA = \angle VET + \angle SEA;$$
$$\therefore \angle VET = \angle ETB + \angle EAB - \angle SEA = p + P - s;$$
$$\therefore \angle VEM_1 = \angle VET + \angle TEM_1 = p + P - s + m;$$
and $\angle VEM_2 = \angle VET - \angle TEM_2 = p + P - s - m;$

For the solar eclipses we have, from the triangle tEA,
$$\angle EtB - \angle EAB = \angle tEA = \angle SEt - \angle SEA$$
$$\therefore \angle SEt = \angle EtB - \angle EAB + \angle SEA = p - P + s.$$
$$\therefore \angle SEm_1 = p - P + s + m,$$
and $\qquad \angle SEm_2 = p - P + s - m.$

Lastly, from the triangle $t'EA$ we have
$$\angle Et'B' - \angle EAB' = \angle AEt' = \angle AES + \angle SEt'.$$
$$\therefore \angle SEt' = \angle B't'E - \angle B'AE - \angle AES = p - P - s.$$
$$\therefore \angle SEm_3 = p - P - s + m.$$

[As an example, the student may show that the greatest latitudes the Moon can have, in order that it may be partially or wholly within the *penumbra* at opposition are $p + s + P + m$ and $p + s + P - m$ respectively.]

*291. Greatest Latitudes of the Moon at Syzygy.— Since S and V are in the ecliptic, it follows that when the Moon is in conjunction or opposition, the plane of the paper in Fig. 89 is perpendicular to the ecliptic. Therefore the angles VEM_1, VEM_2 measure the Moon's latitude at conjunction, and SEm_1, SEm_2, SEm_3 measure its latitude at opposition in the positions represented. The above expressions are, therefore, the greatest possible latitudes at syzygy consistent with eclipses of the kinds named.

Now, taking the mean values we have, roughly,

$$s = 16'; \quad m = 15'; \quad p = 57'; \quad P = 0' 8''.$$

Substituting these values, and collecting the results, we have, roughly, the following limits for the Moon's geocentric latitude, or angular distance from the line of centres:—

(1) *For a* **lunar** *eclipse*, $VEM_1 = p + P - s + m = \mathbf{56'}$;
(2) *For a* **total lunar** *eclipse*, $VEM_2 = p + P - s - m = \mathbf{26'}$;
(3) *For a* **solar** *eclipse*, $SEm_1 = p - P + s + m = \mathbf{88'}$;
(4) *For an* **annular** *eclipse*, $SEm_2 = p - P + s - m = \mathbf{58'}$.

Lastly, taking the Sun at apogee, and the Moon at perigee, we have, $m = 17'$ and $s = 16'$ nearly, whence we have, in the most favourable case,

(4a) *For a* **total solar** *eclipse*, $SEm_3 = p - P - s + m = \mathbf{58'}$.

292. Ecliptic Limits.—From the last results it appears that a lunar eclipse cannot occur unless at the time of opposition the Moon's latitude is less than about 56', and that a solar eclipse cannot occur unless at conjunction the Moon's latitude is less than about 88'. Now the Moon's latitude depends on its position in its orbit relatively to the line of nodes; hence there will be corresponding limits to the Moon's distance from the node consistent with the occurrence of eclipses. These limits are called the **Ecliptic Limits.**

*The ecliptic limits may be computed as follows:—Let the geocentric direction of the Moon's centre be represented on the celestial sphere by M. Let N represent the node, MH a secondary to the ecliptic. [The *ecliptic limit*, strictly speaking, means the limit of NH measured along the ecliptic, and not that of NM.]

Now the limit of latitude MH has been calculated in the last paragraph for the different cases. Let this be denoted by l. Also let I be the inclination of the Moon's orbit to the ecliptic. Then in the spherical triangle NHM, right-angled at H, we have $HM = l$, and $\angle HNM = I$; both of these are known, hence NH can be calculated.

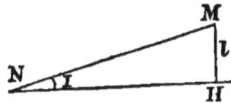

FIG. 90.

For rough purposes it will be sufficient either to treat the small triangle HNM as a plane triangle (Sph. Geom. 24), or to regard MH as approximately the arc of a small circle, whose pole is N. The first method gives

$$l = NH \tan I;$$
$$\therefore NH = l \cot I.$$

Or, adopting the second method, we have (Sph. Geom. 17)

$$l = MH = \angle MNH \times \sin NH = I \sin NH;$$
$$\therefore \sin NH = l/I,$$

whence the ecliptic limit NH is found.

EXAMPLES.

1. To find the Lunar Ecliptic Limit. For a lunar eclipse we have, by § 291, $l = 56'$. Also, $I = 5°$ roughly.

Hence $\sin NH = \dfrac{56}{5 \times 60} = \dfrac{56}{300} = \cdot187$,

$= \sin 11°$ (from table of natural sines)

and the lunar ecliptic limit is about 11°.

2. To find the Solar Ecliptic Limit. For a solar eclipse we have $l = 88'$. Hence, taking $I = 5°$ as before, we have

$$\sin NH = \dfrac{88}{5 \times 60} = \dfrac{88}{300} = \cdot293, \text{ roughly,}$$

$= \sin 17°$, roughly,

and the solar ecliptic limit is about 17°.

293. Major and Minor Ecliptic Limits.—Owing to the variations in the distances of the Sun and Moon their parallaxes and angular semi-diameters are not quite constant. Hence the exact limits of the Moon's latitude l, as calculated by the method of § 291, are subject to small variations.

This alone would render the ecliptic limits variable. But there is another cause of variation in the ecliptic limits, arising from the fact that I, the inclination of the Moon's orbit, is also variable, its greatest and least values being about $5°\ 19'$ and $4°\ 57'$.

The greatest and least values of the limits for each kind of eclipse are called the **Major** and **Minor Ecliptic Limits.**

For an eclipse of the Moon the major and minor ecliptic limits have been calculated to be about $12°\ 5'$ and $9°\ 30'$ respectively at the present time. For an eclipse of the Sun the limits are $18°\ 31'$ and $15°\ 21'$ respectively.

Thus a lunar eclipse **may** take place if the Moon, when full, is within $12°\ 5'$ of a node; and a lunar eclipse **must** take place if the full Moon is within $9°\ 30'$ of a node.

Similarly, a solar eclipse **may** take place if the Moon, when new, is within $18°\ 31'$, and a solar eclipse **must** take place if the new Moon is within $15°\ 21'$ of a node.

The mean values of the lunar and solar ecliptic limits are now $10°\ 47'$ and $16°\ 56'$. But the eccentricity of the Earth's orbit is very slowly decreasing; consequently the major limits are smaller and the minor limits larger than they were, say, a thousand years ago.

294. Synodic Revolution of the Moon's Nodes.—An eclipse is thus only possible at a time when the Sun is within a certain angular distance of the Moon's nodes. Hence the period of revolution of the Moon's nodes, relative to the Sun, marks the recurrence of the intervals of time during which eclipses are possible. This period is called the period of a **synodic revolution of the nodes.**

In § 273 it was stated that the Moon's nodes have a retrograde motion of about $19°$ per annum, more exactly $19°\ 21'$. In one year (365d.) the Sun, therefore, separates from a node by $360° + 19°\ 21'$ or $379·35°$, hence it separates $360°$ in $(360 \times 365\frac{1}{4})/379·35$ days, or about **346·62d.** This, then, is the period of a synodic revolution of the node.

In a synodic lunar month (29½ days), the Sun separates from the line of nodes by an angle

$$379\tfrac{1}{3}° \times 29\tfrac{1}{2} \div 365\tfrac{1}{4}, \text{ or } 30° \ 36',$$

a result which will be required in the next paragraph.

295. To find the Greatest and Least number of Eclipses possible in a Year.—Let the circle in Fig. 91 represent the ecliptic, and let N, n be the Moon's nodes. Take the arcs NL, NL', nl, nl' each equal to the lunar ecliptic limit, and NS, NS'', ns, ns' each equal to the solar ecliptic limit. Then the *least* value of SS' or ss' is twice the *minor solar* ecliptic limit, and is 30° 42', and this is greater than 30° 36', the distance traversed by the Sun relative to the nodes between two new Moons. Hence, at least one new Moon *must* occur while the Sun is travelling over the arc SS', and two *may* occur. Therefore *there must be one, and there may be two eclipses of the Sun, while the Sun is in the neighbourhood of a node.*

Again, the *greatest* value of LL', ll' is double the *major lunar* ecliptic limit, and is, therefore, 24° 10'. This is considerably less than the space passed over by the Sun relative to the nodes between two full Moons. Hence, there cannot be more than one full Moon while the Sun is in the arc LL', and there may be none. Therefore *there cannot be more than one eclipse of the Moon while the Sun is in the neighbourhood of a node, and there may be none at all.*

Fig. 91.

296. **The case most favourable** to the occurrence of eclipses is that in which the Moon is new just after the Sun has come within the solar ecliptic limits, *i.e.*, near S. There will then be an eclipse of the *Sun*.

When the Moon is full (about 14¾ days later) the Sun will be near N, at a point within the lunar ecliptic limits; there will therefore be an eclipse of the *Moon*.

At the following new Moon the Sun will not have reached S'; and there will be a *second* eclipse of the *Sun*.

In six lunations from the first eclipse the Sun will have travelled through just over 180°, and will be within the space ss', near s; there will therefore be a *third* eclipse of the *Sun*.

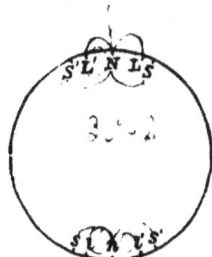

At the next *full* Moon the Sun will be near *n*, and there will be a *second* eclipse of the Moon.

The Sun may just fall within the space *ss'* near *s'* at the next *new* Moon; there will then be a *fourth* eclipse of the Sun.

In twelve lunations from the first eclipse, the Sun will have described about 368°, and will, therefore, be about 8° beyond its first position, and well within the limits *ss'*; there will, therefore, be a *fifth* eclipse of the *Sun*.

About 14¾ days later, at *full* Moon, the Sun will be well within the lunar ecliptic limits *LL'*, and there will be a *third* eclipse of the *Moon*.

All these eclipses occur in 12½ lunations, *i.e.*, 369 days, or a year and four days. We cannot, therefore, have all the eight eclipses in one year, but

There may be as many as seven eclipses in a year, namely, either five solar and two lunar, or four solar and three lunar.

297. The most unfavourable case is that in which the Moon is full just before the Sun reaches the ecliptic limits at *L*.

At new Moon the Sun will be near *N*, and there will be *one solar* eclipse.

At the next full Moon the Sun will have passed *L'*, so that there will be no lunar eclipse. After six lunations the Sun will not have arrived at *l*.

At the next new Moon the Sun will be within the ecliptic limits, and there will be a *second solar* eclipse.

At the next full Moon the Sun will be again just beyond *l'*, and at 12 lunations from first full Moon, the Sun may again not have quite reached *L*.

FIG. 92.

At 12½ lunations there will be a *third solar* eclipse.

The interval between the first and third eclipses will be 12 lunations, or about 354 days. If, therefore, the first eclipse occurs after the 11th day of the year, *i.e.*, January 11, the third will not occur till the following year. Therefore,

The least possible number of eclipses in a year is two. These must both be solar eclipses.

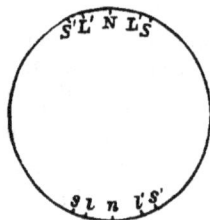

298. The Saros of the Chaldeans.—The period of a synodic revolution of the nodes is (§ 294) approximately 346·62 days. Hence,

19 synodic revolutions of the node take   6585·78 days.
Also 223 lunar months   = 6585·32 days.

It follows that after 6585½ days, or 18 years 11 days, the Moon's nodes will have performed 19 revolutions relative to the Sun, and the Moon will have performed 223 revolutions almost exactly. Hence the Sun and Moon will occupy almost exactly the same position relative to the nodes at the end of this period as at the beginning, and eclipses will therefore recur after this interval.

The period was discovered by observation by the Chaldean astronomers, who called it the **Saros.** By a knowledge of it they were usually able to predict eclipses. Indeed, in the records of eclipses handed down to us in the form of cuneiform inscriptions, they invariably stated whether the circumstances accorded with prediction by the Saros or not.

A "synodic revolution of the Moon's apsides," or the period in which the Sun performs a complete revolution relative to the Moon's apse line, occupies 411·74 days. Hence sixteen such revolutions occupy 6587·87 days, or about two days longer than the Saros. Therefore the Moon's line of apsides also returns to very nearly the same position relative to the Sun and Moon. Hence, the solar eclipses, as they recur, will be nearly of the same kind (total or annular) in each Saros. The whole number of eclipses in a Saros is about 70. The average of all eclipses from B.C. 1207 to A.D. 2162 shows that there are 20 solar eclipses to 13 lunar.

The present values of the mean solar and lunar ecliptic limits, 16° 56', and 10° 47', are in the ratio of 31 : 18 very nearly. This ratio gives, on the whole, a higher average proportion of solar eclipses to lunar than that given above. It must, however, be remembered that all the angles used in calculating the limits are subject to gradual changes. Consequently the numbers of eclipses in that period are subject to very gradual variation; after a large number of Saroses have recurred, the order of eclipses in each will have changed.

*Section III.— *Occultations—Places at which a Solar Eclipse is visible.*

299. Occultations.—When the Moon's disc passes in front of a star or planet, the Moon is said to **occult** it.

An occultation evidently takes place whenever the apparent angular distance of the Moon's centre from the star becomes less than the Moon's angular semi-diameter. As the apparent position of the Moon is affected by parallax, the circumstances of an occultation are different at different places on the Earth's surface.

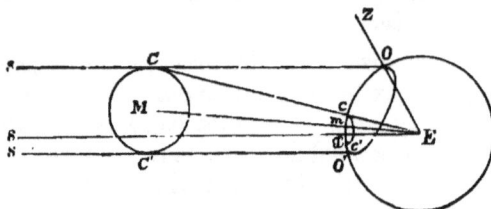

Fig. 93.

Let *m* denote Moon's angular semi-diameter, *p* its horizontal parallax. In the figure, let E and M be the centres of the Earth and Moon, and let sC, sC' represent the parallel rays coming from a star, and grazing the Moon's disc. These rays cut the Earth's surface along a curve OO', and it is evident that only to observers at points within this curve is the star hidden by the Moon's disc. Let EC, Es, EM, EC' cut the Earth's surface in c, x, m, c'; the rays EC, EC' cut the Earth's surface in a small circle cc', whose angular radius $mEc = MEC = m$. Let d be the geocentric angular distance SEM between the Moon's centre and the star.

Then the angle $ECO = $ angle subtended by the Earth's radius EO at C;

$\qquad\qquad = $ parallax of C when viewed from O;

$\qquad\qquad = p \sin COZ \,(\S\ 249);$

$\qquad\qquad = p \sin OEx \text{ (by parallels).}$

But $\qquad\qquad ECO = CEs\,;$

$\qquad\qquad = $ angle subtended by $cx\,;$

$\therefore \quad \sin OEx = \dfrac{\text{angle } cx}{p}.$

Hence we have the following construction for the curve separating those points on the Earth's surface at which the occultation is visible at a given instant from those at which the star is not occulted. Taking the sublunar point m as pole, describe a circle cc' on the terrestrial globe, with the Moon's angular semi-diameter (m) as radius. Through the substellar point x draw any great circle, cutting this small circle in any point c. Measure along it an arc cO such that $\sin cO$ is always the same multiple $\left(\dfrac{1}{p} \right)$ of mc. The locus of the points O, thus determined, is the curve required.

Half of the circle cc' consists of points under the advancing limb of the Moon; hence, over the portion of the curve OO' corresponding to this half-circle, the occultation is just beginning. At points on the other half of cc' the Moon's limb is receding; hence over the other portion of OO' the star is reappearing from behind the Moon's disc.

Since the greatest and least values of cx in any position are $d+m$ and $d-m$, it is evident that the greatest value of d for which an occultation can take is when

$$d - m = p; \quad d = m + p.$$

300. **Occultation of a Planet.**—If s be a planet, the lines Es, Os can no longer be regarded as rigorously parallel; but the angle between them, EsO,

= angle subtended at s by the Earth's radius EO

= parallactic correction at O (§ 248)

= $P \sin ZOs$ (§ 249) = $P \sin OEx$ very nearly.

As before, $ECO = p \sin OEx$. But $ECO = EsO + CEs$;

$$\therefore p \sin OEx = P \sin OEx + cx; \quad \sin OEx = \frac{cx}{p-P}.$$

With this exception, the construction is the same as for a star.

If the planet be so large that we must take account of its angular diameter, the method of the next paragraph must be used.

301. Eclipse of the Sun.—There is a total eclipse of the Sun, provided the Moon's disc completely covers the Sun's; this occurs if the Moon's angular semi-diameter (m) is larger than the Sun's (s), and the apparent angular distance between the Sun's and Moon's centres (as seen from any point at which the eclipse is visible) is less than $m-s$. Hence, if the Moon's angular semi-diameter were reduced to $m-s$, the Sun's centre would then be occulted. Hence the points O, whose locus encloses the places from which the eclipse is visible, can be found as follows :—

With centre m the sublunar point, and angular radius $m-s$, describe a circle. Through the subsolar point x draw any arc of a great circle xc, cutting the circle in c, and take O, on xc produced, such that

$$\sin xO = \frac{xc}{p-P}.$$

For an annular eclipse $m < s$, and the apparent angular distance between the centres is $s-m$; hence the same construction is followed, save that $s-m$ is the angular radius of the small circle first described. For a partial solar eclipse, the angular radius is $s+m$.

When a planet has a sensible disc, the beginning of its occultation may be compared to a partial eclipse of the Sun; and the planet is entirely occulted when the conditions are satisfied corresponding to those for a total eclipse.

EXAMPLE.—Supposing the centres of the Earth, Moon, and Sun to be *in a straight line* and the Moon's and Sun's semi-diameters to be exactly 17′ and 16′, to find the angular radii of the circles on the Earth over which the eclipse is total and partial respectively, taking the relative horizontal parallax as 57′.

At those points at which the eclipse is total, the apparent angular distance between the centres, as displaced by parallax, must be not greater than 17′ − 16′, or 1′. Hence, since the centres are in a line with the Earth's centre, the parallactic displacement must be not greater than 1′. Hence, if z be the Sun's zenith distance at the boundary, then 57′ sin z = 1′; ∴ sin z = $\frac{1}{57}$, or approximately circular measure of $z = \frac{1}{57}$. But a radian contains about 57°; ∴ $\frac{1}{57}$ of a radian = 1° approx. Hence the eclipse is total over a circle of angular radius 1° about the sub-solar point.

Similarly, the eclipse is partial if 57′ sin z < 16′ + 17′, or 33′, or sin z < $\frac{33}{57}$, or ·58. From a table of natural sines, we find that $\sin^{-1}·58 = 35\frac{1}{2}$° roughly; therefore the angular radius is $35\frac{1}{2}$°.

EXAMPLES ON ECLIPSES GENERALLY.

1. To find (roughly) the maximum duration of an eclipse of the Moon, and the maximum duration of totality.

From § 291 we see that a lunar eclipse will continue as long as the Moon's angular distance from the line of centres of the Earth and Sun is less than 58', and the eclipse will continue total while the angular distance is less than 26'. Hence, the maximum duration of the eclipse is the time taken by the Moon to describe $2 \times 58'$, or 116', and the maximum duration of totality is the time taken to describe $2 \times 26'$, or 52'.

Now the Moon describes $360°$ (relative to the direction of the Sun) in the synodic month, $29\frac{1}{2}$ days. Therefore, the times taken to describe 116' and 52' respectively are

$$\frac{29\frac{1}{2} \times 116}{360 \times 60} \quad \text{and} \quad \frac{29\frac{1}{2} \times 52}{360 \times 60} \text{ days,}$$

i.e. 3h. 48m. and 1h. 42m.,

and these are the maximum durations of the eclipse and of totality. The eclipse of Nov. 15, 1891, lasted 3h. 28m., and was total for 1h. 23m.

2. To calculate roughly the velocity with which the Moon's shadow travels over the Earth. (Sun's distance = 93,000,000 miles.)

The radius of the Moon's orbit being about 240,000 miles, its circumference is about 1,508,000 miles. Relative to the line of centres, the Moon describes the circumference in a synodic month, *i.e.*, about $29\frac{1}{2}$ days. Hence its relative velocity is about $1,508,000 \div 29\frac{1}{2}$, or 51,000 miles per day, *i.e.*, 2,100 miles per hour. If q denote the point where the middle of the shadow reaches the Earth (Fig. 88), and if the Earth's surface at q is perpendicular to Sq, we have

velocity of q : vel. of $M = Sq : SM$

$= 93,000,000 : 93,000,000 - 240,000 = 1·0026$ nearly.

Hence the velocity of the shadow at q = vel. of M very nearly
$= 2,100$ miles an hour.

To find the velocity of the shadow relative to places on the Earth, we must subtract the velocity of the Earth's diurnal motion. This, at the Earth's equator, is about 1,040 miles an hour. Hence, if the Earth's surface and the shadow are moving in the same direction, the relative velocity is about 1,060 miles an hour.

3. To find the maximum duration of totality of the eclipse of the example on page 234, neglecting the obliquity of the ecliptic.

The angular radius of the shadow being 1°, or about $69\frac{1}{2}$ miles, its diameter is 139 miles. The obliquity of the ecliptic being neglected, the eclipse is central at a point on the equator, and the shadow and the Earth are therefore moving in the same direction with relative velocity 1,060 miles an hour (by Question 2). The greatest duration of totality is the time taken by the shadow to travel over a distance equal to its diameter, *i.e.*, 139 miles, and is therefore $139 \times 60/1060$ minutes, *i.e.*, 7·9 minutes (roughly).

ASTRON. R

EXAMPLES.—IX.

1. If a total lunar eclipse occur at the summer solstice, and at the middle of the eclipse the Moon is seen in the zenith, find the latitude of the place of observation.

2. If there is a total eclipse of the Moon on March 21, will the year be favourable for observing the phenomenon of the Harvest Moon?

3. Having given the dimensions and distances of the Sun and Moon, show how to find the diameter of the umbra where it meets the Earth's surface.

4. Calculate (roughly) the totality of a solar eclipse, viewed from the Equator at the Equinox, supposing

Moon's diameter 2,160 miles, Sun's diameter 400 times Moon's;
Distance of Moon from Earth 222,000 miles;
Distance of Sun from Earth 92,000,000 miles.

5. If S is the semi-diameter of the Sun, and p, P the horizontal parallaxes of the Sun and the Moon at the time of a lunar eclipse, show that to an observer on the Earth the angular radius of the Earth's shadow at the distance of the Moon is $P+p-S$, and that of the penumbra $P+p+S$. Determine, also, the length of the shadow.

6. If the distance of the Moon from the centre of the Earth is taken to be 60 times the Earth's radius, the angular diameter of the Sun to be half a degree, and the synodic period of the Sun and Moon to be 30 days, show that the greatest time which can be occupied by the centre of the Moon in passing through the umbra of the Earth's shadow is about three hours, and explain how this method might be employed to find the Sun's parallax.

7. If the distance of the Moon were diminished to 30 times the Earth's radius, what would be the time occupied in passing through the shadow?

8. Determine what length of the axis of the Earth's shadow is absolutely dark, having given that the horizontal refraction is about 35'; and account for the copper colour often seen on the Moon when eclipsed.

9. What kind of eclipse is most suitable for the determination of longitude, and why?

10. What would be the greatest possible inclination of the plane of the Moon's orbit to the ecliptic, that there might be a partial eclipse at each conjunction?
(The greatest distance of the Moon = 60 × Earth's radius.)

EXAMINATION PAPER.—IX.

1. What is the cause of eclipses of the Sun, and of the Moon? Why is a solar eclipse visible over so small a portion, and a lunar eclipse over so large a portion of the Earth?

2. Account for the phenomenon called a Lunar Eclipse. Show that it begins and ends at the same instant at all places from which it is visible.

3. Explain briefly the manner in which a solar eclipse passes over the Earth.

4. Explain clearly how an annular eclipse of the Sun is produced. Why are there no annular eclipses of the Moon? Explain why solar eclipses are sometimes total and sometimes annular.

5. Explain why, though there are, on the whole, more eclipses of the Sun than of the Moon, many more of the latter than of the former are visible at Greenwich.

6. Define *umbra* and *penumbra*. Calculate the lengths of the cones of shadow (umbra) cast by the Earth and Moon, and find the breadth of the Earth's umbra at the distance of the Moon.

7. Define and roughly calculate the solar and lunar *ecliptic limits*. What is the greatest number of lunar eclipses which can occur in a year? What is the least number of solar eclipses which can occur in the same interval?

8. What is the *Saros?* State its length, and why it has to be an exact multiple of the synodic period of the Moon and nearly a multiple of that of the node.

9. Do occultations of a star by the Moon occur at the same instant at all observatories?

10. Show how to find at what point (if any) of the Earth's surface a solar eclipse will be central.

CHAPTER X.

THE PLANETS.

SECTION I.—*General Outline of the Solar System.*

302. The name **planet**, or "wanderer," was applied by the Greeks to designate all those celestial bodies, except comets and meteors, which changed their position relative to the stars, independently of the diurnal motion ; these included the Sun and Moon. At present, however, only those bodies are called **planets** which move in orbits about the Sun. The Sun itself is considered to be a **star**, while the Earth is classed among the planets, and the Moon, which follows the Earth in its annual path, and has an orbital motion about the Earth, is described, along with similar bodies which revolve about other planets, as a **satellite** or **secondary.**

303. **The Sun,** ☉, is distinguished by its immense size and mass. It forms the centre of the solar system, for, in spite of the great distances of some of the furthest planets, the centre of mass of the whole system always lies very near the Sun. The Sun resembles the other fixed stars in being self-luminous.

Its diameter is 110 times that of the Earth, or nearly twice as great as the diameter of the Moon's orbit about the Earth.

From observing the apparent motion of the spots or cavities which are usually seen on the Sun's disc, it is inferred that the Sun rotates on its axis in the sidereal period of about 25 days.

304. Bode's Law.—The distances of the planets from the Sun have been observed to be approximately connected by a remarkable law known as *Bode's Law*. This law is purely empirical, that is, it is merely a result of observation, and it has not as yet been proved to be a consequence of any known physical principle. Moreover, it is only *roughly* true, giving, as it does, a result far too great for the furthest planet Neptune.

The law is given by the following rule : Write down the series of numbers

0, 3, 6, 12, 24, 48, 96, 192, 384,

each number (after the second) being double the previous one. Now add 4 to every term ; thus we obtain

4, 7, 10, 16, 28, 52, 100, 196, 388.

These numbers represent fairly closely the relative distances of the various planets from the Sun, the distance of the Earth (the third in the series) being taken as 10.

The planets all revolve round the Sun in the same direction as the Earth. Their motion is, therefore, *direct*.

305. Mercury, $\bar{\varphi}$, is the planet nearest the Sun, its distance on the above scale being represented by 4. It is characterized by its small size, the great eccentricity of its elliptical orbit, amounting to about $\frac{1}{5}$, and the great inclination of the orbit to the ecliptic, namely, about 7°. The sidereal period of revolution round the Sun is about 88 of our days.

Thus, Mercury's greatest and least distances from the Sun are in the ratio of $\quad 1+\frac{1}{5} : 1-\frac{1}{5}\quad$ (*cf.* § 149),
or $\qquad\qquad 3 : 2.$

Professor Schiaparelli, of Milan, has found that Mercury rotates on its axis once in a sidereal period of revolution; consequently it always turns nearly the same face to the Sun, like the Moon does to the Earth (§ 276).

Owing, however, to the great·eccentricity of the orbit, the "libration in longitude" is much greater than that of the Moon, amounting to 47°. Consequently, rather over *one quarter* of the whole surface is turned alternately towards and away from the Sun, *three-eighths* is always illuminated, and three-eighths is always dark.

306. Venus, ♀, is the next planet, its mean distance from the Sun being represented by about 7 (really 7·2). Its orbit is very nearly circular, and is inclined to the ecliptic at an angle of about 3° 23′.

Venus revolves about the Sun in a period of 224 days.

307. The Earth, ⊕, comes next, its mean distance being represented by 10, and its orbit very nearly circular (eccentricity $= \frac{1}{60}$). Its period of revolution in the ecliptic is $365\frac{1}{4}$ days, and its period of rotation is a sidereal day, or 23h. 56m. mean time. It is the nearest planet to the Sun having a *satellite* (the Moon, ☽), which revolves about it in $27\frac{1}{3}$ days.

308. Mars, ♂, is at a mean distance represented roughly by 16, or more accurately by 15·2. Its orbit is inclined at less than 2° to the ecliptic, and is an ellipse of eccentricity about $\frac{1}{11}$. It revolves about the Sun in a sidereal period of about 686 days, and rotates on its axis in about 24h. 37m.

Mars has two very small *satellites*, which revolve about it in the periods $7\frac{1}{2}$ and $30\frac{1}{4}$ hours, roughly. The appearance which would be presented by the inner satellite, if observed from Mars, is rather interesting. As it revolves much faster than Mars, it would be seen to rise in the *west* and set in the *east twice* during the night. The outer satellite would appear to revolve slowly in the opposite direction—from east to west. The inner satellite is eclipsed often at opposition, and would appear to transit the Sun's disc often at conjunction.

309. The Asteroids. — The next conspicuous planet, Jupiter, is at a distance represented by 52 ; but, according to Bode's law, there should be a planet at the distance 28. It was for a long time thought that no planet existed at this distance, but the gap was filled, at the beginning of the century, by the discovery of a number of small planets, to which the name of **Asteroids, or Minor Planets,** was given. Since that time a few new asteroids have been discovered almost every year, the total number found up to October 15, 1891, being 321. It is probable that this number will be very largely increased by stellar photography.

The largest asteroid, **Vesta,** is just visible to the naked eye when in opposition ; and the length of its diameter is between ·1 and ·2 of that of the Moon. Among the others **Juno, Ceres, Pallas,** and **Astræa** are the most conspicuous telescopic objects. Many of the smaller asteroids are less than ten miles in diameter, and are probably simply masses of rock flying round and round the Sun.

The periodic times of revolution of the asteroids vary considerably, but their average is about 1,600 days. The orbits are in many cases very oval, the eccentricity of one (*Polyhymnia*) being over ⅓, and they are often inclined at considerable angles to the ecliptic, the inclination in the case of *Pallas* amounting to nearly 35°, while that of *Juno* is 13°.

The planets outside the asteroid belt are distinguished from those hitherto described by their far greater dimensions and masses, and by their smaller densities. In this respect they resemble the Sun. They are also supposed to be at high temperatures, though not hot enough to emit light.

310. **Jupiter,** ♃, is at a mean distance almost exactly represented by 52. It revolves round the Sun in a period of twelve years, in an orbit nearly circular and inclined at only $1\frac{1}{3}°$ to the ecliptic.

The diameter of Jupiter is about eleven times that of the Earth, and through a telescope the disc is seen to be encircled with a series of *belts* or *streaks* parallel to its equator. On account of their variability, these are supposed to be due to *belts of clouds* in the atmosphere of the planet.

Jupiter is now known to have five *satellites*. The four outer ones are interesting as being the first celestial bodies discovered with the telescope by its inventor Galileo (A.D. 1610). A fairly powerful opera glass will just show them. The outermost of all revolves in an ellipse of considerable eccentricity inclined to the ecliptic plane at about 8°, its period being about 10d. 17h. The three next revolve in orbits nearly circular, and in the ecliptic, in periods of 7d. 4h., 3d. 13¼h., and 1d. 18½h. The fifth or innermost satellite has only just been discovered (1892) by Mr. Barnard with the great Lick telescope; it revolves in a period of nearly 12h., at a mean distance of 70,000 miles from the surface, or 113,000 miles from the centre of Jupiter. Jupiter's satellites are frequently *eclipsed* by passing into the shadow cast by Jupiter, or *occulted* when Jupiter comes between them and the Earth.

311.. **Saturn**, ♄ , is at a mean distance from the Sun of 95⅓, taking the Earth's distance as 10. This is rather less than the distance given by Bode's Law. The periodic time of revolution is 29½ years. The orbit is nearly circular, and inclined to the ecliptic at an angle of 2½°.

Saturn's rings are among the most wonderful objects revealed by the telescope. They appear to be three flat annular discs of extreme thinness, lying in a plane inclined to the ecliptic at an angle of about 28°, and extending to a distance rather greater than the radius of the planet; the middle ring is by far the brightest, while the inner ring is very faint. When the Earth is in the plane of the rings they are seen edgewise, and, owing to their very small thickness, they then become invisible except in the best telescopes.

It is probable that the rings consist of a large number of small satellites or meteors. It is certain that they do not consist of a continuous mass of solid or liquid matter. The surface of the planet itself is encircled with belts similar to those on Jupiter.

In addition to the rings, Saturn has at least eight *satellites*, all situated outside the rings. The seven nearest move in planes nearly coinciding with that of the rings, while the orbit of the eighth is inclined to it at an angle of 10°. The sixth satellite is by far the largest, having a probable diameter not far short of that of the planet Mars. The seventh has been observed, like our moon, always to turn the same side towards the planet. The distances of the satellites from Saturn range from 3 to 60 times the planet's semi-diameter, and the corresponding periods range from 22½ h. to 79 d.

312. **Uranus**, ♅, at mean distance 192, revolves in an approximately circular orbit, nearly coinciding with the ecliptic, in a period of 84 years. It was discovered in 1781 by Sir William Herschel, who named it the *Georgium Sidus* in honour of the king.

Uranus is attended by four satellites at least, and these possess the remarkable peculiarity of revolving in a plane *nearly perpendicular* to the ecliptic and in a *retrograde* direction. In fact, the plane of their orbits makes an angle of 82° with the ecliptic. Their periods are 2½d., 4d., 8⅔d.. and 13½d. roughly.

313. Neptune, ♆.—The position of this planet was predicted in 1846 almost simultaneously by Adams and Leverrier, from the observed effects of its attraction on the orbital motion of Uranus. It was first actually *seen* by Galle, of Berlin, in September, 1846, very close to the position which had been computed beforehand. It has a mean distance 300 (being considerably less than that which it would have according to Bode's Law), and it revolves in its orbit in about 164 years.

Neptune has one *satellite* moving in a *retrograde* direction in a plane inclined to the ecliptic at about 35°.

The discovery of Neptune will be treated more fully in the chapter on Perturbations.

314. Tabular View of the Solar System.—For convenient reference, the mean distances of the planets, measured in terms of the Earth's mean distance as the unit, and their periodic times, are given below, together with the inclinations and eccentricities of the orbits, and the numbers of their satellites.

Name of Planet.	Mean Dist. of Planet. / Mean Dist. of Earth.	Periodic Time.		Inclination of Orbit.	Eccentricity of Orbit.	No. of Satellites
		days	= years	° ′		
Mercury, ☿	0·38	88	0·24	7 0	·206	—
Venus, ♀	0·72	224	0·62	3 23	·007	—
Earth, ⊕	1·00	365	1·00	0 0	·017	1
Mars, ♂	1·52	687	1·88	1 51	·093	2
Ceres, ⚳	2·77	1,681	4·60	10 37	·076	—
Jupiter, ♃	5·20	4,332	11·86	1 19	·048	5
Saturn, ♄	9·54	10,759	29·46	2 30	·056	8 & 3 rings
Uranus, ♅	19·18	30,687	84·02	0 46	·046	4
Neptune, ♆	30·05	60,181	164·78	1 47	·009	1

SECTION 11.—*Synodic and Sidereal Periods—Description of Motion in Elongation of Planets as seen from the Earth— Phases.*

315. Inferior and Superior Planets.—Definitions. —In describing the motions of the planets relative to the Earth, it is convenient to divide the planets into two classes, inferior and superior planets.

An **inferior** planet is one which is nearer to the Sun than the Earth; Mercury and Venus are the two inferior planets.

A **superior** planet is one which is further from the Sun than the Earth : all the planets except Mercury and Venus are superior.

The angle of **elongation** is the difference between the geocentric (§ 156) longitude of the planet and that of the Sun. It has the same meaning as in the case of the Moon (§ 259).

We shall now describe the changes in elongation of the inferior and superior planets, as seen from the Earth. It appears from the preceding section that

(i.) The planets all revolve round the Sun in the same direction;

(ii.) The planets which are nearer the Sun travel at a greater speed than those which are more remote.

The second fact can be easily verified from comparing the distances and periods of the planets given in the previous section. Even if we take into account the fact that the more distant ones have further to travel, we shall still find that they take longer to travel over the same distance.

In order to further simplify the descriptions we shall assume that the planets all revolve uniformly in circles, about the Sun as centre, in the plane of the ecliptic. These assumptions are only roughly true, on account of the small eccentricities of the orbits and their small inclinations to the ecliptic; hence our results will only agree roughly with observation.

316. Changes in Elongation of an Inferior Planet. —Let E be the Earth, V an inferior planet moving in the orbit $A\,UB\,U'$ about S the Sun. Since SV revolves more rapidly about S than SE, the motion of V relative to E, as it would appear from S, is *direct*,

SV separates from SE at a rate which is the difference of the rates at which E, V revolve in their orbits. The changes in the positions of the planet relative to the Sun are therefore the same as if E were at rest and V revolved with an angular velocity equal to the excess of the angular velocity of the planet over that of the Earth.

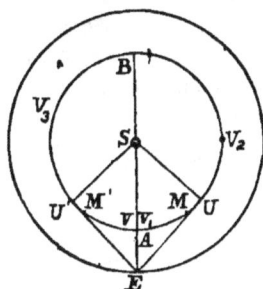

FIG. 94.

Let the line ES meet the orbit of V in A and B. When V is at A or B it has the same longitude as S, and if the planet actually moved in the ecliptic it would be in front of the Sun at A, behind the Sun at B. In reality, owing to the inclination of the orbits, this but rarely happens.

At A, the planet is said to be in **inferior conjunction** with the Sun; it has the same longitude and is nearer the Earth. At B the planet is said to be in **superior conjunction** with the Sun; it has the same longitude but is further away. If we consider the appearances which would be presented on the Sun, the planet is in "heliocentric conjunction" with the Earth at A and in "heliocentric opposition" at B.

After inferior conjunction at A, the planet is seen on the westward side of the Sun, as at V_1. The elongation SEV gradually increases till the planet reaches a point U such that EU is a tangent to the orbit. The planet is then at its **greatest elongation**, the angle SEU being a maximum.

Subsequently, as at V_2, the elongation diminishes, and the planet approaches the Sun, until superior conjunction occurs, as at B. The planet then separates from the Sun, reappearing on the opposite (eastern) side, as at V_3, attains its maximum elongation at U', and finally comes round again to inferior conjunction at A.

The time between two consecutive conjunctions of the same kind (superior or inferior) is called the **synodic period** of the planet (*cf.* § 259), and is the period in which SV separates from SE through 360°.

317. To find (roughly) the Ratio of the Distance from the Sun of an *Inferior* Planet to that of the Earth, it is only necessary to observe the planet's greatest elongation. For if U, E (Fig. 95) represent the planet and Earth at the instant of greatest elongation, the angle EUS is a right angle, and therefore

$$\sin SEU = \frac{SU}{SE} \; ;$$

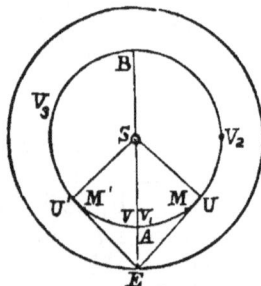

FIG. 95.

that is,

$$\frac{\textbf{Distance of planet}}{\textbf{Distance of Earth}} = \textbf{sine of greatest elongation.}$$

This method is, however, much modified by the fact that the real orbits are not circles, but ellipses.

EXAMPLE 1.—Given that the greatest elongation of Venus is 45°, find its distance from the Sun, that of the Earth being 93,000,000 miles.

Here distance of Venus = 93,000,000 sin 45° = 93,000,000 × $\sqrt{\frac{1}{2}}$
= 93,000,000 × ·70711 = 65,760,000 miles.

EXAMPLE 2.—Taking the Earth's distance as unity, to find the distance of Mercury, having given that Mercury's greatest elongation is $22\frac{1}{2}$°.

The distance of Mercury = 1 × sin $22\frac{1}{2}$° = $\sqrt{\{\frac{1}{2}(1 - \cos 45°)\}}$
= $\frac{1}{2}\sqrt{(2 - \sqrt{2})}$ = ·38268.

318. Changes in Elongation of a Superior Planet.—
Let us now compare the apparent motion of the superior planet
J with that of Sun. Since it revolves about the Sun in the
same direction as the Earth does, but more slowly, the line
SJ will move, relative to SE, in the opposite or retrograde
direction. Hence, in considering the changes in the position
of the planet relative to the Sun, we may regard SE as a
fixed line, and J must then revolve about S in the circle
$ARBT$ with a retrograde motion, i.e., in the same direction
as the hands of a watch.*

At A the planet is in **opposition** with the Sun, and its
elongation is 180°. At B it is in **conjunction,** and its
elongation is 0°. If, however, we were to refer the directions
of the Earth and planet to the Sun, the planet would be in
heliocentric conjunction with the Earth at A, and in heliocentric opposition at B.

The planet is nearest the Earth at A, and since its orbital
velocity is constant, its relative angular velocity is then
greatest, and the elongation SEJ is decreasing at its most
rapid rate. As the planet moves round from opposition A to

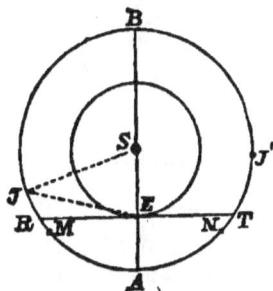

FIG. 96.

conjunction B, the elongation SEJ decreases continuously
from 180° to 0°.

At R the elongation is 90°, and the planet is said to be in
quadrature.

---

* As a simple illustration, both the hour and minute hands of a
watch revolve in the same directions, but the minute hand goes
faster and leaves the hour hand behind. Hence the hour hand
separates from the minute hand in the opposite direction to that in
which both are moving.

At conjunction, B, the elongation is 0°; and we may also consider it to be 360°. As the planet revolves from B to A, the elongation (measured round in the direction BRA) decreases from 360° to 180°.

At T the elongation is 270°, and the planet is again said to be in quadrature.

At A the elongation is again 180°, the planet being once more in opposition. After this the elongation decreases from 180° to 0° as before, as the planet's relative position changes from A through R to B.

The cycle of changes recurs in the **synodic period,** *i.e.*, the period between two successive conjunctions or oppositions. We see that the elongation decreases continually from 360° to 0° as the planet revolves from conjunction round to conjunction, and there is no greatest elongation.

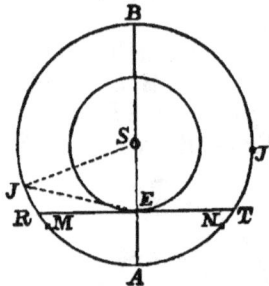

Fig. 97.

319. **To compare (roughly) the Distance of a** *Superior* **Planet with that of the Earth.**—Here there is no greatest elongation, and therefore we must resort to another method.

Let the planet's elongation SEJ (Fig. 97) be observed at any instant, the interval of time which has elapsed since the planet was in opposition being also observed. Let this interval be t, and let S denote the length of the planet's synodic period. Then, in time S the angle JSE increases from 0° to 360°; therefore, if we assume the change to take place uniformly, the angle JSE at time t after conjunction is $= 360° \times t / S$

Hence, JSE is known. Also JES has been observed, and SJE ($= 180° - JES - JSE$) is therefore also known.

Therefore we have, by plane trigonometry,

$$\frac{\textbf{Distance of Planet}}{\textbf{Distance of Earth}} = \frac{SJ}{SE} = \frac{\sin SEJ}{\sin SJE}.$$

which determines the ratio of the distances required.

This method is also applicable to the inferior planets. It is, however, not exact, owing to the fact that the planetary motions are not really uniform (see § 327).

*320. It is not necessary to observe the instant of conjunction or opposition. If S is known, two observations of the elongation and the elapsed time are sufficient to determine the ratio of the distances. The requisite formulæ are more complicated, but they only involve plane trigonometry. We, therefore, leave their investigation as an exercise to the more advanced student.

EXAMPLE.—To calculate the distance of Saturn in terms of that of the Earth, having given that 94 days after opposition the elongation of Saturn was 84° 17', and that the synodic period is 376 days. Given also tan 5° 43' = ·1.

Let the Sun, Earth, and Saturn be denoted by S, E, J. In 376 days $\angle JSE$ increases from 0° to 360°.

∴ in 94 days after opposition $\angle JSE = 90°$;

also, by hypothesis, $\angle JES = 84°$ 17'.

$$\therefore \frac{\text{Distance of Saturn}}{\text{Distance of Earth}} = \frac{SJ}{SE} = \tan SEJ = \tan 84° \ 17''$$

$$= \cot 5° \ 43' = \frac{1}{\cdot 1} = 10.$$

Therefore the distance of Saturn, as calculated from the given data, is 10 times that of the Earth.

321. The synodic period of an inferior planet may be found very readily by determining the time between two transits of the planet across the Sun's disc and counting the number of revolutions in the interval.

For a superior planet this is not possible, and we must, instead, find the interval between two epochs at which the planet has the same elongation.

322. Relations between the Synodic and Sidereal Periods.—The relation between the synodic and sidereal periods is almost exactly the same as in the case of the Moon, the only difference being that the planets revolve about the Sun and not about the Earth.

The **sidereal period** of a planet is the time of the planet's revolution in its orbit about the Sun relative to the stars. The **synodic period** is the interval between two conjunctions with the Earth relative to the Sun. It is the time in which the planet makes one whole revolution as compared with the line joining the Earth to the Sun.

Let S be the planet's synodic period,

P its sidereal period,

Y the length of a year, that is, the Earth's sidereal period, all the periods being supposed measured in days.

Then, in one day,

the angle described by the planet about the Sun $= 360°/P$,
the angle described by the Earth $= 360°/Y$,
and the angle through which their heliocentric
   directions have separated $= 360°/S$.

If the planet be inferior, it revolves more rapidly than the Earth, and $360°/S$ represents the angle gained by the *planet* in one day.

$$\therefore \quad \frac{360°}{S} = \frac{360°}{P} - \frac{360°}{Y};$$

$$\text{or} \quad \frac{1}{S} = \frac{1}{P} - \frac{1}{Y}. \quad \dots\dots\dots\dots(\text{i.}).$$

If the planet be superior, it revolves more slowly than the Earth, and $360°/S$ is the angle gained by the *Earth* in one day.

$$\therefore \quad \frac{360°}{S} = \frac{360°}{Y} - \frac{360°}{P};$$

$$\text{or} \quad \frac{1}{S} = \frac{1}{Y} - \frac{1}{P}.$$

From these relations, the sidereal period can be found if the synodic period is known, and *vice versâ*.

323. Phases of the Planets.—As the planets derive their light from the Sun, they must, like the Moon, pass through different phases depending on the proportion of their illuminated surface which is turned towards the Earth.

Phases of an Inferior Planet.—An inferior planet V will evidently be new at inferior conjunction A, dichotomized like the Moon at its third quarter at greatest elongation U; full at superior conjunction B, dichotomised like the Moon at first quarter when it again comes to greatest elongation at U'. Thus, like the Moon, it will undergo all the possible different phases in the course of a synodic revolution.

There is, however, one important difference. As the planet revolves from A to B its distance from the Earth increases, and its angular diameter therefore decreases. Thus the planet appears largest when new and smallest when full, and the variations in the planet's brightness due to the differences of phase are, to a great extent, counterbalanced by the changes in the planet's distance. For this reason, Venus alters very little in its brightness (as seen by the naked eye) during the course of its synodical revolution.

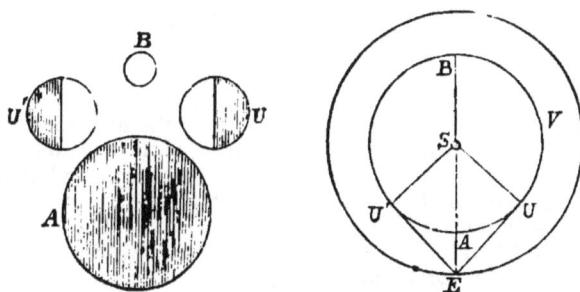

FIG. 98.

The phase is determined by the angle SVE, and this is the angle of elongation of the Earth as it would appear from the planet. The illuminated portion of the visible surface of the planet at V is proportional to $180° - SVE$, and the proportion of the apparent area of the disc which is illuminated varies as $1 + \cos SVE$ or $2 \cos^2 \frac{1}{2} SVE$. (*Cf.* § 263).

The phases of Venus are easily seen through a telescope.

ASTRON. s

324. Phases of a Superior Planet.—For a superior planet J the angle SJE never exceeds a certain value. It is greatest when $SEJ = 90°$, being then the greatest elongation of the Earth as it would appear from the planet. Hence the planet is always nearly full, being only slightly gibbous, and the phase is most marked at quadrature.

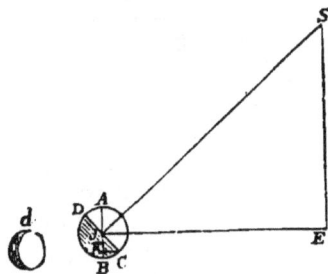

FIG. 99.

The gibbosity of Mars, though small, is readily visible at quadrature, about one-eighth of the planet's disc being obscured. The other superior planets are, however, at a distance from the Sun so much greater than that of the Earth that they always appear very approximately full.

325. The "Phases" of Saturn's Rings are due to an entirely different cause. The plane of the rings, like the plane of the Earth's equator, is fixed in direction, and inclined to the ecliptic at an angle of about 28°. Hence, during the course of the planet's sidereal revolution, the Sun passes alternately to the north and south sides of the rings (just as in the phenomena of the seasons on our Earth, the Sun is alternately N. and S. of the equator). The Earth, which, relatively to Saturn, is a small distance from the Sun, also passes alternately to the north and south sides of the rings, and we see the rings first on one side and then on the other. At the instant of transition the rings are seen edgewise, and are almost invisible.

Unless Saturn is in opposition at this instant, the Sun and Earth do not cross the plane of the rings simultaneously, and between their passages there is a short interval during which the Sun and Earth are on opposite sides of the plane; and the *unilluminated* side of the rings is turned towards the Earth. The last "disappearances" of the rings occurred in Sept., 1891—May, 1892, but they occur twice in each sidereal period, or once about every 15 years.

Other interesting appearances are presented by the shadows thrown by the planet on the rings and by the rings on the planet.

SECTION III.—*Kepler's Laws of Planetary Motion.*

326. Kepler's Three Laws. — We have already seen that the orbits of most of the planets are nearly circular, their distances from the Sun being nearly constant and their motions being nearly uniform. A far closer approximation to the truth is the hypothesis held for a long time by Tycho Brahe and other astronomers, namely, that each planet revolved in a circle whose centre was at a small distance from the Sun, and described equal angles in equal intervals of time about a point found by drawing a straight line from the Sun's centre to the centre of the circle and producing it for an equal distance beyond the latter point.

The true laws which govern the motion of the planets were discovered by the Danish astronomer Kepler, in connection with his great work on the planet Mars (*De Motibus Stellae Martis*). After nine years' incessant labour the first and second of the following laws were discovered, and shortly afterwards the third.

I. Every planet moves in an ellipse, with the Sun in one of the foci.

II. The straight line drawn from the centre of the Sun to the centre of the planet (the planet's "radius vector") sweeps out equal areas in equal times.

III. The squares of the periodic times of the several planets are proportional to the cubes of their mean distances from the Sun.

These laws are known as **Kepler's Three Laws.** We have already proved that the first two laws hold in the case of the Earth. The third law is also found to hold good for the Earth as well as the other planets, and this fact alone affords strong evidence that *the Earth is a planet*

By the **mean distance** of a planet is meant the arithmetic mean between the planet's greatest and least distances from the Sun. If p, a (Fig. 100) be the planet's positions at perihelion and aphelion (*i.e.*, when nearest and furthest from the Sun respectively), the planet's mean distance $= \frac{1}{2}(Sp + Sa) = \frac{1}{2}pa = \frac{1}{2}$ (major axis of ellipse described) (§ 147).

The periodic times are, of course, the *sidereal periods*. Hence the third law is a relation between the sidereal periods and the major axes of the orbits.

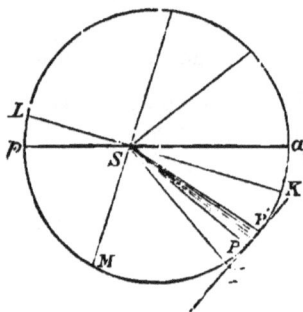

FIG. 100.

327. Verification of Kepler's First and Second Laws.

—We will now roughly sketch the principle of the methods by which Kepler determined the orbit of Mars, and thus proved his First and Second Laws. A verification of the laws in the case of the Earth has already been given, and we have shown (§ 145) how to determine exactly the position of the Earth at any given time; we may regard this, therefore, as known. We may also suppose the length of the sidereal period of Mars to be known, for the average length of the synodic period may be found, as in § 261, and the sidereal period may be deduced by the formulæ of § 322.

Let the direction of the planet be observed when it is at any point M in its orbit, the Earth's position being E. When the planet has returned again to M after a sidereal revolution, the Earth will *not* have returned to the same place in its orbit, but will be in a different position, say F. Let now the planet's new direction FM be observed.*

From knowing the Earth's motion, we know SE, SF and the angle ESF. From the observations of the two directions of M we know the angles SEM and SFM. These data are sufficient to enable us to solve the quadrilateral $SEMF$.†

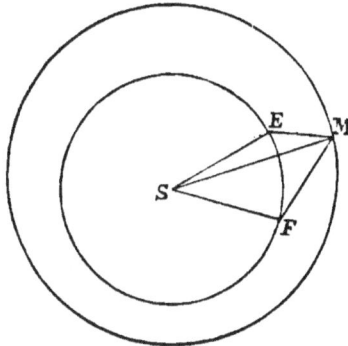

FIG. 101.

We can thus determine SM and the angle ESM, whence the distance and direction of M from the Sun are found. Similarly, any other position of Mars in its orbit can be found by two observations of the planet's sidereal period separated by the interval of the planet's sidereal revolution. In this way, by a series of observations of Mars, extending over two sidereal periods, the planet's direction and distance relative to the Sun can be determined daily, and the whole orbit can thus be plotted out.

This method was that actually adopted by Kepler, except that he had not previously determined the Earth's motion, and believed that it could be accurately represented by Tycho Brahe's hypothesis. This approximation was close enough, for the Earth's orbit is very nearly a circle, and that of Mars, which he was deducing, is very much more eccentric.

----

* For simplicity we suppose Mars to move in the ecliptic plane. The methods require some modification when the inclination of the orbits is taken into account, but the general principle is the same.

† For join EF. In $\triangle SEF$ we know SE, SF and $\angle ESF$. Hence we find EF, $\angle SEF$, $\angle SFE$. Hence $\angle FEM\ (= SEM-SEF)$ and $\angle EFM\ (= SFM-SFE)$ are known. With these and EF solve $\triangle MEF$ and find EM, EF. Lastly, in $\triangle SEM$ we know SE, EM, and $\angle SEM$, and thus we find SM and $\angle ESM$.

328. Verification of Kepler's Third Law.—Kepler's Third Law can be verified much more easily, especially if we make the approximate assumption that the planets revolve uniformly in circles about the Sun as centre. The sidereal periods of the different planets can be found by observing the *average* length of the synodic period (the actual length of any synodic period is not quite constant, owing to the planet not revolving with exactly uniform velocity) and applying the equations of § 322. The distance of the planet may be compared with that of the Earth, either by observing the greatest elongation (§ 317) in the case of an inferior planet, or by the method of § 319. It is then easy to verify the relation between the mean distances and periodic times of the several planets.

In the table of § 314, the student will have little difficulty in verifying (especially if a table of logarithms be employed) that the square of the ratio of the periodic time of the planet to the year (or periodic time of the Earth) is in every case equal to the cube of the ratio of the planet's mean distance to that of the Earth.* The data being only approximate, however, the law can only be verified as approximately true, although it is in reality accurate.

Owing to the importance of Kepler's Third Law, we append the following examples as illustrations.

EXAMPLES.

1. Given that the mean distance of Mars is 1·52 times that of the Earth, to find the sidereal period of Mars.

Let T be the sidereal period of Mars in days. Then, by Kepler's Third Law,

$$\left\{ \frac{T}{365\tfrac{1}{4}} \right\}^2 = \{1\cdot52\}^3 = 3\cdot5118;$$

$$\therefore T = 365\tfrac{1}{4} \times \sqrt{(3\cdot5118)} = 365\tfrac{1}{4} \times 1\cdot874 = 684\cdot5.$$

Hence, from the given data, the period of Mars is 1·874 of a year, or 684·5 days.

Had we taken the more accurate value of the relative distance, viz., 1·5237, we should have found for the period the correct value, namely, 687 days.

---

* In other words, 2 log (period in years) = 3 log (distance in terms of Earth's distance).

2. The *synodic* period of Jupiter being 399 days, to find its distance from the Sun, having given that the Earth's mean distance is 92 million miles.

Let T be the sidereal period of Jupiter. Then, by § 322,

$$\frac{1}{T} = \frac{1}{365\frac{1}{4}} - \frac{1}{399} = \frac{33\frac{3}{4}}{365\frac{1}{4} \times 399}.$$

$$\therefore T = \frac{399}{33\frac{1}{4}} \times 365\frac{1}{4} \text{ days} = \frac{399}{33\frac{1}{4}} \text{ years}$$

$$= 11\cdot82, \text{ or nearly 12 years.}$$

Let a be the distance of Jupiter in millions of miles. Then, by Kepler's Third Law,

$$\left(\frac{a}{92} \right)^3 = \left(\frac{12}{1} \right)^2 = 144.$$

$$\therefore a = 92 \times \sqrt[3]{(144)} = 92 \times 5\cdot24 = 482 \,;$$

that is, Jupiter's distance is **482** millions of miles.

By taking $T = 11\cdot82$ and the Earth's distance as $92\cdot04$, we should have found the more accurate value $477\cdot6$ for Jupiter's distance in millions of miles.

329. Satellites.—The motions of the satellites about any planet are found to obey the same laws as those which Kepler investigated for the orbits of the planets. For example, the Moon's orbit about the Earth is an ellipse, and (except so far as affected by perturbations) satisfies both of Kepler's First and Second Laws. When a number of satellites are revolving round a common primary (*i.e.*, planet) as is the case with Jupiter, the squares of their periodic times are found, in every case, to be proportional to the cubes of their mean distances from the planet.*

EXAMPLE.—To compare (roughly) the mean distances of its two satellites from Mars. The periodic times are $30\frac{1}{4}$h. and $7\frac{1}{2}$h. respectively, and these are in the ratio (nearly) of 4 to 1.

Hence the mean distances are as $4^{\frac{2}{3}}$: 1, or $\sqrt[3]{16}$: 1.

Now, $2\sqrt[3]{16} = \sqrt[3]{128} = 5$ very nearly (since $5^3 = 125$). Hence the mean distances are very nearly in the ratio of **5 to 2**.

---

* Of course the relation does not hold between the periodic times and mean distances of satellites revolving round *different* planets, nor between those of a satellite and those of a planet.

SECTION IV.—*Motions Relative to Stars—Stationary Points.*

330. **Direct and Retrograde Motion.** — We have described (§§ 316–318) the motion of a planet relative to the Sun. In considering its motion *relative to the stars* we must take account of the Earth's motion.

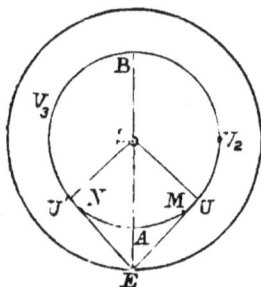

FIG. 102.

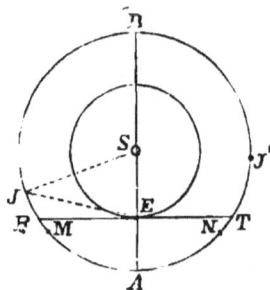

FIG. 103.

An inferior planet moves more swiftly than the Earth. Hence at inferior conjunction the line AE (Fig. 102) joining them is moving in the direction of the hands of a watch. The planet therefore appears to move *retrograde*. At greatest elongation (U, U') the planet's *own* motion is in the line joining it to the Earth, and hence produces no change in its direction; but the Earth's *direct* motion causes the line EU or EU' to turn about U or U' with a rotation *contrary* to that of the hands of a watch; and therefore the apparent motion is *direct*. Over the whole portion UBU' of the relative orbit both the Earth's motion and the planet's combine to make the planet's apparent motion *direct*. There must, therefore, be two positions, M between A and U and N between U' and A, at which the motion is checked and reversed. At these two positions the planet is said to be **stationary**.

A superior planet moves *slower* than the Earth; hence at opposition the line EA (Fig. 103) joining them is turning in the direction of the hands of a watch. The planet therefore appears to move *retrograde*. At quadrature (R, T) the Earth is moving along RET; hence its motion produces no change in the planet's direction. Hence the planet's direct motion about the Sun makes its apparent motion also *direct*. In all parts of the arc RBT the orbital velocities of Earth and planet conspire to produce direct motion. Hence the planet is **stationary** at M, between A and R, and at N between T and A.

In both cases the longitude increases from M to N and decreases from N to M; hence it is a maximum at N and a minimum at M. After a complete synodic revolution the planet's elongation is the same as at the beginning, and the Sun's longitude has been increased; therefore the planet's longitude has also *increased*. Hence the *direct* preponderates over the *retrograde* motion.

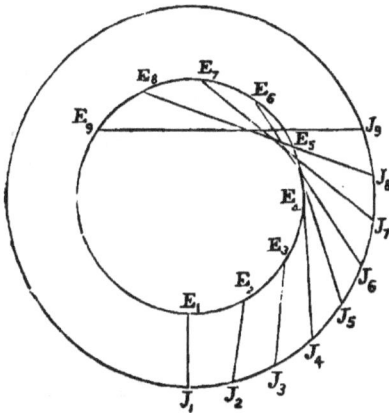

Fig. 104.

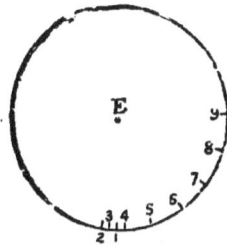

Fig. 105.

331. **Alternative explanation.**—We may also proceed as follows. Let E, J represent two planets at heliocentric conjunction. Let E_1, E_2, E_3, ..., J_1, J_2, J_3, ..., be their successive positions after a series of equal intervals. To find the apparent motion of J among the stars, as seen from E, take any point E, and let $E1$, $E2$, $E3$, ... (Fig. 105) be parallel respectively to E_1J_1, E_2J_2, E_3J_3, ... Then the points 1, 2, 3, ... represent J's direction as seen from E at a series of equal intervals, starting from opposition.

Again, if $J1$, $J2$, $J3$ be taken parallel to J_1E_1, J_2E_2, ...
(Fig. 108), the points 1, 2 now represent E's direction as
seen from J.

We observe from Figs. 107, 108 that the relative motion is
retrograde from 1 to 2, and becomes direct near 3. At the
instant at which this takes place, either planet must be
stationary, relative to the other. Since J_4E_4 is nearly a tangent to E's orbit, E is near its greatest elongation, and J
is near quadrature at the positions 4; hence, E appears
stationary from J between inferior conjunction and greatest
elongation; and J appears stationary between opposition and
quadrature.

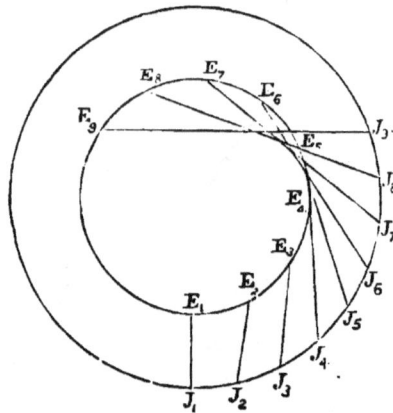

FIG. 106.

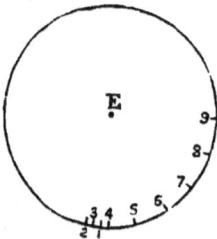

FIG. 107.

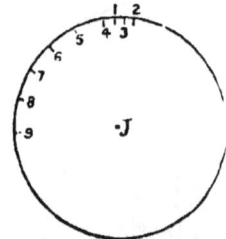

FIG. 108.

We notice that $J1$, $J2$, ... are parallel to $E1$, $E2$, but
measured in opposite directions, showing that the motion of
E relative to J is the same (direct, stationary, or retrograde)
as that of J relative to E.

332. Effects of Motion in Latitude. — Hitherto we have supposed the planet to move in the ecliptic. When, however, the small inclination of the orbit to the ecliptic is taken into account, it is evident that the planet's latitude is subject to periodic fluctuations.

The points of intersection of the planet's orbit with the ecliptic are (as in the case of the Moon) called the **Nodes.** Whenever the planet is at a node its latitude is zero ; and this happens twice in every *sidereal* period of revolution.

A planet is stationary when its longitude is a maximum or minimum, but unless its latitude should happen to be a maximum at the same time, the planet does not remain actually at rest. When the change from direct to retrograde motion, and *vice versâ*, is combined with the variations in latitude, the effect is to make the planet describe a zigzag curve, sometimes containing one or two loops, called "**loops of retrogression.**" This is readily verified by observation.

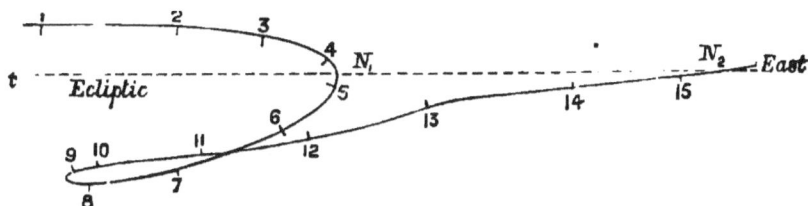

FIG. 109.

Fig. 109 is an example of the path of Venus in the neighbourhood of its stationary points, the numbers representing its positions at a series of intervals of ten days. Here, the planet is stationary close to the node N_1 between 4 and 5, and it describes a loop in the neighbourhood of the stationary point near 9, where its motion changes from retrograde to direct.

The student will find it an instructive exercise to trace out the path of any planet in the neighbourhood of its retrograde motion, using the values of its decl. and R.A., at intervals of a few days, as tabulated in the Nautical or *Whitaker's* Almanack.

333. To find the condition that two planets may be stationary as seen from one another, assuming the orbits circular and in one plane. — Let P, Q be the positions of the planets at any instant; P', Q' their positions after a very short interval of time.

Then, if PQ and $P'Q'$ are parallel, the direction of either planet, as seen from the other, is the same at the beginning and end of the interval; that is, P is stationary as seen from Q, and Q is stationary as seen from P.

Let u, v represent the orbital velocities of the planets P, Q; a, b the radii SP, SQ respectively.

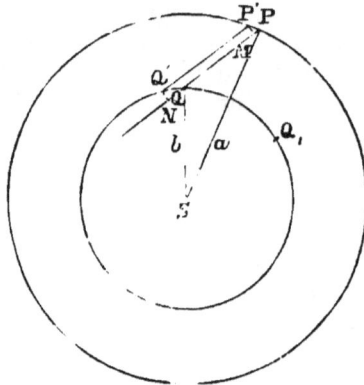

FIG. 110.

Draw $P'M$, $Q'N$ perpendicular to PQ. Then, in the stationary position, we must have $P'M = Q'N$.

But PP', QQ', being the arcs described by the two planets in the same interval, are proportional to the velocities u, v. Therefore $P'M$, $Q'N$ are proportional to the component velocities of the planets perpendicular to PQ. These component velocities must, therefore, be equal, and we have

$$u \sin P'PM = v \sin Q'QN.$$

Whence, since $P'P$ is perpendicular to SP and $Q'Q$ to SQ,

$$u \cos SPQ = v \cos SQN = -v \cos SQP \quad \ldots\ldots(\text{i.}),$$

and this is the condition that the planets may be stationary relative to one another.

*334. To find the angle between the radii vectores in the station-ary position, and the period during which a planet's motion is retrograde.—By projecting SQ, QP on SP, we have

$$a = b \cos PSQ + PQ \cos SPQ.$$

Similarly $\qquad b = a \cos PSQ + PQ \cos SQP.$

$$\therefore \ \cos SPQ : \cos SQP = a - b \cos PSQ : b - a \cos PSQ.$$

Whence, by (i.), $\ u(a - b \cos PSQ) + v(b - a \cos PSQ) = 0;$

$$\therefore \ \cos PSQ = \frac{au + bv}{av + bu} \quad \dots\dots\dots\dots\dots\dots(ii.).$$

By means of Kepler's Third Law, we can express the ratio of u to v in terms of a and b. For if T_1, T_2 denote the periodic times, then evidently $\qquad uT_1 = 2\pi a; \quad vT_2 = 2\pi b;$

$$\therefore \ u : v = aT_2 : bT_1.$$

But $\qquad\qquad T_1 : T_2 = a^{\frac{3}{2}} : b^{\frac{3}{2}};$

$$\therefore \ u : v = \sqrt{b} : \sqrt{a}.$$

Substituting in (ii.), we have

$$\cos PSQ = \frac{a\sqrt{b} + b\sqrt{a}}{a\sqrt{a} + b\sqrt{b}} = \frac{(ab)^{\frac{1}{2}}(a^{\frac{1}{2}} + b^{\frac{1}{2}})}{a^{\frac{3}{2}} + b^{\frac{3}{2}}} = \frac{\sqrt{(ab)}}{a - \sqrt{(ab)} + b}.$$

[From this result it may be easily deduced that

$$\tan \tfrac{1}{2} PSQ = \left(\frac{1 - \cos PSQ}{1 + \cos PSQ}\right)^{\frac{1}{2}} = \frac{\sqrt{b} - \sqrt{a}}{\sqrt{(a + b)}}\right].$$

In the above investigation PSQ is the angle through which SQ separates from SP between heliocentric conjunction and the station-ary point. Hence, since $\angle PSQ$ increases from $0°$ to $360°$ in the synodic period S, the time taken from conjunction to the stationary point

$$= S \times \frac{\angle PSQ}{360}.$$

If $\angle PSQ_1 = \angle PSQ$, there is another stationary point before con-junction, when the planets are in the relative positions P, Q. Hence, the interval between the two stationary positions is twice the time taken by the planets to separate through $\angle PSQ$, and is therefore

$$= 2S \times \frac{PSQ}{360°} = S \times \frac{PSQ}{180°}.$$

This represents the interval during which the motion of either planet, as seen from the other, is retrograde. During the remainder of the synodic period the motion is direct, and the time of direct motion is therefore

$$= S - S \times \frac{PSQ}{180°} = S \times \frac{180° - PSQ}{180°}.$$

SECTION V.—*Axial Rotations of Sun and Planets.*

335. **The Period of Rotation of the Sun** can be found by
observing the passage of sunspots across the disc. These spots, by
the way, are very easily exhibited with any small telescope by
focussing an image of the Sun on to a piece of white paper placed
a few inches in front of the eye-glass—for to look straight at the
Sun would cause blindness. As the Sun's axis of rotation is nearly
perpendicular to the ecliptic, the rotation of the spots is seen in
perspective, and makes them appear to move nearly in straight
lines across the disc. From this observed apparent motion (as
projected on the celestial sphere in a manner similar to that
explained in § 263) their actual motion in circles about the Sun's
axis is readily determined. For example, if a spot moves from the
centre of the disc to the middle point of its radius, we may readily
see that the angle turned through $= \sin^{-1}\frac{1}{2} = 30°$.

The spots are observed to return to the same position in about
$27\frac{1}{4}$ days, and this is their *synodic period* of rotation relative to the
Earth. Call it S, and let T be the time of a sidereal rotation, Y the
length of the year. Then, as in the case of an inferior planet
(§ 322), we may show that

$$\frac{1}{S} = \frac{1}{T} - \frac{1}{Y} \; ; \qquad \therefore \quad \frac{1}{T} = \frac{1}{27\frac{1}{4}} + \frac{1}{365\frac{1}{4}} \; ;$$

whence the true period of rotation $T = 25\frac{1}{4}$ days (roughly).

It has been observed that spots near the Sun's equator rotate
rather faster than those near the poles. This proves the Sun's surface
to be in a fluid condition, for no rigid body could rotate in this way.

336. **Periods of Rotation of Planets.**—The rotation period of a
superior planet is easily found by observing the motions of the
markings across its disc near opposition, allowance being made for
the motions of the Earth and planet. The surface of Mars has well-
defined markings, which give the period 24h. 37m. The principal
mark on Jupiter is a great red spot amid his southern belts, which
rotates in the period of 9h. 56m. Saturn rotates in 10h. 14m.

For an *inferior planet*, the period is more difficult to observe.
There is still some uncertainty as to whether Venus rotates in about
23h. 21m., or whether, like Mercury, it always turns the same face
to the Sun. There are no well-defined markings, and, as the
greatest elongation is only 45°, Venus can only be seen for part of
the night as an evening or morning star, and in the most favourable
positions only a portion of the disc is illuminated. Moreover,
refraction, modified by air-currents, prevents the planet from being
seen distinctly when near the horizon. If the same markings are
seen on the disc of a planet on consecutive nights, they may either
have remained turned towards the Earth, or they may have rotated
through 360° during the day; hence the difficulty of deciding between
the two alternative hypotheses. Before the researches of Schiaparelli (§ 305), it was believed that Mercury also rotated in about 24h.

EXAMPLES.—X.

1. The Earth revolves round the Sun in 365·25 days, and Venus in 224·7 days. Find the time between two successive conjunctions of Venus.

2. If Venus and the Sun rise in succession at the same point of the horizon on the 1st of June, determine roughly Venus' elongation.

3. Find the ratio of the apparent areas of the illuminated portions of the disc of Venus when dichotomized and when full, taking Venus' distance from the Sun to be $\frac{8}{11}$ of that of the Earth.

4. Mars rotates on his axis once in 24 hours, and the periods of the sidereal revolutions of his two satellites are $7\frac{1}{2}$ hours and 30 hours respectively. Find the time between consecutive transits over the meridian of any place on Mars of the two satellites respectively.

5. A small satellite is eclipsed at every opposition. Find an expression for the greatest inclination which its orbit can have to the plane of the ecliptic.

6. If the periodic time of Saturn be 30 years, and the mean distance of Neptune 2,760 millions of miles, find (roughly) the mean distance of Saturn and the periodic time of Neptune. (Earth's mean distance is 92 millions of miles.)

7. If the synodic period of revolution of an inferior planet were a year, what would be its sidereal period, and what would be its mean distance from the Sun according to Kepler's Third Law?

8. Jupiter's solar distance is 5·2 times the Earth's solar distance; find the length of time between two conjunctions of the Earth and Jupiter.

9. Saturn's mean distance from the Sun is nine times the Earth's mean distance. Find how long the motion is retrograde, having given $\cos^{-1}\frac{3}{7} = 65°$.

10. Show that if the planets further from the Sun were to move with greater velocity in their orbits than the nearer ones, there would be no stationary points, the relative motion among the stars being always *direct*. What would be the corresponding phenomenon if the velocities of two planets were *equal*?

EXAMINATION PAPER.—X.

1. Explain the apparent motion of a superior planet. Illustrate by figures.

2. Describe the apparent course among the stars of an inferior planet as seen from the Earth, and the changes in appearance which the planet undergoes.

3. Define the *sidereal* and *synodic period* of a superior or inferior planet, and find the relation between them. Calculate the synodic period of a superior planet whose period of revolution is thirty years.

4. How is it that Venus alters so little in apparent magnitude (as seen by the naked eye) in her journey round the Sun? Why does not Jupiter exhibit any perceptible phases?

5. State Bode's Law connecting the mean distances of the various planets from the Sun.

6. Prove that the time of most rapid approach of an inferior planet to the Earth is when its elongation is greatest, and that the velocity of approach is then that under which it would describe its orbit in the synodic period of the Earth and the planet. Give the corresponding results for a superior planet. (The orbits are to be taken circular and in the same plane.)

7. What is meant by *stationary points* in the apparent motion of a planet? Prove that, if a planet Q is stationary as seen from P, then P will be stationary as seen from Q.

8. State Kepler's Three Laws, and, assuming the orbits of the Earth and Venus to be circular, show how the Third Law might be verified by observations of the greatest elongation and synodic period of Venus.

9. Find the periods during which Venus is an evening star and a morning star respectively, being given that the mean distance of Venus from the Sun is ·72 of that of the Earth.

10. Having given that there will be a full Moon on the 5th of June, that Mercury and Venus are both evening stars near their greatest elongations, that Mars changed from an evening to a morning star about the vernal equinox, and that Jupiter was in opposition to the Sun on April 21st, draw a figure of the configuration of these heavenly bodies on May 1st. (All these bodies may be supposed to move in one plane.)

CHAPTER XI.

THE DISTANCES OF THE SUN AND STARS.

SECTION I.—*Introduction—Determination of the Sun's Parallax by Observations of a Superior Planet at Opposition.**

337. In Chapter VIII., Section I., we explained the nature of the correction known as parallax, and showed how to find the distance of a celestial body from the Earth in terms of its parallax. We also described two methods of finding the parallax of the Moon or of a planet in opposition—the first by meridian observations at two stations, one in the northern and the other in the southern hemisphere (§ 252); the second by micrometric observations made at a single observatory shortly after the time of rising and shortly before the time of setting of the planet or observed body (§ 254).

In both methods the position of the body is compared with that of neighbouring stars. This is impossible in the case of the Sun, for the intensity of the Sun's rays necessitates the use of darkened glasses in observations of the Sun, and these render all near stars invisible.

Of course the star could theoretically be dispensed with in the method of § 252, but only (as there explained) at a great sacrifice of accuracy; and if a star is used which crosses the meridian at night, the temperature of the air has changed considerably, and the corrections for refraction are therefore quite different, besides which other errors are introduced by the change of temperature of the instrument.

---

* The student will find it of great advantage to revise Section I. of Chapter VIII. before commencing the present Section.

In § 264 we described a method, due to Aristarchus, in which the ratio of the Sun's to the Moon's distance was determined by observing the Moon's elongation when dichotomized, but this method was rejected, owing to the irregular boundary of the illuminated part of the disc, and the consequent impossibility of observing the instant of dichotomy.

338. **Classification of Methods.**—The principal *practicable* methods of finding the Sun's distance may be conveniently classified as follows :—

A. *Geometrical Methods.*

(1) By observations of the parallax of a superior planet at opposition (Section I.).

(2) By observations of a transit of the inferior planet Venus (Section II.).

B. *Optical Methods* (Section IV.).

(3) By the eclipses of Jupiter's satellites (Roemer's Method).
(4) By the aberration of light.

C. *Gravitational Methods* (Chapter XIV., Section IV.).

(5) By perturbations of Venus or Mars.
(6) By lunar and solar inequalities.

339. **To find the Sun's Parallax by Observation of the Parallax of Mars.**—By observing the parallax of Mars when in opposition, the Sun's parallax can readily be found. For the observed parallax determines the distance of Mars from the Earth, and this is the **difference** of the distances of the Sun from the Earth and Mars respectively. The **ratio** of their mean distances may be found, if we assume Kepler's Third Law (§ 326), by comparing the sidereal period of Mars with the sidereal year, and is therefore known. Hence the distance of either planet from the Sun may readily be found, and the Sun's parallax thus determined.

The parallax of Mars in opposition may be observed by either of the methods described in Chapter VIII., Section I. The method of § 252 (by meridian observations at two stations) was employed by E. J. Stone in 1865. The observations were made at Greenwich and at the Cape, and the Sun's parallax was computed as 8·943″. The method of § 254 (by observations at a single observatory) was employed by Gill at Ascension Island in 1879, and the result was 8·783″.

EXAMPLE.

If the parallax of Mars when in opposition be 14″, to find the Sun's parallax, assuming the distances of the Sun from the Earth and Mars to be in the ratio of 10 : 16.

The distance of the Earth from Mars in opposition is the difference of the Sun's distances from the two planets. Hence

Distance of Earth from Mars : Distance of Earth from Sun

$$= 16 - 10 : 10 = 3 : 5.$$

But the parallax of a body is inversely proportional to its distance (§ 230).

$$\therefore \text{ Parallax of Sun : Parallax of Mars} = 3 : 5;$$

$$\therefore \text{ Sun's parallax} = \frac{3 \times 14''}{5} = 8\cdot4''.$$

*340. Effect of Eccentricities of Orbits.—Owing to the eccentricities of the orbits of the Earth and Mars, their distances from the Sun when in opposition will not in general be equal to their mean distances, and therefore their ratio will differ from that given by Kepler's Third Law. But, by the method of § 145, the Earth's distance at any time may be compared with its mean distance, and similarly, since the eccentricity of the orbit of Mars and the position of its apse line are known, it is easy to determine the ratio of Mars' distance at opposition to its mean distance, and thus to compare its distance with that of the Earth.

341. Sun's Parallax by Observations on the Asteroids and on Venus.—The Sun's parallax may also be found by observing the parallax of one of the asteroids when in opposition, the method being identical with that employed in the case of Mars. In this way Galle, by meridian observations of the parallax of Flora at opposition in 1873, computed the Sun's parallax at 8·873″, and Lindsay and Gill, by observing the parallax of Juno in 1877, found the value 8·765″.

The next planet, Jupiter, is too distant to be utilized in this way. Its parallax at opposition is less than a quarter of the Sun's parallax, and is too small to be observed with sufficient accuracy.

The Sun's parallax might also be found by an observation of Venus near its greatest elongation. The ratio of its distance to the Sun's might be calculated and its parallax found by the method of § 252, and that of the Sun deduced. The method of § 254 could not be employed, because one of the observations would have to be made in full sunshine.

EXAMPLES.

1. Having given that the *greatest possible* parallax of Mars when in opposition is 21·08″, to find the Sun's mean parallax, the eccentricities of the orbits of the Earth and Mars being $\frac{1}{60}$ and $\frac{1}{17}$ respectively, and the periodic time of Mars being 1·88 of a year.

The parallax of Mars is greatest when Mars is nearest the Earth; hence the greatest possible value occurs when, at opposition, Mars is in perihelion and the Earth is at aphelion.

Let r, r' denote the *mean distances* of the Earth and Mars from the Sun respectively. By Kepler's Third Law we have

$$\frac{r'^3}{r^3} = \frac{(1\cdot88)^2}{1^2}; \quad \therefore \frac{r'}{r} = (1\cdot88)^{\frac{2}{3}} = 1\cdot523.$$

(The calculation is most easily performed with a table of logarithms.)

But since the Earth is in aphelion, its distance from the Sun at the time of observation is greater than its mean distance by $\frac{1}{60}$, and is therefore

$$= r\left(1 + \tfrac{1}{60}\right) = 1\cdot017\, r.$$

Also the distance of Mars from the Sun at perihelion

$$= r'\left(1 - \tfrac{1}{17}\right) = \left(1 - \tfrac{1}{17}\right) \times 1\cdot523\, r$$
$$= (1\cdot523 - \cdot090)\, r = 1\cdot433\, r.$$

Hence the least distance of Mars from the Earth at opposition

$$= \cdot416\, r.$$

Therefore, since r is the Sun's *mean* distance from the Earth, we have

Observed parallax of Mars : *mean* parallax of Sun $= 1 : \cdot416$;

$\therefore$ Sun's mean parallax $= 21\cdot08'' \times \cdot416 = 8\cdot77''$.

2. To find the Earth's mean distance from the Sun, and its distances at perihelion and aphelion, taking the Sun's parallax as 8·79″.

If a denote the Earth's equatorial radius, we have, approximately,

$$r = \frac{a}{\sin 8\cdot79''} = \frac{a}{\text{circ. meas. of } 8\cdot79''} = a \times \frac{206{,}265}{8\cdot79}.$$

Taking $a = 3963\cdot3$, this gives

r (Earth's mean solar distance) = **93,002,000 miles,**
correct to the nearest thousand miles.

Also, perihelion distance from Sun $= 93{,}002{,}000 \times \left(1 - \tfrac{1}{60}\right)$
$$= 93{,}002{,}000 - 1{,}550{,}000 = \textbf{91,452,000 miles,}$$

and aphelion distance $= 93{,}002{,}000 \times \left(1 + \tfrac{1}{60}\right)$
$$= 93{,}002{,}000 + 1{,}550{,}000 = \textbf{94,552,000 miles.}$$

Section II.—*Transits of Inferior Planets.*

342. When Venus is very near the ecliptic at inferior con-junction, it passes in front of the Sun's disc, appearing like a black dot on the Sun. Now the circumstances of such a transit are different at different places, for although both the Sun and planet are displaced by parallax, their displace-ments are different, and their relative directions are therefore not the same. Now the *ratio* of the parallaxes of the Sun and planet at conjunction can be calculated from comparing their periodic times, or from the ratio of their distances, as determined by observations of the planet's greatest elonga-tion or otherwise. Hence, by comparing the circumstances of the transit at different places, it becomes possible to deter-mine the parallaxes of both the Sun and planet.

The various methods of finding the Sun's parallax from observing transits of Venus may be classified as follows:—

(i.) By simultaneous observations of the relative position of the planet at different stations, either by micrometric mea-surements, or from photographs.

(ii.) *Delisle's method*, by comparing the times of the *begin-ning or end* of the transit at stations in different *longitudes*.

(iii.) *Halley's method*, by comparing the *durations* of the transit at stations in different *latitudes*.

Of these methods Halley's is the earliest, Delisle's the next.

343. First Method.—Let P and p be the horizontal parallaxes of the Sun and of Venus respectively at the time of transit. Then, at a place where the planet's zenith distance is z, its direction is depressed by parallax through an angle $p \sin z$ (§ 249); also the Sun is depressed through $P \sin z$.* Hence the planet appears to be brought nearer to the Sun's lower limb by an angle $(p-P) \sin z$.

If, now, the positions of the planet relative to the Sun's disc be simultaneously observed at any two or more different places, and the Sun's zenith distances be also determined, the difference of parallaxes $p-P$ can be readily found. Thus, if one of the stations be chosen where the Sun is

---

* Strictly speaking, this should be $P \sin z_1$, where z_1 is the Z.D. of the Sun's centre, but z_1 is very nearly equal to z, and no sensible error is introduced by taking z instead of z_1.

vertical, and another where the Sun is on the horizon, the relative displacement will be zero at the former station, and $p - P$ at the latter. Hence, the two directions of the planet relative to the Sun will be inclined at an angle $p - P$. If two stations are at opposite ends of a diameter of the Earth, the angular distance between the relative positions will be $2(p - P)$. Hence, in either case, $p - P$ can be readily found.

Let now r' and r denote the distances of Venus and the Earth from the Sun respectively. Then, if T be the ratio of the sidereal period of Venus to a year, we have, by Kepler's Third Law (assuming the orbits circular),

$$r'/r = T^{\frac{2}{3}},$$

whence the ratio of r' to r is found. Also, since Venus is in conjunction, its distance from the Earth is $= r - r'$. Therefore

$$p : P = r : r - r',$$

and $$\frac{P}{p - P} = \frac{r - r'}{r'} = \frac{r}{r'} - 1.$$

Whence, since the ratio of r to r is known, and $P - p$ has been observed, the Sun's horizontal parallax P may be found.

We have roughly (by Bode's Law) $r' = \frac{7}{10} r$, and therefore

$$P = \frac{3}{7} (p - P).$$

Hence the displacement of Venus on the Sun's disc at a place where its zenith distance is z, is about $\frac{1}{3} P \sin z$.

The apparent position of Venus on the Sun's disc may be observed either by measuring the planet's distance from the edge of the disc with a micrometer or heliometer, or by taking a photograph of the Sun. But the photographic method, though easier, does not give such accurate results.

For, to obtain P correct to $0.01''$, it would be necessary to find $2(p - P)$ correct to $\frac{1}{3} \times 0.01''$, or about $0.05''$. Since the Sun's diameter is $32'$, the greatest possible difference of positions would be only $\dfrac{1}{20 \times 32 \times 60}$, or $\dfrac{1}{37400}$,

of the Sun's diameter. It is difficult to obtain a good photograph of the Sun more than $4\frac{1}{2}$ inches in diameter, and it would, therefore, be necessary to measure the planet's position correct to $\frac{1}{8300}$ of an inch, a degree of accuracy unattainable in practice. The slightest distortion or imperfection in the photographic plate would render the observations worthless.

344. Delisle's Method.—In this method, the Sun's parallax is determined by observing the difference between the times at which the transit begins or ends at different places. Let A, B be two stations near the Earth's equator in widely different longitudes, say at the ends of the diameter of the Earth, and in the plane containing UV, the path of Venus' relative motion. Draw AUL and BVL, touching the Sun in L and cutting the path of Venus in U, V. Then, when Venus reaches U the transit begins at A, the planet appearing to enter the Sun's disc at L, and when Venus is at V the transit begins at B. In the interval between the times of commencement of the transit as seen from A and B, the planet moves through the angle ULV or ALB about the Sun relative to the Earth, and this angle, being the angle subtended at the Sun by the Earth's diameter AB, is *twice the Sun's parallax*.

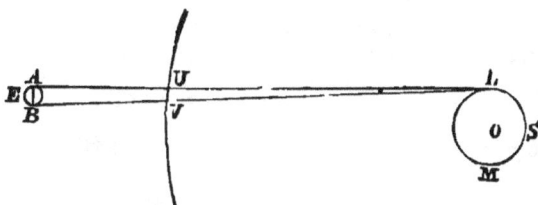

FIG. 111.

But the rate of relative angular motion of Venus is known, being 360° in a synodic period. Hence the angle ULV, described in the observed interval, is known, and the Sun's parallax is thus found.

In a similar way, the Sun's parallax may be determined by observing the interval between the times at which the transit ends at two stations A, B. We should have to draw two tangents from A, B to the opposite side of the Sun (M). As before, the angle described by Venus in the observed interval is twice the Sun's parallax.

In employing Delisle's method, the observed times of ingress or egress must be the *Greenwich* times, or must be reckoned from an epoch common to both observers. For this reason the difference of longitudes of the two stations must be accurately known. In the following example the observed interval 690s. corresponds to 8·86″ of parallax, and it follows that an error of 1s. in the estimated interval would give rise to an error of just over 0·01″ in the computed parallax. Hence if the interval of time be estimated correct to the nearest second, the parallax will be correct to two decimals of a second.

In practice it would be difficult to make observations from the extremity of a diameter of the Earth, but the method is readily modified so as to be applicable when the stations are not so favourably situated.

EXAMPLE.

Given that the synodic period is 584 days, and that the difference between the times of ending of a transit, as seen from opposite ends of a diameter of the Earth, is 11m. 30s., to find the Sun's parallax.

In 584 days Venus revolves through 360° about the Sun relative to the Earth; therefore its angular motion per minute

$$= \frac{360 \times 60 \times 60}{584 \times 24 \times 60} \text{ seconds} = 1\text{·}541''.$$

Therefore in 11½m. Venus describes an angle 1·541″ ∠ 11½ = 17·72″.
This angle is twice the Sun's parallax;

∴　Sun's parallax = 8·86″.

345. **Halley's Method.** — The method now to be described was invented by Dr. Halley in 1716, and was first put into use at the transits in 1761 and 1769. In Halley's method the *times of duration* of the transits are observed from two stations *A*, *B*, one in north and the other in south latitude, in a plane as nearly as possible *perpendicular* to the ecliptic, or, more strictly, to the relative path of Venus. Take this plane as the plane of the paper in Fig. 112, and suppose also (for the purpose of simplifying the explanation) that *A*, *B* are at the ends of a diameter of the Earth. Let *LM* be the diameter of the Sun's disc perpendicular to the line of centres, and let the directions of Venus *AV*, *BV*, when produced, meet the disc in *a*, *b*. Then *a*, *b* are the relative positions of Venus as seen at conjunction from *A* and *B*.

In Fig. 113 the Sun's disc is represented as seen from the Earth; a, b are the positions of Venus as seen on the disc from A, B, projected on LM, in Fig. 112, and PQR, $P'Q'R'$ are the apparent paths of Venus as it appears to cross the disc at B and A respectively.

As in § 343, the angular measure of the arc ab or QQ' measures the sum of the displacement of Venus due to relative parallax at A and B, and this, in the circumstances here considered, is twice the difference of the parallaxes of the Sun and Venus.

FIG. 112.

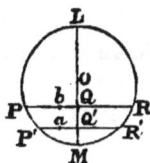

FIG. 113.

Now the observed times of duration of the transit at A and B are the times taken to describe the chords $P'Q'R'$ and PQR respectively. Knowing the synodic period of Venus and the ratio of its distances from the Sun and Earth, the rate at which Venus travels across the Sun's face can be found. Hence, the angular lengths of the chords PQR, $P'Q'R'$ can be found. Also the Sun's angular diameter LM is known. Hence the angular distances OQ, OQ', QQ' can be calculated, for we have (very approximately)

$$OQ^2 = OM^2 - PQ^2, \quad OQ'^2 = OM^2 - P'Q'^2,$$

and

$$QQ' = OQ' - OQ.$$

Hence QQ' is known, and therefore the difference of parallaxes of Venus and the Sun is found; whence the Sun's parallax may be found as in § 343.

*346. Or if AB be known in miles, the length of ab in miles can be found from the proportion $ab : AB = Va : VA$, and then, the angle aAb being known (being the angular measure of QQ'), we can find the Sun's distance in miles, for we have

$$\text{circular measure of } \angle aAb = \frac{ab}{aA}; \quad \text{whence}$$

$$\text{Sun's distance } Aa \text{ (in miles)} = \frac{\text{length } ab \text{ (in miles)}}{\text{circular measure of } \angle aAb}.$$

The working of Halley's method will be made much clearer by a careful study of the following numerical examples. The student should copy Figs. 112 and 113.

EXAMPLES.

1. To find the angular rate at which Venus moves across the Sun's disc.

Let S, E, V denote the Sun, Earth, and Venus respectively (Fig. 112).

From the example of § 344, SV separates from SE with relative angular velocity, about S, of $1\cdot54''$ per minute, or $1' 32\cdot4''$ per hour.

But Venus is nearer the Earth than the Sun in the ratio $28 : 72$ (roughly). And we have

angular velocity of EV : ang. vel. of SV

$$= \frac{1}{EV} : \frac{1}{SV} = 72 : 28 = 18 : 7.$$

Therefore EV separates from ES with angular velocity

$$= \frac{18}{7} \times 1' 32\cdot4'' \text{ per hour} = 3' 57\cdot6'' \text{ per hour}$$

$$= 4'' \text{ per minute very nearly.}$$

2. Neglecting the motion of the observatory due to the Earth's rotation, find the position on the Sun's disc of the chord PR, traversed by the planet, in order that the transit may take four hours.

Draw the figures as in § 345.

In four hours Venus moves $4 \times 3' 58'$, or very nearly $16'$ relative to the Sun (by Ex. 1); ∴ the chord PR must measure $13'$. Hence PR is equal to the Sun's angular semi-diameter OP.

Therefore, PR is a side of a regular inscribed hexagon in the Sun, and $\angle MOP = 30°$.

3. If, at A, B, at opposite ends of a diameter of the Earth perpendicular to the plane of the ecliptic, the durations of transit are 3h. 21m. and 4h. respectively, to find the Sun's parallax.

Here the arc PR takes 39m. longer to describe than $P'R'$. Hence it is longer by $39 \times 4''$, or $156''$. Draw $R'K$ perpendicular to PR. Then, $\qquad KR = \frac{1}{2}(PR - P'R') = \frac{1}{2} \times 156'' = 78''$.

Now, by Example 2,
$$\angle MOR = 60°.$$
And RR', being very small, is approximately a straight line perpendicular to OR; $\therefore$ $R'RK = 30°$ approximately. Hence
$$Q'Q = R'K = RK \tan 30° = RK\sqrt{\tfrac{1}{3}} = \tfrac{18}{3}\sqrt{3}'' = 45'' \text{ nearly}.$$
But angular measure of $Q'Q$: twice Sun's parallax
$$= SV : EV = 18 : 7;$$
$$\therefore \text{ twice Sun's parallax} = 45'' \times \tfrac{7}{18} = 17\cdot50'';$$
$$\therefore \text{ Sun's parallax} = 8\cdot75''.$$

4. A transit of Venus was observed from two stations selected as favourably as possible, one in N. the other in S. latitude, the zenith distances of the planet being $53° 8'$ (sin $53° 8' = \cdot8$) and $30°$ respectively. Given that the times occupied by the planet in passing across the disc were 4h. 52m. and 4h. 30m., to find the Sun's parallax, assuming the distances of Venus and the Earth from the Sun to be in the ratio of 18 : 25 and neglecting the rotation of the Earth.

Venus moves nearly $4''$ per minute relative to the Sun; hence in 4h. 30m. it moves through $18'$.
In 4h. 52m. it moves through $19' 28''$;
$\therefore$ in Fig. 113, $P'Q' = 18' \times \frac{1}{2}$ $= 9'$,
$\qquad\qquad PQ = 19' 28'' \times \frac{1}{2}$ $= 9\cdot73'$,
and the Sun's semi-diameter SP $= 16'$ nearly;
$\therefore$ $SQ = \sqrt{SP^2 - PQ^2} = \sqrt{256 - 94\cdot67}$ $= 12\cdot70'$;
$\qquad SQ' = \sqrt{SP'^2 - P'Q'^2} = \sqrt{256 - 81}$ $= 13\cdot23'$;
$\qquad\qquad\qquad \therefore QQ'$ $= \cdot53' = 31\cdot8''$.

Now, if A and B be well chosen, QQ' is the sum of the relative displacements of Venus at the two stations. Let P be the Sun's parallax, p that of Venus; then we have
$$QQ' = (p - P)(\sin z + \sin z') = (p - P) \times (\sin 30° + \sin 53° 8')$$
$$= (p - P) \times (\cdot5 + \cdot8) = (p - P) \times 1\cdot3;$$
$$\therefore p - P = \frac{31\cdot8''}{1\cdot3} = 24\cdot5''.$$

Again, $P : p = \dfrac{1}{ES} : \dfrac{1}{EV} = EV : ES = 7 : 25;$
$$\therefore P : p - P = 7 : 18;$$
$$\therefore P = 24\cdot5'' \times \tfrac{7}{18} = 9\cdot5''.$$

Hence, with the given data, the Sun's parallax is $9\cdot5''$.

347. Difficulties of Observing the Duration of a Transit.—In Examples 3, 4, above, the observed differences of duration were 39m. and 20m. respectively. An error of one second in the estimated durations of transit would give rise to an error of less than 0·1 per cent., and if we could be sure of observing the durations to within a second, the Sun's parallax could be found correct to two decimal places. But in practice it is extremely difficult to estimate the times of beginning and ending of a transit, even to the nearest second.

For in the first place, Venus, when seen through the telescope, is not a mere point, but a disc of finite dimensions, its angular diameter at conjunction being about 67″, or one-thirtieth of the diameter of the Sun. Hence its passage across the edge of the disc from external to internal contact occupies an interval which is never less than about 17s. (See Example on page 279.)

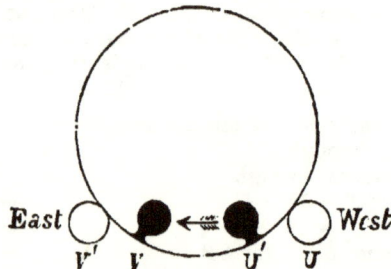

East   West

V′   V   U′   U

Fig. 114.

Now, it is impossible to observe the first external contact (*U*) of Venus with the Sun, because the planet is invisible until it has cut off a perceptible portion from the edge of the Sun's disc, and by that time it has advanced considerably beyond the point of contact. The last external contact (*V*) at the end of the transit is also difficult (though rather less so) to observe, for a similar reason.

For this reason, the internal contacts *U′*, *V*. are alone observed, and a correction is applied for the angular semi-diameter of Venus.

But in observing the first **internal** contact *U′*, when the planet's disc separates from the edge of the Sun, another difficulty, in the form of an optical illusion, makes itself manifest.

Instead of remaining truly circular, the planet's disc appears to become elongated towards the edge of the Sun, and remains for some time connected with the edge by a narrow neck called the **" black drop."** This breaks suddenly at last, but not until the planet has separated some distance from the Sun's edge.* Even if the "black drop" be remedied, the atmosphere surrounding the planet Venus renders the contacts uncertain and ill-defined.

It is worthy of notice that in Delisle's method the times of ingress and egress at both stations are equally affected by the "black drop" appearance, and therefore it has no effect on the computation, provided that both observers take the same stage of the phenomenon for the observed time of ingress.

EXAMPLE.

Having given that the angular diameter of Venus at conjunction is 67″, to find the interval between external and internal contact (i.) when Venus passes across the centre of the Sun's disc, (ii.) in the circumstances of Example 2, § 346.

(i.) Between external and internal contacts the planet moves through a distance equal to its angular diameter; therefore, since its rate of motion is 4″ per second, the time occupied = 67 ÷ 4s. = **17s.** very nearly.

(ii.) Here the planet is 67″ nearer the centre at internal than at external contact. Now the planet's direction of motion UV is inclined at angle 60° to the radius through the centre of the disc (Fig. 114). Hence the planet's component relative velocity along the radius is 4″ cos 60° per second, and therefore the interval required, in seconds,

$$= \frac{67}{4 \cos 60°} = \frac{67}{2}$$
$$= 33.5s.$$

348. Recent Determinations of the Parallax of the Sun.—Professor Arthur Auwers, the well-known Berlin astronomer, has recently (December 11, 1891) completed the calculations based on the observations in Germany of the transit of Venus in 1882. He finds that the parallax of the Sun is **8·800** seconds, with an error of 0·03 of a second at most. From the old observations of the transits of 1761 and 1769, Prof. Newcomb has lately computed the parallax at **8·79″**.

---

* The "black drop" may be illustrated by holding two globes in the sunshine, at *different* distances from a white screen, and moving them until their shadows nearly touch.

349. Advantages and Disadvantages of Halley's and Delisle's Methods.

—In Halley's method the observed data are the *intervals* of time occupied by Venus in crossing the Sun's disc at the two stations. It is not necessary to know the *actual* times of the transit; hence neither the Greenwich time nor the longitude of the observatories need be known. In Delisle's method it is essential that the Greenwich times of the observations should be known with great accuracy, but it is not necessary to observe *both* the beginning and end of the transit at the two stations. Still, if these be both observed, we have two independent data for calculating the parallax, which afford some test of the accuracy of the computations.

On the other hand, Delisle's method possesses the advantage that the places of observation must be near the Earth's equator, and it may therefore be possible to select the stations nearly at opposite ends of a diameter of the Earth, and thus to get the greatest effect of parallax, while in Halley's method it is necessary that the stations shall be in as high latitudes as possible, and, owing to the practical difficulties of taking observations near the poles, the greatest effect of parallax cannot be utilized.

Delisle's method is most easily employed if the transit is nearly *central*, *i.e.*, if Venus passes nearly across the *centre* of the Sun's disc. This condition is fatal to the success of Halley's method; here the best results are obtained when Venus transits near the *edge* of the disc.

For in Fig. 113 (page 275) we have

$$OQ'^2 - OQ^2 = QP^2 - Q'P'^2,$$

or

$$QQ' = \frac{PR - P'R'}{2} \times \frac{QP + Q'P'}{OQ + OQ'}.$$

Hence the effect on QQ' of a small error in the computed length of PR or $P'R'$ will be least when $QP + Q'P'$ is smallest and $OQ + O'Q'$ is largest, a condition satisfied when the transit takes place near the edge M of the disc.

On the other hand, for a nearly central transit, OQ, $O'Q'$ would be small, and very slight errors in the estimated lengths of PR, $P'R'$ would produce such large errors in the computed displacement QQ' as to render the method practically worthless.

The transits of 1874 and 1882 were both favourable to the use of Halley's method.

*350. **To determine the frequency of Transits of Venus.**—Since the Sun's angular semi-diameter is about $16'$, a transit of Venus only occurs when the angular distance between the centres of the Sun and Venus, as seen from some place on the Earth, is $16'$. Hence, neglecting the effects of the relative parallax ($P-p = 23''$ by Ex. 3. § 346, and this is small compared with $16'$), Venus must be at an angular distance (SEV) $< 16'$ from the ecliptic at the time of conjunction. Hence the planet's *heliocentric* latitude ESV must be less than $16' \times EV/SV$, that is $16' \times \frac{7}{18}$, or about $6'$. Now the orbit of Venus is inclined to the ecliptic at about $3° 23'$, or $203'$. Hence, by a method similar to that of § 292, we see that the planet must be at a distance from the node of not more than about $\sin^{-1}\frac{6}{203} = \sin^{-1}\frac{1}{34}$ (roughly) $= 1° 42'$, in order that a transit may take place. The smallness of this limit alone shows that transits of Venus are of rare occurrence.

Now, a synodic period of Venus contains about 584 days, that is, 1·599, or, more accurately, 1·598662 of a year. Hence five synodic revolutions occupy almost exactly eight years, the difference only amounting to $\frac{1}{150}$ of a year. This difference corresponds to an arc of $\frac{360}{150}$, or $2° 24'$ on the ecliptic. This arc is much less than the *double* arc $3° 24'$ within which transits take place. Hence it frequently happens that, eight years after one transit has taken place, the Sun and Venus are again at conjunction within the necessary limits, and another transit occurs near the same node. But after sixteen years, conjunction will occur at $4° 48'$ from its first position; this is greater than $3° 24'$; hence there cannot be more than two transits near the same node at intervals of eight years. And if a transit should be *central*, occurring almost exactly *at* the node, the conjunctions occurring eight years before and after would fall outside the required limits, and no second transit would then take place in eight years.

Again, it may be shown that $1·598662 \times 147 = 235·003$. Hence 147 synodic periods of Venus occupy almost exactly 235 years, the difference being only ·003 of a year. Thus a transit of Venus may recur at the same node at an interval of 235 years. And it is possible to prove that there is no intermediate interval between 8 and 235 years at which transits recur at the same node.

If the orbits of the Earth and Venus were circular, a transit at one node would be followed by one at the opposite node in $113\frac{1}{2}$ or $121\frac{1}{2}$ years. For

$$1{\cdot}598662 \times 71 = 113\frac{1}{2} + {\cdot}005 \; ; \; 1{\cdot}598662 \times 76 = 121\frac{1}{2} - {\cdot}002.$$

But this result is modified by the eccentricities of the orbits (which now cause a difference of nearly a day in the times taken by the Earth to describe the two halves into which its orbit is divided by the line of nodes).

In reality it is found that the intervals between transits of Venus occur at present in the following order :—

$$8, \; 105\frac{1}{2}; \; 8, \; 121\frac{1}{2}; \; 8, \; 105\frac{1}{2}; \; 8, \; 121\frac{1}{2}.$$

Transits have occurred, and are about to occur, in 1761, 1769, **1874, 1882,** 2004, 2012 (the thick and thin type being used to distinguish the two different nodes).

*351. **Transits of Mercury** occur much more frequently than transits of Venus. For although the orbit of Mercury is inclined to the ecliptic at about twice as great an angle as that of Venus, this cause is more than compensated for by the greater proximity of the planet to the Sun; and since the synodic period of Mercury is only about $\frac{1}{4}$ of that of Venus, conjunctions occur five times as often, so that we should *ceteris paribus* expect five times as many transits. By a method similar to that employed for Venus it is found that transits occur at the same node at intervals of 7, 13, 33, or 46 years. The next transit will occur in 1894.

Although transits of Mercury thus occur far more often than transits of Venus, they cannot be used to determine the Sun's parallax with such accuracy, for Mercury is so near the Sun that the parallaxes of the two bodies are more nearly equal, and the planet's relative displacement is therefore much smaller than that of Venus. Moreover, Mercury moves much more rapidly across the Sun's disc, giving less time for accurate observations; besides which, owing to the great eccentricity of the orbit, the ratio of Mercury's to the Earth's distance from the Sun cannot be so exactly computed.

SECTION III.—*Annual Parallax, and Distances of the*
Fixed Stars.

352. Annual Parallax, Definition. — By **Annual Parallax** is meant the angle between the directions of a star as seen from different positions of the Earth in its annual orbit round the Sun.

We have several times (§§ 5, 247) mentioned that the fixed stars have no appreciable *geocentric parallax.* Their distances from the Earth are so great that the angle subtended at one of them by a diameter of the Earth is far too small to be observable even with the most accurately constructed instruments. But the diameter of the Earth's annual orbit is about 23,400 times as great as the Earth's diameter, or about 186 million miles (twice the Sun's distance), and this diameter subtends, at certain of the nearest fixed stars, an angle sufficiently great to be measurable, sometimes amounting to between 1″ and 2″.

Now, the Earth, by its annual motion, passes in six months from one end to the other of a diameter of its orbit; hence, by observing the same star at an interval of six months, its displacement due to annual parallax can be measured.

Since the Sun is *fixed,* the position of a star on the celestial sphere is corrected for annual parallax by referring its direction to the *centre of the Sun ;* this is called the star's *heliocentric* direction, as in § 156.

The **correction for annual parallax** is the angle between the geocentric and heliocentric directions of a star. Let S be the Sun, E the Earth, x the star (Fig. 115). Then Ex is the apparent or geocentric direction of the star, Sx its heliocentric direction, and $\angle ExS$ is the correction for annual parallax. This angle is also equal to xEx' where Ex' is parallel to Sx.

We notice that *the correction for annual parallax* (ExS) *is the angular distance of the Earth from the Sun as they would appear if seen by an observer on the star.*

FIG. 115.

ASTRON.   U

353. To find the Correction for Annual Parallax.

Let $r = ES =$ radius of Earth's orbit.

$\quad\quad = Sx =$ distance of star.

$E = \angle SEx =$ angular distance of star from Sun.

$p = \angle ExS =$ annual parallax of star.

By trigonometry we have in the triangle SEx

$$\frac{\sin ExS}{\sin SEx} = \frac{ES}{Sx};$$

whence*　　　　$$\mathbf{sin}\, p = \frac{r}{d}\, \mathbf{sin}\, E \quad\quad\ldots\ldots\ldots\ldots\ldots\ldots(\text{i.}).$$

Fig. 116.

Hence the parallactic correction p is greatest when $E = 90°$. This happens twice a year, and the corresponding positions of the Earth in its orbit are evidently the intersections of the ecliptic with a plane drawn through S perpendicular to Sx. Let this greatest value of p be denoted by P, then P is called the **star's annual parallax**, or simply the star's **parallax.**†

Putting $E = 90°$ in (i.), we have

$$\sin P = \frac{r}{d};$$

and therefore　　　$$\sin p = \sin P . \sin E.$$

---

* Notice the close similarity between the present investigation and that of § 249.

† There is no risk of confusion in the use of the term *parallax* alone, because a star has no geocentric parallax. *The "parallax" of a body means its equatorial horizontal parallax if the body belongs to the solar system. If not, its "parallax" is its annual parallax.*

But the angles P, p are always very small; therefore their sines are very approximately equal to their circular measures. Thus we have approximately

$$P \text{ (in circular measure)} = \frac{r}{d},$$

$$p = P \sin E;$$

and, if P'', p'' denote the numbers of seconds in P, p,

$$P'' = \frac{180 \times 60 \times 60}{\pi} \frac{r}{d} = 206,265 \frac{r}{d} \text{ (approximately)},$$

and
$$p'' = P'' \sin E.$$

354. Relation between the Parallax and Distance of a Star.—If a star's parallax be known, its distance from the Sun is given by the formula

$$P'' = \frac{180 \times 60 \times 60}{\pi} \frac{r}{d} = 206,265 \frac{r}{d};$$

whence
$$d = \frac{180 \times 60 \times 60}{\pi \times P''} r = 206,265 \frac{r}{P''},$$

where r is the Sun's distance from the Earth.

For most purposes r may be taken as 93 million miles.

EXAMPLES.

1. The parallax of *Castor* is $0.2''$; to find its distance. We have
$$d = 206,265 \frac{r}{P''} = \frac{206,265 \times 93,000,000}{0.2}$$
$$= 5 \times 206,265 \times 93,000,000$$
$$= 95,900,000,000,000, \text{ or } 959 \times 10^{11} \text{ miles}$$

approximately. It would be useless to attempt to calculate more figures of the result with the given data, which are only approximate. It is most convenient (besides being shorter) to write the result in the second form.

2. To find the distance of *a Centauri* (i.) in terms of the Sun's distance, (ii.) in miles, taking its parallax to be $0.750''$.

Here
$$d = \frac{206,265}{.75} r = 275,000r$$
$$= 275 \times 10^3 \times 93 \times 10^6 = 25,575 \times 10^9$$
$$= 256 \times 10^{11} \text{ miles approximately.}$$

355. General Effects of Parallax. — Since Ex' is parallel to Sx, it is in the same plane as ES and Ex. Hence the lines ES, Ex, Ex' cut the celestial sphere of E at points S, x, x_0, lying in one great circle, and we have the two following laws : —

(i.) *Parallax displaces the apparent position of a star from its heliocentric position in the direction of the Sun.*

(ii.) *The parallactic displacement of any star at different times varies as the sine of its angular distance from the Sun.*

 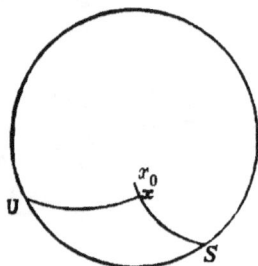

FIG. 117.   FIG. 118.

Let Fig. 118 represent the observer's celestial sphere, S the Sun. Let x be the apparent or geocentric position of the star, whose parallax is P. Draw the great circle Sx and produce it to x_0, making

$$x x_0 = P \sin Sx.$$

Then x_0 represents the star's heliocentric position, and this is its position as corrected for annual parallax.

Conversely, if the star's heliocentric position x_0 is given, we may obtain its geocentric or apparent position x by joining $x_0 S$, and on it taking

$$x_0 x = P \sin Sx = P \sin Sx_0 \text{ very approximately}$$

(for the difference between $P \sin Sx$ and $P \sin Sx_0$ is exceedingly small, and may be neglected).

The terms **Parallax in Latitude** and **Parallax in Longitude** are used to designate the corrections for parallax which must be applied to the celestial latitude and longitude of a star respectively. Similarly, the **parallax in decl.** and **parallax in R.A.** denote the corresponding corrections for the decl. and R.A.

356. To show that any star, owing to parallax, appears to describe an ellipse.

In Fig. 117, Ex' is parallel to the star's heliocentric direction; therefore, x' is fixed, relative to the Earth. Moreover, $x'x = ES$. Hence, as the Sun S appears to revolve about the Earth in a year, the star x will appear as though it revolved in an equal orbit about its heliocentric position x', in a plane parallel to the ecliptic.

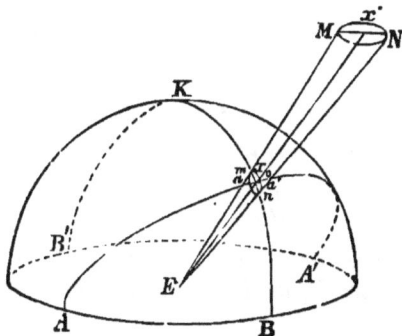

FIG. 119.

Let the circle MN (Fig. 119) represent this path, which the star x appears to describe in consequence of parallax. This circle is viewed obliquely, owing to its plane not being in general perpendicular to Ex'; hence, if mn denote its projection on the celestial sphere, the laws of perspective show that mn is an ellipse. (Appendix, 12.) This small ellipse is the curve described by the star on the celestial sphere during the year.

Particular Cases.—*A star in the ecliptic* moves as if it revolved about its mean position in a circle in the ecliptic plane, hence its projection on the celestial sphere oscillates to and fro in *a straight line* (more accurately a small arc of a great circle) of length $2P$.

For a star in the pole of the ecliptic the circle MN is perpendicular to Ex', hence Ex describes a right cone, and the projection x describes on the celestial sphere a *circle*, of angular radius P, about the pole K.

If the eccentricity of the Earth's orbit be taken into account, the curve MN will be an ellipse instead of a circle, but its projection mn will still be an ellipse.

357. Major and Minor Axes of the Ellipse. —We shall now prove the following properties of the small ellipse described during the course of the year by a star whose parallax is P, and celestial latitude l.

(i.) (A) *The length of the* **semi-axis major** *is P.*

(B) *The major axis is parallel to the ecliptic.*

(C) *When the star is displaced along the major axis it has no parallax in latitude.*

(D) *At these times the Sun's longitude differs from the star's by* 90°.

(ii.) (A) *The length of the* **semi-axis minor** *is $P \sin l$.*

(B) *The minor axis is perpendicular to the ecliptic.*

(C) *When the star is displaced along the minor axis it has no parallax in longitude.*

(D) *At these times the Sun's longitude is either equal to the star's, or differs from it by* 180°.

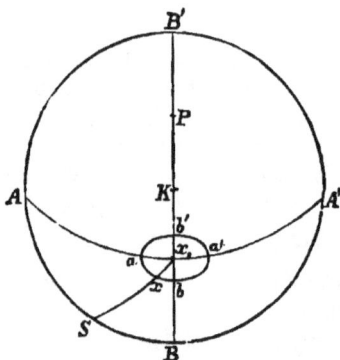

FIG. 120.

On the celestial sphere let x_0 denote the heliocentric position of the star, $ABA'B'$ the ecliptic, K its pole, BKx_0B the secondary to the ecliptic through the star.

Then, if S is the Sun, the star x_0 is displaced to x, where

$$x_0 x = P \sin x_0 S.$$

(i.) The displacement is greatest when $\sin x_0 S$ is greatest, and this happens when

$$\sin x_0 S = 1, \quad x_0 S = 90°.$$

If, therefore, we take A, A' on the ecliptic so that

$$x_0 A = x_0 A' = 90°,$$

A, A' are the corresponding positions of the Sun.

Now A, A' are the poles of BKB' (Sph. Geom., 11, 14, 15), and therefore the great circle $A x_0 A'$ is a secondary to BKB'. Hence, if a, a' denote the displaced positions of the star, aa' is perpendicular to KB, and is therefore, parallel to the ecliptic.

Also, $\qquad x_0 a = x_0 a' = P \sin 90° = P$;

therefore the semi-axis major of the ellipse is P.

Since $AB = A'B = 90°$, the star's longitude (ΥB) differs from the Sun's longitude at A or A' by $90°$.

And since the star is displaced parallel to the ecliptic, its latitude, or angular distance from the ecliptic, is unaltered, and therefore the parallax in latitude is zero.

(ii.) The parallactic displacement is least when $\sin x_0 S$ is least, and this happens when S is at B. For (Sph. Geom., 26) B is the point on the ecliptic nearest to x_0. Also, since

$$\sin x_0 B' = \sin (180° - x_0 B) = \sin x_0 B,$$

it follows that the parallactic displacement is also least when S is at B'.

If, therefore, b, b' be the extremities of the minor axis, the arc bb' is along KB, and is therefore perpendicular to the ecliptic.

Also, $\qquad x_0 b = x_0 b' = P \sin x_0 B = P \sin l$;

therefore the semi-axis minor is $P \sin l$.

When the Sun is at B, it has the same longitude as the star; when at B', the longitudes differ by $180°$.

And since the star is displaced in a direction perpendicular to the ecliptic, its longitude ΥB is unaltered; therefore the parallax in longitude is zero.

The parallax in latitude is evidently equal to the apparent angular displacement of the star resolved parallel to $x_0 K$, and its maximum value is $x_0 b$, or $x_0 b'$. The parallax in longitude is not equal to the star's angular displacement perpendicular to $K x_0$, but to the change of longitude thence resulting, and this is measured by the angle $x \bar{K} x_0$. Hence, in Fig. 120,

(i.) The maximum parallax in latitude $= x_0 b = P \sin l$.

(ii.) The maximum parallax in longitude $= \angle\ x_0 K a$
$= x_0 K a' = x_0 a / \sin\ \bar{K} x_0$ (Sph. Geom. 17) $= P / \cos\ x_0 B$
$= P \sec l$.

358. To determine the Annual Parallax of any Star, the following methods have been employed :—

(i.) The absolute method, by the Transit Circle;

(ii.) Bessel's, or the differential method, by the micrometer or heliometer;

(iii.) The photographic method.

The absolute method consists simply in observing with the Transit Circle the apparent decl. and R.A. of a star at different times in the year. From the small variations in these coordinates it is possible to find the star's parallax.

Although this method has been successfully employed, it possesses many disadvantages. For the observations are considerably affected by errors of adjustment of the Transit Circle and by refraction. Moreover, several other causes give rise to variations in the star's apparent decl. and R.A. during the year. These include aberration (*vide* Section IV.), precession (§ 141), and nutation, all of which produce displacements much larger than those due to parallax.

In § 372 we shall see that when either the latitude or longitude is most affected by parallax it is unaffected by aberration. Hence the best plan is to find the changes in these coordinates when they are respectively most affected by parallax. These changes are $P \sin l$ and $P \sec l$ (§ 357) and from them P may be found.

359. Bessel's Method consists in observing with a micrometer (§ 79) or heliometer (§ 80) the variations in the angular distance and relative position of two optically near stars during the course of a year.

The stars, being nearly in the same direction, are very nearly equally affected by refraction, and we may also mention that the same is true of aberration, precession and nutation. These corrections do not therefore sensibly affect the relative angular distance and positions of the stars. On the other hand, the two stars may be at very different distances from the Earth ; if so, they are differently displaced by parallax, and their angular distance and position undergo variations depending on their *relative* parallax or *difference* of parallax. Hence, by observing these variations during the year the difference of parallax can be found.

This method does not determine the *actual* parallax of either star. But if one of the observed stars is very bright and the other is very faint, it is reasonable to assume that the former is comparatively near the Earth, while the latter is at such a great distance away that its parallax is insensible. Under such circumstances the observed relative parallax is the parallax of the bright star alone. By making comparisons between the bright star and *several* different faint stars in its neighbourhood, this point may be settled.

If a considerable discrepancy is found in the observed relative parallaxes, one or more of the comparison stars must themselves have appreciable parallaxes, but since the vast majority of stars in any neighbourhood are too distant to have a parallax, we shall be able to find the parallax not only of the star originally observed, but of that with which we had first compared it.

The parallax of a star can *never be negative*; if the relative parallax should be found to be negative, we should infer that the comparison star has the greater parallax, and is therefore nearer the Earth.

360. **The Photographic Method** is identical in principle with the last, but instead of observing the relative distances of different stars with a micrometer, portions of the heavens are photographed at different seasons, and the displacements due to parallax are measured at leisure by comparing the positions of any star on the different plates. This method has been used by Dr. Pritchard, of Oxford, and possesses the advantages of great accuracy, combined with convenience.

361. Parallaxes of certain Fixed Stars.—The nearest stars are α *Centauri*, with a parallax of 0·75″, and 61 *Cygni*, with parallax 0·54″. Among others, the following may be mentioned : α *Lyræ*, 0·18″, *Sirius*, 0·2″, *Arcturus*, 0·13″, *Polaris*, 0·07″, α *Aquilæ*, 0·19″. Of these, 61 *Cygni* is by no means bright ; and a companion star to *Sirius* is invisible in all but two or three of the best telescopes. So it is not an *invariable* rule that faint stars are most distant, and have no appreciable parallax ; it is, however, true in the great majority of cases.*

362. Proper Motions.—Binary Stars.—Many stars, instead of being fixed in space, are gradually changing their positions. They are then said to have a **proper motion.** This motion may partly belong to the star, but is also partly an apparent motion, due to the fact that the solar system is itself moving through space in the direction of a point in the constellation *Hercules.* The displacement due to this cause can be allowed for approximately.

Many of these motions, like that of our own Sun, are apparently *progressive ; i.e.,* the star moves with constant velocity and in the same direction. Others are *orbital, i.e.,* the star revolves about some other star, or (more accurately) two stars revolve about their common centre of mass. Such a system of stars is called a **Binary Star.** It is usually seen by the naked eye as a single heavenly body, its components being too near to be distinguished. Frequently a system of stars has itself a progressive motion ; and sometimes an apparently progressive motion may really be an orbital one, with a period so long that the path has not sensibly diverged from a straight line during the short period for which stellar motions have been watched.

A progressive or orbital motion cannot be confounded with the displacement due to annual parallax, for the former is always in the same direction, and the latter has a period differing from a year, while parallax always produces an *annual* variation.

---

* These figures can only be regarded as very rough approximations, for considerable discrepancies exist between the values found by different methods.

SECTION IV.—*The Aberration of Light.*

363. Velocity of Light. — We now come to certain methods of finding the Sun's distance which depend on the fact that light is propagated through space with a large but measurable velocity.

The velocity of light has been measured by laboratory experiments in two different ways, invented by two French physicists, Fizeau and Foucault. For the description of these the reader is referred to Wallace Stewart's *Text Book of Light*, Chapter IX.* The experiments give the velocity of light in air; the velocity in vacuo can be obtained by multiplying this by the index of refraction of air.† The latter quantity may be found either by direct experiment or from the coefficient of astronomical refraction (see § 183).

In 1876, Cornu, by employing Fizeau's method, found the velocity of light *in vacuo* to be 300,400,000 metres per second. Still more recently, Michelson, by a modification of Foucault's method, has found the velocity to be **299,860,000** metres, or **186,330** miles per second; this may be taken as the most probable value.

364. Roemer's Method.—The Equation of Light.— In the last chapter we stated that Jupiter has four satellites, which revolve very nearly in the plane of the planet's orbit. Consequently a satellite passes through the shadow cast by Jupiter once in nearly every revolution, and is then eclipsed, as is our Moon in a lunar eclipse.

Since the orbits and periods of the satellites have been accurately observed, it is possible to predict the recurrence of the eclipses, so that when one eclipse has been observed the times at which subsequent eclipses will begin and end can be computed.

Now, the Danish astronomer Roemer in 1675 observed a remarkable discrepancy between the predicted and the observed times of eclipses. If of two eclipses one happens when Jupiter is near opposition, and the other happens near the planet's superior conjunction, the observed interval

---

* The student will find it useful to read this chapter before commencing the present section.

† Stewart's *Light*, § 41.

between the former and the latter is always *greater* than the
computed interval; similarly the observed interval between
an eclipse near superior conjunction and the next eclipse
near opposition is always *less* than the computed interval.
The eclipses at conjunction are thus always retarded, relatively
to those at opposition, by an interval of time which is observed
to be about 16m. 40s. As explained by Roemer, this apparent
retardation is due to the fact that light travels from Jupiter
to the Earth with finite velocity, and therefore takes 16m.
40s. longer to reach the Earth when the planet is furthest
away at superior conjunction (*B*) than when the planet is
nearest the Earth at opposition (*A*).

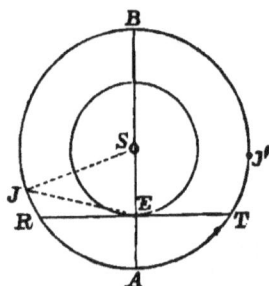

FIG. 121.

The relative retardation is the difference between the times
taken by the light to travel over the distances AE and BE.
But $BE - AE = 2SE$. Therefore *the retardation is twice the
time taken by the light to travel from the Sun to the Earth.*
Taking the retardation as 16m. 40s., we see that light takes
8m. 20s. to travel from the Sun to the Earth.
This interval is sometimes called the "*equation of light.*"
If we know the equation of light and the velocity of light,
we may calculate the Sun's distance. Conversely, if the
Sun's distance and the equation of light are known, the
velocity of light can be determined.
Knowing the Sun's distance, the Sun's *parallax* can be
computed, as in Chapter VIII., Section I. The present
method differs from those described in Sections I., II., in
that it gives the distance instead of the parallax of the Sun.

EXAMPLE 1.—To find the Sun's distance, having given that the velocity of light is 186,330 miles per second, and that eclipses of Jupiter's satellites which occur when the planet is furthest from the Earth, are retarded 16m. 40s. relatively to those which occur when the planet is nearest.

Here the time taken by light to pass over a diameter of the Earth's orbit is 16m. 40s.; therefore light travels from the Sun to the Earth in 8m. 20s., or 500 seconds.

$$\therefore \text{ the Sun's distance} = 186.330 \times 500 \text{ miles}$$
$$= 93,165,000 \text{ miles.}$$

EXAMPLE 2.—Taking the value of the Sun's distance calculated in the preceding example, the Sun's parallax will be found to be about 8·78″.

365. The Aberration of Light is a displacement of the apparent directions of stars, due to the effect of the Earth's motion on the direction of the *relative* velocity with which their light approaches the earth.

The rays of light emanating from a star travel in straight lines through space* with a velocity of about 186,330 miles per second. We *see* the star when the rays reach our eye, and the appearance presented to us depends solely on how the rays are travelling at that instant. If the Earth were at rest, and there were no refraction, we should see the star in its true direction, because the light would be travelling towards our eyes in a straight line from the star. But in every case the direction in which a star is *seen* is the direction of approach of the light-rays from the star at the instant of their reaching the eye.

Now the velocity of approach is the *relative velocity* of the light with respect to the observer. If the observer is in motion, this relative velocity is partly due to the motion of the light and partly due to the motion of the observer. If the observer happens to be travelling towards or away from the source of light, the only effect of his motion will be to increase or decrease the velocity of approach of the light, without altering its direction, but if he be moving in any other direction, his own motion will alter the direction of the relative velocity of approach, and will therefore alter the direction in which the star is seen.

---

* Of course the rays are refracted when they reach the Earth's atmosphere, but the effects of refraction can be allowed for separately.

Suppose the light to be travelling from a distant star x in the direction xO. Let V be the velocity of light, and let it be represented by the length MO. Suppose also that an observer is travelling along the direction NO with velocity u, represented by the straight line NO. Then, if we regard O as a *fixed point*, the light is approaching O with velocity represented by MO. Also since the observer is approaching O with velocity represented by NO, the point O is approaching the observer N with an equal and opposite velocity represented therefore by ON. Hence the whole relative velocity with which the light is travelling towards the observer is the resultant of the velocities represented by MO and ON.

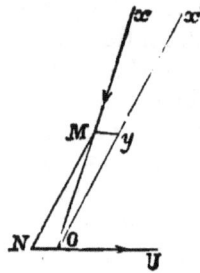

F<small>IG</small>. 122.

By the Triangle of Velocities this resultant velocity is represented in magnitude and direction by MN. Hence MN represents the direction of approach of the light towards the observer's eye. Therefore when the observer has reached O the star is seen in the direction Ox' drawn parallel to NM, although its real direction is Ox.

In consequence, the star appears to be displaced from its true position x to the position x'. This displacement is called the **aberration** of the star, and its amount is, of course, measured by the angle xOx'. This angle is sometimes called the **angle of aberration** or the **aberration error**.

366. **Illustrations of Relative Velocity and Aberration.**—The following simple illustrations may possibly assist the reader in understanding more thoroughly how aberration is produced.

(1) Suppose a shower of rain-drops to be falling perfectly vertically, with a velocity, say, of 40 feet per second. Then, if a man walk through the shower, say with a velocity of 4 feet

per second, the drops will appear to be coming towards him, and
therefore to be falling in a direction inclined to the vertical. Here
the man is moving towards the drops with a horizontal velocity of
4 feet per second, and therefore the drops appear to be coming
towards the man with an equal and opposite horizontal velocity of
4 feet per second.

Their whole relative velocity is the resultant of this horizontal
velocity and the vertical velocity of 40 feet per second with which
the drops are approaching the ground. By the rule for the compo-
sition of velocities, this relative velocity makes an angle $\tan^{-1}\frac{4}{40}$ or
$\tan^{-1}\cdot1$ with the vertical. Hence the man's own motion causes an
apparent displacement of the direction of the rain from the vertical
through an angle $\tan^{-1}\cdot1$. This angle corresponds to the angle of
aberration in the case of light.

(2) Suppose a ship is sailing due south, and that the wind is blowing from due west with an equal velocity. Then to a person on the
ship the wind will appear to be blowing from the south-west, its
southerly component being due to the motion of the ship, which is
approaching the south. In this case the ship's velocity causes the
wind to apparently change from west to south-west, *i.e.*, to turn
through 45°. We might, therefore, consistently say that the
"angle of aberration" of the wind was 45°.

367. Annual and Diurnal Aberration.—A point on
the Earth's surface is moving through space with a velocity
compounded of

(i.) The orbital velocity of the Earth in the ecliptic about
the Sun ;

(ii.) The velocity due to Earth's rotation about the poles.

These give rise to two different kinds of aberration, known
respectively as **annual** and **diurnal aberration.** Now the
Earth's orbital velocity is about $2\pi \times 93{,}000{,}000$ miles per
annum, or rather over 18 miles per second, while the
velocity due to the Earth's rotation at the equator is roughly
$2\pi \times 4000$ miles per day, or 0·3 miles per second. The
former velocity is about $\frac{1}{10000}$ of the velocity of light, and
therefore the annual aberration is a small though measurable
angle. The latter velocity is only $\frac{1}{60}$ as great ; hence the
diurnal aberration is much smaller and less important. For
this reason the term "aberration" always signifies annual
aberration, unless the word "diurnal" is also used. We shall
now consider the effects of annual aberration, leaving diurnal
aberration till the end of this section.

**368. To determine the correction for aberration
on the position of a Star.**—Let Ox be the actual direction
of a star x seen from the Earth at O; OU the direction of the
Earth's orbital motion at the time of observation. On Ox
take OM representing on any scale the velocity of light, and
draw MY parallel to OU, and representing on the same scale
the velocity of the Earth. Then YO represents the relative
velocity of the light in magnitude and direction, so that OYx'
is the direction in which the star x is *seen* (Fig. 123).

[For if ON be drawn parallel and equal to YM, the parallelogram
of velocities $MNOY$ shows that MO, the actual velocity of the light-
rays in space is the resultant of the two velocities YO and NO, or
YO and MY, and therefore YO is the required relative velocity.]

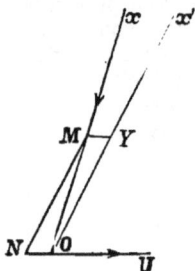

FIG. 123.

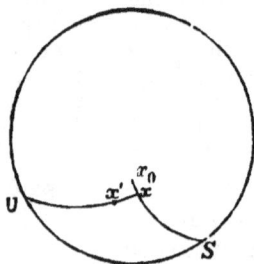

FIG. 124.

Since Ox, Ox', and OU all lie in one plane, it follows, by
representing their directions on the celestial sphere, that *a
star is displaced by aberration along the great circle joining its
true place to the point on the celestial sphere towards which the
Earth is moving.*

The displacement xOx' is called the star's **aberration
error.** Let it be denoted by y, and let

$$u = NO = \text{velocity of Earth,}$$
$$V = MO = \text{velocity of light.}$$

Then the triangle OMY gives

$$\frac{\sin MOY}{\sin MYO} = \frac{MY}{MO} = \frac{u}{V};$$

or $\sin y = \dfrac{u}{V} \sin MYO = \dfrac{u}{V} \sin UOx'$.

The aberration error y is, therefore, greatest when UOx' $= 90°$. Let its value, then, be k. Putting $UOx' = 90°$, we have
$$\sin k = u/V;$$
and $\therefore$ $\sin y = \sin k \sin UOx'$.

The angle UOx' is called the **Earth's Way** of the star, and k is called the **Coefficient of Aberration**. Since a and k are both small, we have, approximately

$$y = k \sin (\text{Earth's way}),$$
$$k \ (\text{in circular measure}) = u/V;$$

and, therefore, if y'', k'' denote the number of seconds in y, k respectively

$$y'' = k'' \sin (\text{Earth's way}),$$
$$k'' = \frac{180 \times 60 \times 60}{\pi} \frac{u}{V}$$
$$= 206{,}265 \times \frac{\text{velocity of Earth}}{\text{velocity of light}}.$$

369. General effect of Aberration on the Celestial Sphere.—Neglecting the eccentricity of the Earth's orbit, the direction of motion of the Earth, in the ecliptic plane, is always perpendicular to the radius vector drawn to the Sun. Hence, on the celestial sphere, the point U, towards which the Earth is moving, is on the ecliptic, at an angular distance 90° behind the Sun. This point is sometimes called the **apex of the Earth's Way**.

Let x' denote the observed position of the star. Draw the great circle $x'U$, and produce it to a point x, such that

$$xx' = k \sin x'U.$$

Then x represents the star's true position, corrected for aberration.

Conversely, if we are given the true position x, we can find the apparent position x' by joining xU and taking

$$xx' = k \sin xU,$$

for it is quite sufficiently approximate to use $k \sin xU$ instead of $k \sin x'U$.

ASTRON. x

We thus have the following laws :—.

(i.) *Aberration produces displacement in the apparent position of a star towards a point U on the ecliptic, distant 90° behind the Sun.*

(ii.) *The amount of the displacement varies as the sine of the* **Earth's Way** *of the star,* i.e., *the star's angular distance from the point U.*

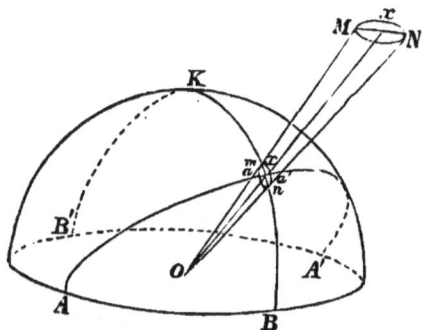 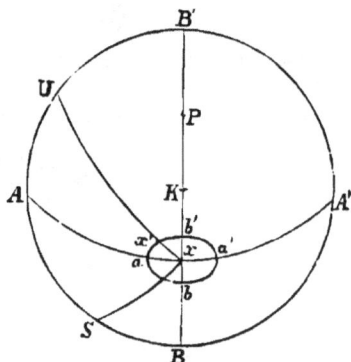

FIG. 125.	FIG. 123.

370. Comparison between Aberration and Annual Parallax.—The student will not fail to notice the close analogy between the corrections for aberration and annual parallax.

The point *U* for the former corresponds to the point *S* for the latter, in determining the direction and magnitude of the displacement. In fact, *the aberration error of a star is exactly the same as its parallactic correction would be three months earlier* (when the Sun was at *U*) *if the star's annual parallax were k.*

There is, however, this important difference that *the annual parallax depends on a star's distance, whilst the constant of aberration k is the same for all stars.*

For *k* depends only on the ratio of the Earth's velocity to the velocity of light, and not on the star's distance. The value of *k* in seconds is about **20·492″**; for rough purposes it may be taken as **20·5″**.

371. To show that the aberration curve of a star is an ellipse.—This result, which follows immediately from the analogy between aberration and parallax, may be proved independently as follows:—On Ox (Fig. 125), the true direction of a star x, take Ox to represent the velocity of light, and xM to represent the Earth's velocity. Then MO meets the celestial sphere in m, the star's apparent position.

As the Earth's direction of motion in the ecliptic varies, while its velocity remains constant, M describes a circle about x as centre in a plane parallel to the ecliptic plane. The projection of this circle on the celestial sphere is an ellipse (*cf.* § 356), and this is the curve traced out by a star during the year in consequence of aberration.

Particular Cases.—*A star in the ecliptic* oscillates to and fro in a *straight line*, or more accurately an arc of a great circle of length $2k$. *A star at the pole of the ecliptic* revolves in a small circle of radius k (*cf.* § 356).

372. Major and Minor Axes of the Aberration Ellipse.—By writing U for S and k for P in the investigation of § 357, we obtain the analogous results relating to the ellipse described by a star in consequence of aberration, namely:—

(i.) (A) *The length of the* **semi-axis major** *is k.*

(B) *The major axis of the ellipse is parallel to the ecliptic.*

(c) *When the star is displaced along the major axis it has no aberration in latitude.*

(D) *At these times the Sun's longitude is either equal to the star's, or differs from it by 180°.**

(ii.) (A) *The length of the* **semi-axis minor** *is k sin l.*

(B) *The minor axis is perpendicular to the ecliptic.*

(c) *When the star is displaced along the minor axis, it has no aberration in longitude.*

(D) *At these times the Sun's longitude differs from the star's by 90°.*

COROLLARY.—The maximum aberration in longitude $= k \sec l$ (*cf.* § 357, ii.).

---

* Note that (i., D) and (ii., D) are the reverse of the corresponding properties in § 357.

*373. **Effect of Eccentricity of Earth's Orbit.**—Owing to the elliptic form of the Earth's orbit the Earth's velocity is not quite uniform, and therefore the coefficient of aberration is subject to small variations during the year. The earth's velocity is greatest at perihelion and least at aphelion. The angular velocities at those times are inversely proportional to the squares of the corresponding distances from the Sun, but the actual (linear) velocities are inversely proportional to the distances themselves, and these are in the ratio of $1-e : 1+e$, or $1-\frac{1}{60} : 1+\frac{1}{60}$ (§ 149). Since the coefficient of aberration is proportional to the Earth's velocity, its greatest and least values are therefore in the ratio of $61 : 59$, and are respectively $\frac{61}{60}$ and $\frac{59}{60}$ of its mean value.

Moreover, the direction of the Earth's motion is not always exactly perpendicular to the line joining it to the Sun, hence the "apex of the Earth's way," towards which a star is displaced, may be distant a little more or a little less than 90° from the Sun at different seasons.

The aberration curve is still an ellipse. The student who has read the more advanced parts of particle dynamics may know that the curve MN, traced out by M, is in this case the "hodograph" of the Earth's orbital motion. It is also known, in the case of elliptic motion, such as the Earth's, that this hodograph is a circle, whose centre does not, however, quite coincide with x. Hence the aberration-curve hk is an ellipse.

374. **Discovery of Aberration.**—Aberration was discovered by Bradley, in 1725, in the course of a series of observations made with a zenith sector on the star γ *Draconis* for the purpose of discovering its annual parallax. The star's latitude was observed to undergo small periodic variations during the course of the year, and these differed from the variations due to annual parallax in the fact that *the displacement in latitude was greatest when the Sun's longitude differed from that of the stars by* 90°; that is, at the time *when the parallax in latitude should be zero* (§ 357, i., c.). The fact that the phenomenon recurred annually led Bradley to suppose that it was intimately connected with the Earth's motion about the Sun, and he was thus led to adopt the explanation which we have given above, It will be seen that the peculiarity which led Bradley to discard annual parallax as an explanation is quite in harmony with the results of § 372.

375. **To Determine the Constant of Aberration by Observation.**—The constant k can best be found by observing different stars with a zenith sector or transit circle, as in the direct method of finding a star's parallax (§ 358).

The differential method of § 359 cannot be used, because the coefficient of aberration is the same for all stars. But aberration is much larger than parallax (the coefficient of aberration being 20·49″, while the greatest stellar parallax is < 1″), and can therefore be found directly with greater accuracy. Of course it is necessary to make corrections for refraction and precession. The former correction is the most liable to uncertainty, as it varies slightly according to atmospheric conditions. But, as all stars have the same constant of aberration, a star may be selected which transits near the zenith, and is therefore but little affected by refraction.

This condition was secured by Bradley when he observed the star γ *Draconis*. The star is very favourable in another respect, for its longitude is very nearly 270°. It therefore lies very nearly in the " solstitial colure," its declination circle passing nearly through the pole of the ecliptic.

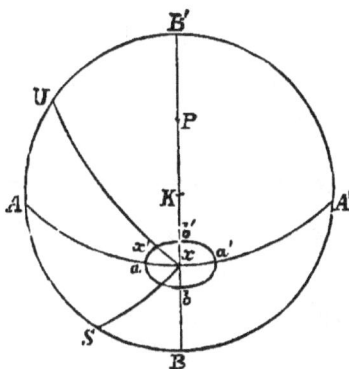

FIG. 127.

At the vernal equinox, the star's longitude is less than the Sun's by 90°, and it is therefore displaced away from the poles of the ecliptic and equator through a distance $k''\sin l$, its declination being therefore decreased by $k''\sin l$. At the autumnal equinox its declination is increased by $k''\sin l$.

Hence the difference of the apparent declinations $= 2k''\sin l$, and this is also the difference of the star's apparent meridian zenith distances. By observing these, k'' may be found, l being of course known.

The value of k'' is very approximately **20·493″**.

376. Relation between the Coefficient of Aberration and the Equation of Light.—We have seen (§ 368) that

$$k'' = \frac{180 \times 60 \times 60}{\pi} \frac{u}{V} \quad\dots\dots\dots\dots\text{(i.),}$$

where k'' is the coefficient of aberration in seconds, u the velocity of the Earth, V that of light, both of which we will suppose measured in miles per second.

Now let r represent the radius of the Earth's orbit (supposed circular) in miles. Then in one sidereal year, or $365\frac{1}{4}$ days, the Earth travels round its orbit through a distance $2\pi r$ miles. Hence the Earth's velocity in miles per second is

$$u = \frac{2\pi r}{365\frac{1}{4} \times 24 \times 60 \times 60}.$$

Substituting in (i.), we have

$$k'' = \frac{15}{365\frac{1}{4}} \frac{r}{V}.$$

But r/V is the time taken by the light to travel from the Sun to the Earth, measured in seconds, or the "equation of light." Hence,

The coefficient of aberration in seconds

$$= \frac{15}{365\frac{1}{4}} \times \textbf{number of seconds taken by Sun's light to reach Earth.}$$

Thus, by observing the retardation of the eclipses of Jupiter's satellites at superior conjunction, the coefficient of aberration can be found independently of the methods of § 375, the number of days ($365\frac{1}{4}$) in the sidereal year being of course known.

The close agreement between the values found thus and by direct observation affords the strongest evidence in support of Bradley's explanation of aberration.

EXAMPLE.—To find the coefficient of aberration in seconds, having given that light takes 8m. 20s. to travel from the Sun to the Earth.

Here the required coefficient of aberration

$$k'' = \frac{15 \times 500}{365\frac{1}{4}} = \frac{7500}{365 \cdot 25} = 20 \cdot 534''.$$

377. To find the time taken by the light from a star to reach the Earth.—It is sometimes convenient to estimate the distance of a star by the number of years which the light from it takes to reach the Earth. This may be determined from a knowledge of the star's parallax, and of the coefficient of aberration, without knowing either the Sun's distance or the velocity of light.

Let the parallax of a star be $= P''$ in seconds $= P$ radians, and let the coefficient of aberration $= k''$ seconds $= k$ radians.

Then, if r, d be the Earth's and star's distances from the Sun, we have

$$P = \frac{r}{d}, \qquad k = \frac{\text{velocity of Earth}}{\text{velocity of light}}.$$

Now, in one year, the Earth travels over a distance $2\pi r$;

$\therefore$ in one year light travels a distance $\dfrac{2\pi r}{k}$;

$\therefore$ the number of years taken by light to travel from the star (distance d) to the Earth

$$= d \div \left(\frac{2\pi r}{k}\right) = \frac{dk}{2\pi r} = \frac{k}{2\pi P} = \frac{k''}{2\pi P''}.$$

The distance travelled by light in a year is sometimes called a " **light-year.**" Hence,

The product of a star's parallax and its distance in light-years is equal to the coefficient of aberration divided by 2π.

EXAMPLES.—1. To find how long the light would take to reach us from a star having a parallax $0.1''$.

The required time, in years,

$$= \frac{1}{2\pi} \frac{k''}{0.1} = \frac{10 \times 20.49 \times 7}{2 \times 22} \text{ approximately}$$

$$= 32.6.$$

2. To find the time taken by the light from the nearest star, *a Centauri*, taking its parallax as $0.75''$.

The parallax is 7.5 times that of the star in the last question, therefore its distance is $10/75$ as great, and the time taken by the light

$$= \frac{32.6}{7.5} = 4.35 \text{ years.}$$

378. Relation between the Coefficient of Aberration, the Sun's Parallax, and the Velocity of Light. — It follows from § 376 that if the coefficient of aberration k'' be determined by observation, the fraction r/V is also known, independently of observations of the eclipses of Jupiter's satellites. And if V, the velocity of light, be determined experimentally by the method of Foucault or Fizeau, the Sun's distance r can be found. Thus the Sun's parallax can be calculated from the coefficient of aberration and the velocity of light. And generally, if, of the four quantities, Sun's parallax, coefficient of aberration, velocity of light, and length of sidereal year in days, any three are observed, the value of the fourth may be deduced from them.

In this manner Foucault, by his determination of the velocity of light, in 1862, found the Sun's parallax to be 8·86″. Cornu, by experiments in 1874 and 1877, combined with the values for k'' determined by Struve, obtained the values 8·83″ and 8·80″ respectively. Michelson's experiments make the parallax 8·793″.

EXAMPLE.—If the velocity of light = 186,000 miles per second and the Earth's radius (a) = 3,960 miles, to prove that the product of the Sun's parallax and the coefficient of aberration, both measured in seconds, is 180·35.

The Sun's parallax $P'' = \dfrac{180 \times 60 \times 60}{\pi} \dfrac{a}{r}$,

and

$$k'' = \frac{15}{365\frac{1}{4}} \frac{r}{V} = \frac{60}{1461} \frac{r}{V};$$

$$\therefore\ P''k'' = \frac{180 \times 60 \times 60 \times 60}{1461\pi} \frac{a}{V} = \frac{206265 \times 60}{1461} \cdot \frac{3960}{186000}$$

$$= 180·35.$$

379. Planetary Aberration.—The direction of any planet is affected by aberration, which is due partly to the motion of the Earth, and partly to that of the planet itself.

For, during the time occupied by the light in travelling from a planet to the Earth, the planet itself will have moved from the position which it occupied when the light left it.

We shall, however, show that *the direction in which a planet is seen at any instant was the actual direction of the planet relative to the Earth at the instant previously when the light left the planet.*

Let t be the time required by the light to travel from the planet to the Earth. Let P, Q be the positions of the planet and Earth at any instant; P', Q' their positions after an interval t.

The light which leaves the planet when at P reaches the Earth when it has arrived at Q'; the direction of the actual motion of the light is, therefore, along PQ'. But PQ' and QQ' are the spaces passed over by the light and the Earth

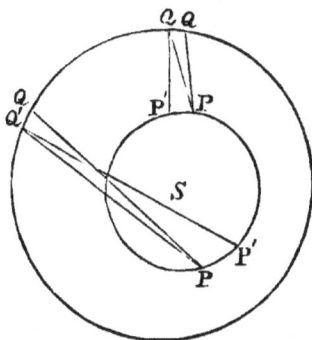

FIG. 128.

respectively in the time t (and QQ' is so small an arc that it may be regarded as a straight line). Therefore

$$QQ' : PQ' = \text{velocity of Earth : velocity of light.}$$

Hence it follows from § 368 that the line PQ represents the direction of relative velocity of the light with respect to the Earth. Therefore, when the Earth is at Q' the planet is seen in a direction parallel to PQ, and its apparent direction is exactly what its real direction was at a time t previously.

The same is true in the case of the Sun or a comet, or any other body, provided that the time taken by the light from the body to reach the Earth is so small that the Earth's motion does not change sensibly in direction in the interval.

The **aberration** of the planet at any instant is the angle between the apparent direction PQ and the actual direction $P'Q'$.

EXAMPLE.—To find the effect of aberration on the positions of (i.) the Sun, (ii.) Saturn in opposition, taking its distance from the Sun to be 9½ times the Earth's.

(i.) The light takes 8m. 20s. to travel from the Sun to the Earth therefore the Sun's apparent coordinates at any instant are its actual coordinates 8m. 20s. previously. Thus, its apparent decl. and R.A. at noon are its true decl. and R.A. at 23h. 51m. 40s., or 11h. 51m. 40s. A.M.

Now the Sun describes 360° in longitude in 365¼ days. Hence, in 500 seconds it describes 20·492″, and the Sun's aberration in longitude is 20·492″. This is otherwise evident from the fact that the Earth's way of the Sun is 90°; and it is at rest, consequently its aberration $= k$.

(ii.) The distance of Saturn from the Earth at opposition is $= 9½-1$, or 8½ times the Sun's distance. Light travels over this distance in 8m. 20s. × 8½ $= 500 × 8½$s. $= 1$h. 10m. 50s. Therefore, the apparent coordinates are the actual coordinates 1h. 10m. 50s. previously.

Thus the observed decl. and R.A. at midnight (12h. 0m. 0s.) are the true decl. and R.A. at 10h. 49m. 10s.

380. **Diurnal Aberration** is due to the effect of the Earth's diurnal rotation about the poles on the relative velocity of light.

As the Earth revolves from west to east, the portion of the motion of an observer due to this diurnal rotation is in the direction of the east point E of the horizon.

The effect of diurnal aberration can thus be investigated by methods precisely similar to those of § 368, E taking the place of U.*

Hence, every star x is displaced by diurnal aberration towards the east point E. And if x' be its displaced position, then

the displacement $xx' = A \sin xE$,

where

Circular measure of $A = \dfrac{\text{velocity of observer}}{\text{velocity of light}}.$

---

* The student will find it useful to go through the various steps of §§ 368–371, considering the diurnal motion.

Taking a for the Earth's radius, V for the velocity of light, let the observer's latitude be l.

In a sidereal day (86164·1 mean seconds) the Earth's rotation carries the observer round a small circle, whose distance from the Earth's axis is $a \cos l$, and whose circumference is, therefore, $2\pi a \cos l$. Hence, the observer's velocity

$$= \frac{2\pi a \cos l}{86164 \cdot 1} \text{ miles per second};$$

$$\therefore \text{ circular measure of } A = \frac{2\pi a \cos l}{86164 \cdot 1 \times V};$$

$\therefore A''$ (number of seconds in A)

$$= \frac{180 \times 60 \times 60}{\pi} \times \frac{2\pi a \cos l}{86164 \cdot 1\, V}$$

$$= \frac{15 a \cos l}{V} \text{ approximately.}$$

Thus, the coefficient of diurnal aberration varies as the cosine of the latitude. If K'' denote the coefficient of diurnal aberration at the equator in seconds, we therefore, have

$$K'' = \frac{15a}{V} = \frac{15 \times 3963}{186,000} = 0 \cdot 32'',$$

$$A'' = K'' \cos l = 0 \cdot 32'' \cos l.$$

* Effect of Diurnal Aberration on Meridian Observations.

The correction for diurnal aberration is greatest when the star is 90° from the east point, i.e., is on the meridian. In this case, the displacement is perpendicular to the meridian, and is equal to A''.

The star's meridian altitude is thus unaffected, but its time of transit is somewhat retarded at upper culmination, and (for a circumpolar star) accelerated at lower culmination, since the star appears on the meridian, when it is really A'' west of the meridian. The effect of diurnal aberration on the time of transit is thus equivalent to that of a small collimation error A'' in the Transit Circle.

For a star on the equator, seen from the Earth's equator, the retardation of the time of transit would be $\frac{1}{15} K''$ seconds, $= \frac{1}{50}$ of a second nearly, and it would be difficult to observe such a small interval.

381. To determine the Coefficient of Diurnal Aberration by Observations of the Azimuths of Stars when on the Horizon.

When a star is rising or setting it is evidently displaced by diurnal aberration along the horizon towards the east point. Consider two stars, one of which rises S. of E., and the other N. of E. It is evident that their rising points are drawn towards one another. But the stars set S. of W. and N. of W., and their displacements are still towards the E. point; hence, their setting points are separated away from one another. And, if the stars, at rising and setting, be carefully observed with an altazimuth, the difference between their azimuths at setting will exceed that between their azimuths at rising by an amount proportional to the diurnal aberration. From this, the coefficient of diurnal aberration may be found.

The azimuths are unaltered by refraction (§ 184), but the times of rising and setting are slightly altered by refraction. If the co-efficient of refraction be the same at both observations, however, the acceleration in rising will be equal to the retardation at setting, and the refraction will increase the azimuths at rising and setting by the same amount; thus the data will be unaffected. If the temperature of the air has changed considerably between rising and setting, it is only necessary to make the observations at equal intervals before and after the stars transit.

382. Relation between the Coefficients of Aberration and the Sun's Parallax.—We have evidently

$$\frac{k'''}{k''} = \frac{\text{velocity of diurnal motion at equator}}{\text{velocity of Earth's orbital motion}}.$$

But the velocities in miles, per sidereal day, are $2\pi a$ and $2\pi r/366\frac{1}{4}$;

$$\therefore \frac{k'''}{k''} = 366\frac{1}{4} \times \frac{a}{r} = 366\frac{1}{4} \times \begin{array}{c}\text{(Sun's parallax in circular}\\ \text{measure)}.\end{array}$$

This gives the coefficient of diurnal aberration at the equator in terms of the coefficient of annual aberration and the Sun's parallax. Conversely, if it were possible to observe the coefficient of diurnal aberration accurately, we should thus have another way of finding the Sun's parallax.

But the smallness of the diurnal aberration renders it impossible to obtain good results by this method.

EXAMPLES.—XI.

1. Prove that cosec 8·76″ = 23546 approximately, and thence that the distance of the Sun is nearly 81 million geographical miles, the angle 8·76″ being the Sun's parallax, and a geographical mile subtending 1′ at the Earth's centre.

2. Find the Sun's diameter in miles, taking the Sun's parallax as 8·8″, its angular diameter as 32′, and the Earth's radius as 3,960 miles.

3. A spot at the centre of the Sun's disc is observed to subtend an angle of 5″. What is its absolute diameter?

4. Show, by means of a diagram, that the general effect of the Earth's diurnal rotation is to shorten the duration of a transit of Venus, and that this circumstance might be used to find the Sun's parallax.

5. Supposing the equator, ecliptic, and orbit of Venus all to lie in one plane, and that a transit of Venus would last eight hours, at a point on the Earth's equator, if the Earth were without rotation ; show that, if the Sun is vertically overhead at the middle of the transit, the duration is diminished by about 9m. 55½s. owing to the Earth's rotation, taking the Sun's parallax to be 8·8″, and the synodic period of Venus to be 586 days.

6. If the annual parallax be 2″, determine the distance of the star, taking the Sun's distance to be 90,000,000 miles. Hence, deduce the distance of a star whose parallax is 0·2″.

7. Find, roughly, the distance of a star whose parallax is 0·5″, given that the Sun's parallax is 9″, and the Earth's radius is 4000 miles.

8. The parallax of 61 *Cygni* is 0·5″, and its proper motion, perpendicular to the line of sight, is 5″ a year ; compare its velocity in that direction with that of the Earth in its orbit round the Sun.

9. Account for the following phenomena : (i.) all stars in the ecliptic oscillate in a straight line about their mean places in the course of the year ; (ii.) two very near stars in the ecliptic appear to approach and recede from one another in the course of the year.

10. Suppose the velocity of light to be the same as the velocity of the Earth round the Sun. Discuss the effect on the Pole Star as seen by an observer at the North Pole throughout the year.

11. Sound travels with a velocity 1,100 feet per second. Determine the aberration produced in the apparent direction of sound to a person in a railway train travelling at sixty miles an hour, if the source of sound be exactly in front of one of the windows of the carriage.

12. Show that, in consequence of aberration, the fixed stars whose latitude is l appear to describe ellipses whose eccentricity is cos l.

13. How must a star be situated so as to have no displacement due to (i.) aberration, (ii.) parallax? Where must a star be so that the effect may be the greatest?

14. On what stars is the effect of aberration or parallax to make them appear to describe (i.) circles, (ii.) straight lines?

15. Show that the effect of annual parallax on the position of a star may be represented by imagining the star to move in an orbit equal and parallel to the Earth's orbit, and that the effect of aberration may be represented by imagining it to revolve in a circle whose radius is equal to the distance traversed by the Earth while the light is travelling from the star.

16. Supposing the star η *Virginis* to be situated (as it nearly is) at the first point of Libra, find the direction and magnitude of its displacement due to aberration about the 21st day of every month of the year, taking the coefficient of aberration to be 20·5″. When is its aberration greatest?

17. At the solstices show that a star on the equator has no aberration in declination. If its R.A. be 22h., show that its time of transit is retarded at the summer and accelerated at the winter solstice by ·68 of a second.

18. If the coefficient of aberration be 20″, and an error of 2,000 miles a second be made in determining the velocity of light, find, in miles, the consequent error in the value of the Sun's mean distance as computed from these data.

19. Show that when a planet is stationary its position is unaffected by aberration.

20. Taking the Earth's radius as 4,000, velocity of light 186,000 miles per second, show that the coefficient of diurnal aberration at the equator is about one-third of a second.

MISCELLANEOUS QUESTIONS.

1. Explain the following terms:—*asteroid*, *libration*, *lunation* *parallax*, *perihelion*, *planet's elongation*, *right ascension*, *synodical period*, *syzygies*, *zenith*.

2. Given that the R.A. of Orion's belt is 80°, show by a figure its position at different hours of the night about March 21 and September 23.

3. Prove that the number of minutes in the dip is equal to the number of nautical miles in the distance of the visible horizon.

4. Show how to determine the latitude of a place by meridional observations on a circumpolar star, taking into account the refraction error.

5. Show how to find longitude from lunar distances. The cleared lunar distance of a star at 8h. 30m. local mean time is 15°0′45″, and the tabular distances are 15°0′0″ at 6h. and 15°1′30″ at 9h. of Greenwich mean time. Find the longitude.

6. At what time of the year can the waning moon best be seen ?

7. On July 21 at 2 A.M. the Moon is on the meridian. What is the age of the Moon? Indicate the position on the celestial sphere of a star whose declination is 0 and whose R.A. is 30°.

8. Taking the distance of Venus from the Sun to be ¾ of that of the Earth, find the ratio of the planet's angular diameters at superior and inferior conjunction and greatest elongation, and draw a series of diagrams showing the changes in the planet's appearance during a synodic period, as seen through a telescope under the same magnifying power.

9. Defining a *lunar day* as the interval between two consecutive transits of the Moon across the meridian, find its mean length in (i.) mean solar, and (ii.) sidereal units.

10. At what season is the aberration of a star least whose R.A. is 90° and whose declination is 60° ?

11. Show that the constant of aberration can be determined by observation of Jupiter's satellites, without a knowledge of the radius of the Earth's orbit.

12. How is it possible to calculate separately the aberration—the constant of aberration being supposed unknown—annual parallax, and proper motion of a star, from a long series of observations of the apparent place of a star ?

EXAMINATION PAPER.—XI.

1. Why is the method for finding the Moon's parallax not available in the case of the Sun? Show how the determination of the parallax of Mars leads to the determination of the Sun's parallax.

2. Show how the Sun's parallax can be found by comparing the times of *commencement* or of *termination* of a transit of Venus at two stations not far from the Earth's equator.

3. Show how the Sun's parallax can be found by comparing the *durations* of a transit of Venus at two stations in high N. and S. latitudes. Why is this method not available when the transit is *central?*

4. Distinguish between solar and stellar parallax. Towards what point does a star seem to be displaced by heliocentric parallax? Find an expression for the displacement.

5. Describe Bessel's method of determining the annual parallax of a fixed star.

6. How might the Sun's parallax be determined by observations of the eclipses of Jupiter's satellites?

7. Explain the *aberration of light*, and investigate the direction and magnitude of the displacement which it produces on the apparent position of a star.

8. Show that owing to aberration a star in the pole of the ecliptic appears to describe a circle, and that a star in the ecliptic appears to oscillate to and fro in a straight line during the course of the year.

9. Show how the velocity of light may be determined from the aberration of a star when the Sun's mean distance is known.

10. Investigate the general effects of *diurnal aberration* due to the Earth's rotation about its axis. In what direction are stars displaced by diurnal aberration? Show that the coefficient of diurnal aberration at a place in latitude l is $K \cos l$, where K is the coefficient at the equator.

DYNAMICAL ASTRONOMY.

CHAPTER XII.

THE ROTATION OF THE EARTH.

383. Introductory.—In the preceding chapters we have shown how the motions of the celestial bodies can be determined by actual observation, and we have also explained certain resulting phenomena. But no use has yet been made of the principles of dynamics; consequently we have been unable to investigate the *causes* of the various motions. In particular, while we have assumed that the diurnal rotation of the stars is an appearance due to the Earth's rotation, we have not as yet given any definite proof that this is the only possible explanation.

The ancient Greeks accounted for the motions of the solar system by means of the *Theory of Epicycles*, according to which each planet moved as if it were at the end of a system of jointed rods rotating with uniform but different angular velocities. Suppose AB, BC, CD to be three rods jointed together at B, C. Let A be fixed; let AB revolve uniformly about A; let BC revolve with a different angular velocity about B; and let CD revolve with another different angular velocity about C. Then, by properly choosing the lengths and angular velocities of the rods, the motion of D, relative to A, may be made nearly to represent the motion, relative to the Earth, of a planet.

Copernicus (A.D. 1500 *circ.*) was the first astronomer who explained the motions of the solar system on the theory that the diurnal motion is due to the Earth's rotation, and that the Earth is one of the planets which revolve round the Sun. This theory was adopted by **Kepler** (A.D. 1609 *circ.*) whose laws of planetary motion have already been mentioned (§ 326).

These laws were, however, unexplained until their true cause was found by **Newton** (A.D. 1687) by his discovery of the law of gravitation.

384. **Arguments in Favour of the Earth's Rotation.**—Without appealing to dynamical principles, the probability of the Earth's rotation about its axis (§ 87) may be inferred from the following considerations :—

(i.) If the Earth were at rest, we should have to imagine the Sun and stars to be revolving about it with inconceivably great velocities.　If the Earth rotates, the velocity of a point on its equator is somewhere about 1,050 miles an hour.　But since the Sun's distance is about 24,000 times the Earth's radius, the alternative hypothesis would require the Sun—a body whose diameter is nearly 110 times as great as that of the Earth—to be moving with a velocity 24,000 times as great, or about 25,000,000 miles an hour ; while most of the fixed stars are at such distances from the Earth that they would have to move with velocities vastly greater than the velocity of light.　It is inconceivable that such should be the case.

(ii.) The diurnal rotations all take place about the pole, and are all performed in the same period—a sidereal day. This uniformity is a natural consequence of the Earth's rotation, but if the Earth were at rest, it could only be explained by supposing the stars to be rigidly connected in some manner or other.　Were such a connection to exist it would be difficult to explain the proper motions of certain fixed stars, and the independent motions of the Sun, Moon, and planets.

(iii.) By observing the motion of the spots on the Sun at different intervals, it is found that the Sun rotates on its axis. Moreover, similar rotations may be observed in the planets ; thus, Mars is known to rotate in a period of nearly 24 hours.　There is, therefore, nothing unreasonable in supposing that the Earth also rotates once in a sidereal day.

(iv.) The phenomenon of *diurnal aberration* affords a proof of the Earth's rotation.　Were it not for the difficulty of its observation, this proof alone would be conclusive.

We may mention that diurnal parallax could be equally well accounted for if the celestial bodies revolved round the Earth; not so, however, diurnal aberration.

385. Dynamical Proofs of the Earth's Rotation.— The following is a list of the methods by which the Earth's rotation is proved from dynamical considerations :—

(1) The eastward deviation of falling bodies.

(2) Foucault's pendulum experiment.

(3) Foucault's experiments with a gyroscope.

(4) Experiments on the deviation of projectiles.

(5) Observations of ocean currents and trade winds.

(6) Experiments on the differences in the acceleration of gravity in different latitudes, due to the Earth's centrifugal force, as observed by counting the oscillations of a pendulum ; combined with

(7) Observations of the figure of the Earth.

386. The Eastward Deviation of Falling Bodies.— If the Earth is rotating about its polar axis, those points which are furthest from the Earth's axis move with greater velocity than those which are nearer the axis. Hence the top of a high tower moves with slightly greater velocity than the base. If, then, a stone be dropped from the top of the tower, its eastward horizontal velocity, due to the Earth's rotation, is greater than that of the Earth below, and it falls to the east of the vertical through its point of projection. The same is true when a body is dropped down a mine. This eastward deviation, though small, has been observed, and affords a proof of the Earth's rotation.

Consider, for example, a tower of height h at the equator. If a be the Earth's equatorial radius, the base travels over a distance $2\pi a$ in a sidereal day, owing to the Earth's rotation, while the top of the tower describes $2\pi(a+h)$ per sidereal day. Thus, the velocity at the top exceeds that at the bottom by $2\pi h$ per sidereal day.

If h be measured in feet, the difference of velocities is $\pi h/3600$ *inches* per sidereal second, and is sufficiently great to cause a small but perceptible deviation when a body is let fall from a high tower.

The earliest experiments were too rough to show this deviation, and were, therefore, used as evidence *against*, instead of *for*, the Earth's rotation. The deviation can only be observed in experiments conducted with very great care, and it is very difficult to measure. Its amount is largely modified by the resistance of the air and other causes, and therefore differs considerably from that by theory.

387. Foucault's Pendulum Experiment.—In 1851, M. Foucault invented an experiment by which the Earth's rotation is very clearly shown. A pendulum is formed of a large metal ball suspended by a fine wire from the roof of a high building, and is set in motion by being drawn on one side and suddenly released; it then oscillates to and fro in a vertical plane. If now the pendulum be sufficiently long and heavy to continue vibrating for a considerable length of time, the plane of oscillation is observed to very gradually change its direction relative to the surrounding objects, by turning slowly round from left to right at a place in the northern hemisphere, or in the reverse direction in the southern. If the experiment is performed in latitude l, the plane of oscillation appears to rotate through $15° \times \sin l$ in a sidereal hour, $360° \sin l$ in a sidereal day, or $360°$ in $\operatorname{cosec} l$ sidereal days. This apparent rotation is accounted for by the Earth's rotation, as follows.

(i.) Let us first imagine the experiment to be performed at the **north pole** of the Earth. Let the pendulum AB be vibrating about A in the arc BB' in the plane of the paper. The only forces acting on the bob are the tension of the string BA and the weight of the bob acting vertically downwards; both are in the plane of the paper. The Earth's rotation about its axis CA produces no forces on the bob. Hence there is nothing whatever to alter the direction of the plane of oscillation; this plane therefore remains fixed in space. But the Earth is not fixed in space; it turns from west to east, making a complete *direct* revolution in a sidereal

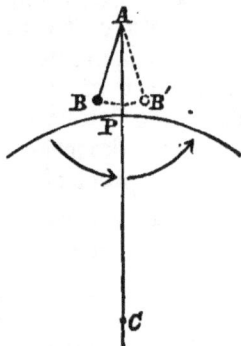

Fig. 129.

day. Hence the plane of the pendulum's oscillation appears, to an observer not conscious of his own motion, as though it rotated once in a sidereal day, in the reverse or *retrograde* direction (east to west). If, however, he were to compare the plane of oscillation not with the Earth but with the stars, whose directions are actually fixed in space, he would see that it always retained the same position relatively to them.

Since, then, the pendulum at the pole of the Earth appears to follow the stars, it evidently appears to rotate in the same direction as the hands of a watch at the north pole, and in the direction opposite to the hands of a watch at the south pole.

(ii.) Next suppose the experiment performed at the Earth's equator. If the bob be set swinging in the plane of the equator, take this as the plane of the paper (Fig. 130). The direction of the vertical AQC is now rotating about an axis through C perpendicular to the plane of the paper; hence it always remains in that plane. Hence there is nothing whatever to turn the plane of oscillation of the pendulum out of the plane of the Earth's equator. It therefore continues always to pass through the east and west points, and there is *no* apparent rotation of the plane of oscillation.

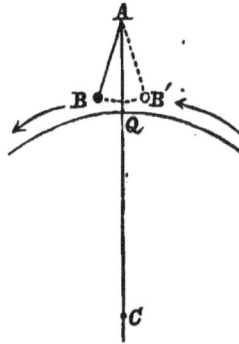

FIG. 130.

If the pendulum do not swing in the plane of the equator, the explanation is much more complicated. As the Earth rotates, the direction of gravity performs a *direct* revolution in a sidereal day. Hence, relative to the point of support, gravity is gradually and continuously turning the bob westwards, in such a way as to keep its mean position always pointed towards the centre of the Earth. When the bob is south of its position of equilibrium, this westward bias tends to turn the plane of oscillation in the clockwise direction, but when the bob is north of the mean position, the westward bias has an equal tendency to turn the plane in the reverse direction. Consequently the two effects counteract one another, and therefore produce no apparent rotation of the plane of oscillation relative to surrounding objects.

(iii.) Lastly, consider the case of an observer O in latitude l (Fig. 131). Let n denote the angular velocity with which the Earth is rotating about its polar axis CP. It is a well-known theorem in Rigid Dynamics that an angular velocity of rotation *about* any line may be resolved into components *about* any two other lines, by the parallelogram law, in just the same way as a linear velocity or a force *along* that line; this theorem is called the Parallelogram of Angular Velocities. Applying it to the angular velocity n about CP, we may resolve it into two components—

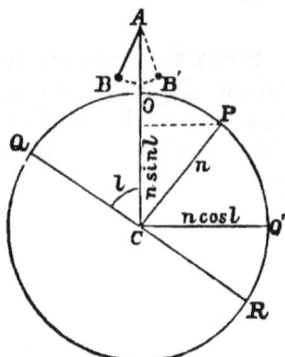

FIG 131.

$$n \cos PCO \text{ or } n \sin l \text{ about } CO,$$

and

$$n \sin PCO \text{ or } n \cos l \text{ about a line } CO' \text{ perpendicular to } CO,$$

and we may consider the effects of the two angular velocities separately.

As in case (i.), the component $n \sin l$ causes the Earth to turn about CO, without altering the direction in space of the plane of oscillation; this plane, therefore, appears to rotate *relatively* in the reverse or retrograde direction, with angular velocity $n \sin l$. As in case (ii.), the angular velocity $n \cos l$ about CO' produces no apparent rotation of the plane of oscillation relative to the Earth. Hence the plane of oscillation appears to revolve, relative to the Earth, with retrograde angular velocity $n \sin l$.

But the angular velocity $n = 15°$ per sidereal hour
$= 360°$ per sidereal day.

Therefore the plane of oscillation turns through

$$15° \sin l \text{ per sidereal hour} = 360° \sin l \text{ per sidereal day},$$

and its period of rotation $= \dfrac{360°}{n \sin l}$

$= \operatorname{cosec} l \text{ sidereal days.}$

388. The Gyroscope or Gyrostat is another apparatus used by Foucault to prove the Earth's rotation. It is simply a large spinning-top, or, more correctly, a heavy revolving wheel M (Fig. 132), whose axis of rotation AB is supported by a framework, so that it can turn about its centre of gravity in any manner. Thus, by turning the wheel and the inner frame $ACBD$ about the bearings CD, and then turning the outer frame $DECF$ about the bearings EF, the axis AB (like the telescope in an altazimuth or equatorial) can be pointed in any desired direction. The three axes AB, CD, EF all pass through the centre of gravity of the top; hence its weight is entirely supported, and does not tend to turn it in any way; and the bearings A, B, C, D, E, F are very light, and so constructed that their friction may be as small as possible. The top may be spun by a string in the usual way, and it continues to spin for a long time.

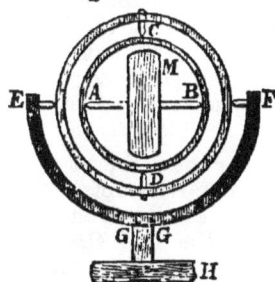

Fig. 132.

When a symmetrical body, such as the wheel M, is revolving rapidly about its axis of figure, and is not acted on by any force or couple, it is evident that no change of motion can take place, and therefore the axis of rotation AB must remain fixed in direction. This is the case with the gyroscope, for, from the mode in which the weight of the wheel is supported, there is no force tending to turn it round.

When the experiment is performed it is observed that the axis AB follows the stars in their diurnal motion; if pointed to any star, it always continues to point to that star, its position relative to the Earth changing with that of the star. Hence it is inferred that the directions of the stars are fixed in space, and that the diurnal motion is not due to them, but to the rotation of the Earth.

389. If while the gyroscope is spinning rapidly any attempt be made to alter the direction of the axis of rotation AB by pushing it in any direction, a very great resistance will be experienced, and the axis will only move with great difficulty. This shows that the small friction at the pivots CD, EF can have but little effect in turning the axis of the top, and therefore the gyroscope spins as if it were practically free, as long as its angular velocity remains considerable.

The following additional experiments with the gyroscope can be also used to prove the Earth's rotation.

Experiment 1.—Let the hoop $CEDF$ be steadily rotated about the line EF. The line AB is no longer free to take up any position, for the pivots C and D obviously force it always to be in a plane through EF and perpendicular to plane $CEDF$. Hence the axis of rotation is no longer able to maintain always the same position, unless that position coincides with EF. The result is that the axis gradually turns about CD till it *does* coincide with EF, the direction of rotation of the wheel being the same as that in which frame is forced to revolve. It will then have no further tendency to change its place. Of course we suppose the hoop turned so quickly that the effect of the slow motion of the Earth is imperceptible.

FIG. 133.

Experiment 2.—We may now repeat Experiment 1, using the Earth's rotation. Let the framework $CEDF$ be fixed in a horizontal position, the line CD being held pointed due east and west. The axis AB is then free to turn in the plane of the meridian. Now, owing to the Earth's rotation, the framework carrying CD is turning about the Earth's polar axis, and this causes the top to turn *till its axis points to the celestial poles.* The result of experiment agrees with theory, thus affording a further proof of the Earth's rotation about the poles.

Experiment 3.—Let the framework $CEDF$ be clamped in a vertical plane. The axis AB can then turn in a horizontal plane, but it cannot point to the pole. It will, however, try to point in a direction differing as little as possible from the direction of the Earth's axis, and will therefore turn till it points due north and south. This has also been verified by actual observation.

Experiments 3 and 2, if performed with a sufficiently perfect gyroscope, would enable us to find the north point, and then to find the celestial pole, and thus determine the latitude without observing any stars. By means of Foucault's pendulum experiment we could also (theoretically) determine the latitude.

390. **The Deviation of Projectiles.**—If we suppose a cannon ball to be fired in any direction, say from the Earth's North Pole, the ball will travel with uniform horizontal velocity in a vertical plane. But, as the Earth rotates from right to left, the object at which the ball was aimed will be carried round to the left of the plane of projection, and therefore the ball will appear to deviate to the right of its mark. At the South Pole the reverse would be the case, because in consequence of the direction of the vertical being reversed, the Earth would revolve from left to right; hence the ball would deviate to the left of its mark. At the equator no such effect would occur.

The deviation, like that in Foucault's pendulum, depends on the Earth's component angular velocity about a vertical axis at the place of observation, and this component, in latitude l, is $n \sin l$ (§ 387, iii.). Now the Earth rotates about the poles through 15" per sidereal second. Hence, if t be the time of flight measured in sidereal seconds, the deviation is

$$= nt \sin l = 15''. t \sin l,$$

and it is necessary to aim at an angle 15". $t \sin l$ to the *left* of the target in N. lat. l, or 15". $t \sin l$ to the *right* in S. lat. l. The formula is sufficiently approximate even if t be measured in *solar* seconds. It is necessary to allow for this deviation in gunnery—thus affording another proof of the Earth's rotation.

391. **The Trade Winds** are due to a similar cause. The currents of air travelling towards the hotter parts of the Earth at the equator, like the projectiles, undergo a deviation towards the right in the northern hemisphere, and towards the left in the southern. This deviation changes their directions from north and south to north-east and south-east respectively. In a similar manner the Earth's rotation causes a deviation in the ocean currents, making them revolve in a direction opposite to that of the Earth's rotation, which is " counter clockwise " in the N. and " clockwise " in the S. hemisphere. The rotatory motion of the wind in cyclones is also due to the Earth's rotation.

392. Centrifugal Force.—If a body of mass m is revolving in a circle of radius r with uniform velocity v under the action of any forces, it is known that the body has an acceleration v^2/r towards the centre of the circle.* Hence the forces must have a resultant mv^2/r acting towards the centre, and they would be *balanced* by a force mv^2/r acting in the reverse direction, *i.e.*, *outwards* from the centre. This force is called the **centrifugal force.**

Thus, in consequence of its acceleration, the body appears to exert a centrifugal force outwards. If it be attached to the centre of the circle by a string, the pull in the string is mv^2/r. If m be measured in pounds, r in feet, and v in feet per second, then mv^2/r represents the centrifugal force in *poundals*. Similarly, in the centimetre-gramme-second system of units, mv^2/r is the centrifugal force in *dynes*.

If n represent the body's angular velocity in radians per second, $v = nr$, and the centrifugal force is therefore mn^2r.

393. General Effects of the Earth's Centrifugal Force.—If the Earth were at rest the weight of a body would be entirely due to the Earth's attraction. But in consequence of the diurnal rotation the apparent weight is the resultant of the Earth's attraction and the centrifugal force.

Let QOR represent a meridian section of the Earth (Fig. 134). Consider a body of mass m supported at any point O on the Earth's surface. Since the Earth is nearly, but not quite, spherical, the force g_0 of the Earth's attraction on a unit mass is not directed exactly to the Earth's centre, but along a line OK. But, owing to the body's central acceleration along OM, the force which it exerts on the support is not quite equal to the Earth's attraction mg_0, but is compounded of mg_0 acting along OK, and the centrifugal force $m . n^2 . MO$ acting along MO.

On KQ take a point G such that
$$KG : KO = n^2 . MO : g ;$$

---

* See any book on Dynamics.

then, by the triangle of forces, OG is the direction of the resultant force exerted by the body on its support, and this force is the apparent weight of the body. Hence, also OG represents the apparent direction of gravity, or the vertical as indicated by a plumb-line. Producing GO, KO to Z, Z'', we see that *the effect of centrifugal force is to displace the vertical from Z'' towards the nearest pole* (P).

The angle ZGQ measures the (geographical) latitude of the place, and is greater than $Z''KQ$, which would measure the latitude if the Earth were at rest. Hence *the apparent latitude of any place is increased by centrifugal force.*

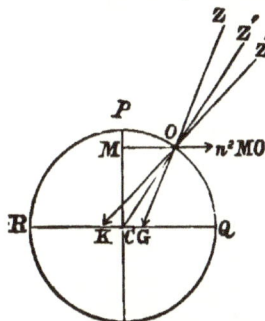

FIG. 134.

Again, if the apparent weight be denoted by mg, we have, by the triangle of forces,

$$g : g_0 = GO : KO;$$

now from the figure it is evident that $GO < KO$, and therefore $g < g_0$. Hence *the apparent weight of a body is diminished by centrifugal force.*

394. Effect on the Acceleration of a Falling Body.—If a body is falling freely towards the Earth near O, the whole acceleration of its motion in space is due to the Earth's attraction, and is g_0, along OK. But the Earth at O has itself an acceleration $n^2 OM$ towards M. Hence the acceleration of the body *relative* to the Earth is the resultant of g_0 along OC, and $n^2 . MO$ along MO, and is therefore g along OG. Hence the body approaches the Earth with acceleration g along OG. *Therefore its* **relative** *acceleration is the acceleration due to its* **apparent weight,** *that is, to the resultant of the Earth's attraction and centrifugal force.*

395. **To find the loss of weight of a body at the equator, due to centrifugal force.**— At the equator centrifugal force is directly opposed to gravity; hence, if a denote the Earth's radius CQ,

$$g = g_0 - n^2 a.$$

Now we have roughly

$$g_0 = 32 \cdot 18 \text{ feet per second per second,}$$

$$a = 3963 \text{ miles} = 3963 \times 5280 \text{ feet,}$$

and $n = 2\pi$ radians per sidereal day

$$= \frac{2\pi}{86164} \text{ radians per } mean \ solar \ second.$$

Hence $n^2 a = \dfrac{3963 \times 5280 \times 4\pi^2}{86164 \times 86164} = \cdot 11127,$

and therefore $\dfrac{n^2 a}{g_0} = \dfrac{\cdot 11127}{32 \cdot 18} = \dfrac{1}{289}$ nearly.

Hence $g = g_0 - \dfrac{1}{289} g_0,$

or the effect of the Earth's rotation is to decrease the weight of a body by about $\dfrac{1}{289}$ *of the whole.*

· For rough calculations it would be sufficient to take $g = 32 \cdot 2$, $a = 3960$ miles, and to neglect the difference between a solar and a sidereal day. This would give $\frac{1}{289}$, as before.

396. **To find approximately the loss of weight of a body and the deviation of the vertical due to centrifugal force in any given latitude.**—

Let $l = QGO =$ astronomical latitude of O; $D = GOK = ZOZ'' =$ deviation of vertical from direction of Earth's attraction, or increase of latitude due to centrifugal force.

We have $OM = CO \cos COM$

$$= a \cos l \text{ approximately;}$$

where a is the Earth's radius, since the Earth is very nearly spherical, and $\angle COM$ is therefore very nearly equal to the latitude l. Therefore centrifugal force per unit mass at O

$$= n^2 . OM = n^2 . a \cos l = \frac{1}{289} g_0 \cos l \text{ (from § 395)}.$$

Resolving along OG, we have, if g_0 be the Earth's attraction per unit mass at O^*,

$$g = g_0 \cos D - n^2 \cdot OM \cos l$$

$$= g_0 - \frac{g_0}{289} \cos^2 l \text{ approximately}$$

(since D is small, and $\therefore \cos D = 1$ nearly).

Hence, *in latitude l, the Earth's rotation diminishes the weight of a body by approximately* $\frac{1}{289} \cos^2 l$ *of itself.*

Resolving perpendicular to OG, we have

$$g_0 \sin D - n^2 OM \sin l = 0 ;$$

$$\therefore \quad \sin D = \frac{n^2 a \cos l \sin l}{g_0}$$

$$= \frac{1}{289} \frac{\sin 2l}{2} .$$

Since d is small, this gives approximately

circular measure of $d = \frac{1}{289} \frac{\sin 2l}{2} ;$

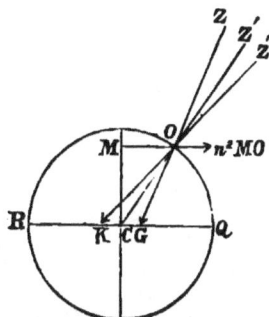

FIG. 135.

$\therefore \quad D''$ (number of seconds in D)

$$= \frac{180 \times 60 \times 60}{289 \times 2\pi} \sin 2l$$

$$= \frac{206265}{578} \sin 2l = 357'' \sin 2l.$$

Hence the deviation $D = \mathbf{5'\ 57''} \cdot \mathbf{\sin 2}l$, and this is the increase of latitude due to centrifugal force.

COROLLARY.—The deviation of the vertical due to centrifugal force is greatest in latitude $45°$ ($\therefore \sin 2l = 1$), and is there $5'\ 57''$.

---

* Since the Earth is not quite spherical, g_0 is not the same at O as at the equator. The difference may be neglected, however, when multiplied by the small constant $\frac{1}{289}$.

397. Figure of the Earth.—In § 114 we stated that the form of the Earth has been observed to be an oblate spheroid. Now it has been proved mathematically that a mass of gravitating liquid-when rotating takes the form of an oblate spheroid whose least diameter is along its axis of rotation. Thus the Earth's form may be accounted for on the theory that the Earth's surface was formerly in a fluid or molten state, and that it then assumed its present form, owing to its diurnal rotation. We thus have another argument in favour of the Earth's rotation; but it is only fair to say that this theory of the Earth's origin has not been satisfactorily demonstrated. It accounts satisfactorily, however, for the form of the surface of the *ocean*.

This theory may be illustrated by the following general considerations. When a mass of liquid is acted on by no forces beyond the attractions of its particles, it is easy to realize that the whole is in equilibrium in a spherical form, being then perfectly symmetrical.

If, however, the fluid be rotating about the axis PCP', the centrifugal force tends to pull the liquid away from this axis and towards the equatorial plane. The liquid would, therefore, fly right off, but its attraction is always trying to pull it back to the spherical form. Hence, the only effect of centrifugal force (which, for the Earth, is small compared with gravity) is to distort the liquid from its spherical form by pulling it out towards the equator ; and it is therefore reasonable to suppose that the fluid will assume a more or less oblate figure, whose equatorial is greater than its polar diameter.

It may also be remarked that the form assumed by the liquid would be such that the effective force of gravity (*i.e.*, the resultant of the attraction and centrifugal force) on the surface would everywhere be perpendicular (*i.e.*, normal) to the surface.

*398. Gravitational Observations.**—If the Earth were a sphere, its attraction g_0 would everywhere tend to its centre, and would be of the same intensity at all points on its surface, while the variations in g, the apparent intensity of gravity, would be entirely due to the Earth's centrifugal force, its value in latitude l being proportional to $1 - \frac{1}{289} \cos^2 l$ (§ 396). By comparing the values of g at different places, we should then be able to demonstrate the Earth's centrifugal force, and hence prove its rotation. But, owing to the Earth's ellipticity, its attraction g_0 does not pass through the centre, except at the poles and equator, and its intensity is not everywhere constant. It is, therefore, important to determine experimentally the values of g at different stations. By allowing for centrifugal force, the corresponding values of the Earth's attraction g_0 can be found, and the variations in its intensity at different places afford a measure of the amount by which the Earth differs from a sphere. We thus have a gravitational method of finding the Earth's ellipticity.

But the Earth's ellipticity can also be determined by direct observation, as explained in Chapter III., Section III. The agreement between the results thus independently obtained furnishes another proof of the Earth's rotation.

In consequence of the Earth's ellipticity it is found (by observation) that the difference in the intensity of gravity between the pole and equator is increased from $\frac{1}{230}$ to $\frac{1}{190}$ of the whole.

399. To compare the Intensity of Gravity at different places.— The intensity of gravity may be measured by the force with which a body of unit mass is drawn towards the Earth. This cannot be measured by weighing a body with a *common balance*, because the weights of the body and of the counterpoise, by means of which it is weighed, are equally affected by variations in the intensity of gravity, and two bodies of equal *mass* will, therefore, balance one another when placed in the scale pans, no matter what be the intensity of gravity. In fact, by weighing a body with weights in the ordinary way, we determine only its *mass*, and not the absolute force with which it is drawn to the Earth.

We might determine the intensity of gravity by means of a "*spring balance*," for the elasticity of the spring does not depend on the intensity of gravity, and therefore the extension of the spring gives an absolute measure of the force with which the body is drawn towards the Earth. If the apparatus were to support a mass of one pound, first at the equator and then at the pole, the force on it would be greater at the latter place by about $\frac{1}{190}$, and this spring would there be extended about $\frac{1}{190}$ more. It would be very difficult to construct a spring balance sufficiently sensitive to show such a small relative difference of weight, but it has been done.

Atwood's machine might be used to find g, but this method is not capable of giving very accurate results.

The most accurate method of finding g is by timing the oscillations of a *pendulum* of known length.

[* A theoretical *simple pendulum*, consisting of a mere heavy particle of no dimensions, suspended by a thread without weight, is of course impossible to realize in practice, but the difficulty is overcome by the use of a pendulum called *Captain Kater's Reversible Pendulum*. This pendulum is a bar which can be made to swing ab ut either of two knife-blades fixed at opposite sides of, but unequal distances from, its centre of gravity, and it is so loaded that the periods of oscillation, when suspended from either knife-edge, are equal. It is then known that the pendulum will swing about either knife-edge in just the same manner as if it were a simple pendulum whose whole mass was concentrated at the *other* knife-edge. The distance between the knife-edges is, therefore, to be regarded as the *length of the pendulum*.]

400. Oscillations of a Simple Pendulum.—In a simple
pendulum, formed of a small heavy particle suspended by a
fine light thread of length l, the period of a complete oscillation
to and fro is

$$t = 2\pi \sqrt{\frac{l}{g}},$$

the time of a single swing or "*beat*" being of course half of
this.

Hence by observing the time of oscillation t and measuring
the length l, the intensity of gravity g can be found.

By the "**seconds pendulum**" is meant a pendulum in
which one *beat* occupies one second, hence a complete
oscillation occupies *two seconds*.

EXAMPLE.—Having given that the length of the seconds pendulum
is 99·39 centimetres, to find g in centimetres per second per second.

$$t = 2\pi\sqrt{l/g} = 2 \text{ seconds, and } l = 99\cdot39 \text{ cm.,}$$
$$\therefore \quad g = \pi^2 l = 99\cdot39 \times (3\cdot1416)^2 = 981.$$

It is often necessary to compare the lengths of two
pendulums whose periods of oscillation are very nearly equal,
to find the effect of small changes in the length of a pendulum
due to variations in temperature, or, in comparing the intensity
of gravity at different places, to find the effect of a small
alteration in the value of g on the period of oscillation and on
the number of oscillations in a given interval. If the differences are small, the calculations may be much simplified by
means of the following methods of approximation.*

**401. To find the change in the time of oscillation of
a pendulum, and in the number of oscillations in a
given interval, due to a small variation in its length
or in the intensity of gravity.**

If t be the time of a complete oscillation of a pendulum of
length l, we have, by § 400,

$$t^2 = 4\pi^2 \frac{l}{g} \quad \dots\dots\dots\dots\dots\dots\dots \text{ (i).}$$

---

* The same results can of course be obtained by means of the
differential calculus.

(i.) Suppose the length increased to l', and let t' be the new period of oscillation. We have

$$t'^2 = 4\pi^2 \frac{l'}{g}.$$

Therefore, by division,

$$\frac{t'^2}{t^2} = \frac{l'}{l},$$

and therefore also

$$\frac{t'^2 - t^2}{t^2} \equiv (t'-t)\frac{t'+t}{t^2} = \frac{l'-l}{l}.$$

These formulæ are exact. But if l' is very nearly equal to l, t' is very nearly equal to t, and therefore, putting $t+t'=2t$, we have approximately

$$2\frac{t'-t}{t} = \frac{l'-l}{l},$$

whence, if t, l be known, the change $t'-t$, consequent on the increase of length $l'-l$, may be readily found approximately without the labour of extracting any square roots.

(ii.) Suppose the intensity of gravity increased to g', the length l being unaltered, and let t' be the new period. Since

$$t'^2 = 4\pi^2 \frac{l}{g'},$$

we have, by division,

$$\frac{t'^2}{t^2} = \frac{g}{g'}$$

and therefore also

$$\frac{t'^2 - t^2}{t^2} \equiv (t'-t)\frac{t'+t}{t^2} = \frac{g-g'}{g'}.$$

But, if t, g are very nearly equal to t', g', this gives approximately

$$2\frac{t'-t}{t} = -\frac{g'-g}{g}.$$

(iii.) If l and g both vary, becoming l' and g', we have, in like manner

$$\frac{t'^2}{t^2} = \frac{l'g}{l\,g'}.$$

Therefore also

$$\frac{t'^2 - t^2}{t^2} = (t'-t)\frac{t'+t}{t^2} = \frac{l'g - lg'}{lg'} = \frac{l'g - l'g'}{lg'} + \frac{l'g' - lg'}{lg'}$$

$$= \frac{l'}{l}\frac{g - g'}{g'} + \frac{l' - l}{l},$$

or approximately, if the variations are small,

$$2\frac{t'-t}{t} = \frac{l'-l}{l} - \frac{g'-g}{g},$$

showing that the effects of the two variations may be considered separately.

(iv.) If n, n' be the number of complete oscillations of the pendulum in a given interval T, and if, in consequence of the change, this number be altered to n', we have

$$nt = n't' = T,$$

$$\therefore \qquad \frac{n'}{n} = \frac{t}{t'},$$

whence

$$\frac{n'-n}{n} = \frac{t-t'}{t'}.$$

If t' is very nearly equal to t, this gives approximately

$$\frac{n'-n}{n} = -\frac{t'-t}{t} = -\tfrac{1}{2}\frac{l'-l}{l} + \tfrac{1}{2}\frac{g'-g}{g},$$

which determines the number of beats gained by the pendulum in the time T, in consequence of the variations, the original number n being supposed known.

EXAMPLE.—To find the number of oscillations gained or lost in an hour by the pendulum of the Example of § 400, supposing (i.) its length increased to 1 metre; (ii.) the acceleration of gravity increased to 982; (iii.) both changes made simultaneously.

(i.) The pendulum beats seconds; therefore it performs 3600 half oscillations or 1800 whole oscillations in an hour. Also $l' = 100.00$ $l' - l = 0.61$, $g' - g = 0$.

Hence, if n' be the new number of oscillations in an hour,

$$\frac{n'-1800}{1800} = -\frac{0\cdot61}{2l} = -\frac{0\cdot61}{2l'}\text{(approx.)} = -\frac{0\cdot61}{200};$$

$$\therefore \quad n'-1800 = -9 \times \cdot61 = -5\cdot49.$$

Hence the pendulum loses nearly $5\frac{1}{2}$ oscillations in an hour, and the number of oscillations is therefore $1794\frac{1}{2}$.

(ii.) Here $$\frac{n'-1800}{1800} = \frac{g'-g}{g} = \frac{982-981}{2\times981} = \frac{1}{2\times981};$$

$$\therefore \quad n'-1800 = \frac{1800}{1962} = \cdot9 = 1 \text{ nearly.}$$

Hence the pendulum gains 1 oscillation in an hour, the total number of oscillations being 1801.

(iii.) Since from the first cause the pendulum loses $5\frac{1}{2}$ oscillations and from the second it gains 1 oscillation, therefore on the whole it loses $5\frac{1}{2}-1$ or $4\frac{1}{2}$ oscillations per hour. It therefore performs $1795\frac{1}{2}$ oscillations or 3591 *beats* per hour.

402. To compare the times of oscillations of two pendulums whose periods are very nearly equal.— If two pendulums of nearly equal periods are simultaneously started swinging in the same direction, the one whose period is a little the shortest will soon begin to swing before the other. After some time it will gain a half oscillation, and the pendulums will then be swinging in opposite directions. After another equal interval, the quicker pendulum will have gained one whole oscillation on the slower, and both will be again swinging together in the same direction. Similarly, every time the quicker pendulum has gained an exact number of complete oscillations on the slower, both will be swinging together in the same direction. Thus, the number of **coincidences,** or the number of times that the two pendulums are together, in any interval, is equal to the number of complete oscillations (to and fro) gained by the quicker pendulum over the slower, *i.e.*, the difference between the numbers of complete oscillations performed by the two pendulums.

Thus, if n, n' be the number of oscillations of the slower and faster pendulums in any given interval, then $n'-n$ is the the number of oscillations gained by the latter, and is, therefore, the number of " coincidences." If either of the numbers n, n' is known, we can, by counting the coincidences, find the other number.

403. To find g, the acceleration of gravity, the simplest plan is to use a Captain Kater's pendulum, the beat of which is very nearly one second. By counting the "coincidences" of the pendulum with the pendulum of a clock regulated to beat seconds during, say, an hour (as shown by the clock) the exact time of oscillation can be found. Moreover, from the number of beats gained or lost, and the observed length of the pendulum, we may calculate the amount by which the length must be increased or decreased in order to make the pendulum beat seconds. The length of the seconds pendulum is thus known, and the value of g can be found.

The reason for using *two pendulums* is that it would be extremely difficult to measure the length of the pendulum of the clock, and it would be equally difficult to find the period of oscillation of a pendulum without comparing it with that of a clock, whose rate can be regulated daily by astronomical observations.

404. To compare the value of g at two different stations, the simplest plan is to determine the number of seconds gained or lost in a day by a clock after it has been taken from one station to the other, the length of the pendulum remaining the same. If n, n' be the number of seconds marked by the clock in a day at the two places, we have exactly

$$\frac{n'^2}{n^2} = \frac{g'}{g},$$

or approximately,

$$\frac{n'-n}{n} = \tfrac{1}{2}\frac{g'-g}{g},$$

whence the ratio of g' to g may be found.

Here there is no necessity to use a Captain Kater's pendulum, because the length of the pendulum is not required; hence the ordinary compensating pendulum of the clock answers the purpose. If a *non-compensating* pendulum were used, it would be necessary to make allowance for any change in the length of the pendulum due to variations in temperature.

EXAMPLES.—XII.

1. A Foucault's pendulum being set vibrating in latitude 30°, show that after one sidereal day it is again vibrating in the same plane. Find the corresponding interval in latitude 45°.

2. If two *conical pendulums* of equal length revolve in opposite directions, describing cones of equal vertical angle, show that at a place in the northern hemisphere the pendulum which revolves in the same direction as the hands of a watch will have the greater apparent angular velocity, and will gain two complete revolutions on the other in the period in which the plane of Foucault's pendulum turns through 360°. Consider, in the first place, the phenomena at the North Pole. Also describe the corresponding phenomena in the southern hemisphere.

3. If a railway is laid along a meridian, and a train is travelling from the equator towards the pole, investigate whether it will exert an eastward or a westward thrust on the rails, and why.

4. A bullet is fired in N. latitude 45°, with a velocity of 1,600 feet per second, at an elevation 45°. Prove that it must be aimed in a vertical plane 12' 30" to the left of the target; and, if this precaution be neglected, calculate how many feet it will deviate to the right.

5. Show that if the Earth were to rotate seventeen times as fast, a body at the equator would have no weight.

6. If the Earth were a homogeneous sphere rotating so fast that bodies at the equator had no weight, show that in any latitude the plumb-line would point to the celestial pole.

7. Would the latitude of Greenwich be increased or decreased by an increase in the speed of the Earth's rotation? If the latitude of a place be 60°, find what would be its latitude if (i.) the Earth were reduced to rest, (ii.) its angular velocity were doubled.

8. Prove that if the Earth were reduced to rest, a pendulum in latitude 45° would gain one oscillation in every 1156, but if the Earth's angular velocity were doubled, it would lose *three* oscillations in 1156.

9. A clock and a chronometer are taken from London to Gibraltar and it is observed that the clock begins to lose, while the chronometer continues to keep correct time. Why is this?

10. Assuming that a body loses $\frac{1}{193}$ of its weight when taken from the poles to the equator, show that a clock which keeps mean time at the equator would keep sidereal time at the poles, with a rate amounting to only a fraction of a second per day.

11. With the data of the last question, show that the Earth's attractions on a unit mass placed at the equator and at the poles are in the ratio of (nearly) 496 : 497.

12. If a railway train is travelling along the equator from east to west, show that it presses on the rails with a force greater than its apparent weight when at rest. If the train is travelling at forty-five geographical miles per hour, and its mass is 144 tons, find (roughly) in pounds the increase in the downward thrust on the rails,

EXAMINATION PAPER.—XII.

1. Give reasons for supposing that the diurnal rotation of th heavens is only an appearance caused by a real rotation of th Earth. Name methods by which it has been claimed that this i proved.

2. Describe the *gyroscope experiment*, and the *gyroscope*.

3. Give any theoretical methods of determining latitude withot observing a heavenly body.

4. Describe Foucault's experiment for exhibiting the Earth rotation ; and find the time of the complete rotation of the plane (vibration of a simple pendulum freely suspended in latitude 60°.

5. Having given that the Earth's circumference is 40,000 kil metres, find the acceleration of a body at the equator due to th Earth's rotation in centimetres per second per second, and takin g_0, the acceleration of gravity, to be 981 of these units, deduce th ratio of centrifugal force to gravity at the equator.

6. What is meant by the *vertical* at any point of the Earth surface ? Supposing the Earth to be a uniform sphere revolvir round a diameter, calculate the deflection of the vertical from th normal to the surface.

7. State what argument is drawn from the Earth's form to suppo the hypothesis of its rotation.

8. Why is it that the intensity of gravity is less at the equat than in higher latitudes ? Show that the alteration in the appare weight of a body due to centrifugal force varies nearly as $\cos^2$ where l is the latitude, and state the ratio of centrifugal force gravity at the equator.

9. If a body is weighed by a spring balance in London and Quito, a difference of weight is observed. Why is this not observec an ordinary pair of scales be used ?

10. Show that an increase in the intensity of gravity will cau a pendulum to swing more rapidly, and *vice versâ*. If the accele tion of gravity be increased by the small fraction $1/r$ of its val show that a pendulum will gain one complete oscillation in every

CHAPTER XIII.

THE LAW OF UNIVERSAL GRAVITATION.

Section I.—*The Earth's Orbital Motion—Kepler's Laws and their Consequences.*

405. Evidence in favour of the Earth's Annual Motion round the Sun.—The theory that the Earth is a planet, and revolves round the Sun, was propounded by Copernicus (*circ.* 1530) and received its most convincing proof, over 150 years later from Newton (A.D. 1687), who accounted for the motions of the Earth and planets as a consequence of the law of universal gravitation. This proof is based on dynamical principles; but the following arguments, based on other considerations, afford independent evidence in favour of the theory that the Earth revolves round the Sun rather than the Sun round the Earth.

(i.) The Sun's diameter is 110 times that of the Earth's, and it is much easier to believe that the smaller body revolves round the larger, than that the larger body revolves round the smaller.

If the dynamical laws of motion be assumed, it is impossible to see how the larger body could revolve round the smaller, unless either its mass and therefore its density were very small indeed, or the smaller one were rigidly fixed in some way.

(ii.) The stationary points, and alternately direct and retrograde motions of the planets, are easily accounted for on the theory that the Earth and planets revolve round the Sun (Chap. X.) in orbits very nearly circular, and it would be impossible to give such a simple explanation of these motions on any other theory. It is true that we might suppose, with Tycho Brahé (*circ.* 1600), that the planets revolve round the Sun as a centre, while that body has an orbital motion round the Earth, but this explanation would be more complicated than that which assumes the Sun to be at rest. And it would be hard to explain how such huge bodies as Jupiter and Saturn could be brought to describe such complex paths.

(iii.) As seen through a telescope, Venus and Mars are found to be very similar to the Earth in their physical characteristics, and their phases show that, like the Earth and Moon, they are not self-luminous. It is, therefore, only natural to suppose that their property of revolving round the Sun is shared by the Earth. Moreover, the Earth's relative distance from the Sun agrees fairly closely with that given by Bode's law ; hence there is a strong analogy between the Earth and the planets.

(iv.) The orbital motion of the Earth is in strict accordance with Kepler's Laws of Planetary Motion. In particular, the relation between the mean distances and periodic times given by Kepler's Third Law (§ 326) is satisfied in the case of the Earth's orbit.

Moreover, a similar relation is observed to hold between the periodic times of Jupiter's satellites and their mean distances from Jupiter. Hence it is probable that the Earth and planets form, like Jupiter's satellites, one system revolving about a common centre. But it is improbable that the Sun and Moon should both revolve about the Earth, for their distances from it and their periods are *not* connected by this relation.

(v.) The changes in the relative positions of two stars during the year in consequence of annual parallax can only be accounted for on the hypothesis either of the Earth's orbital motion, or of a highly improbable rigid connection between all the nearer stars and the Sun, compelling them all to execute an annual orbit of the same size and position.

(vi.) The aberration of light affords the most convincing proof of all. In particular, the relation between the coefficient of aberration and the retardation of the eclipses of Jupiter's satellites has been fully verified by actual observations, and affords incontestible evidence that the phenomenon is actually due to the finite velocity of light, as explained in Chapter XI. And the alternative hypothesis which would account for annual parallax would *not* give rise to aberration, but would produce an entirely different phenomenon. Hence the evidence derived from the aberration of light, unlike the previous evidence, furnishes a conclusive proof, and not merely an argument, in favour of the Earth's orbital motion.

406.—NEWTON'S THEORETICAL DEDUCTIONS FROM KEPLER'S LAWS.

Kepler's Three Laws of planetary motion naturally suggest the following questions :—

(1) What makes the planets move in ellipses ?

(2) Why does the radius vector from the Sun to any planet trace out equal areas in equal times ?

(3) Why are the squares of the periodic times proportiona: to the cubes of the mean distances from the Sun ?

These questions were first answered by Newton about 1687, or nearly sixty years after the death of Kepler. The theoretical interpretation of the Second Law necessarily precedes that of the first; accordingly we now repeat the laws in their new order, together with Newton's interpretations of them.

Kepler's Second Law.—*The radius vector joining each planet to the Sun moves in a plane describing equal areas in equal times.*

NEWTON'S DEDUCTION.— **The force under which a planet describes its orbit always acts along the radius vector in the direction of the Sun's centre.**

Kepler's First Law.—*The planets move in ellipses, having the Sun in one focus.*

NEWTON'S DEDUCTION.—**The force on any planet varies inversely as the square of its distance from the Sun.**

Kepler's Third Law.—*The squares of the periodic times of the several planets are proportional to the cubes of their mean distances from the Sun.*

NEWTON'S DEDUCTION.—**The forces on** *different* **planets vary directly as their masses, and inversely as the squares of their distances from the Sun, or, in other words, the accelerations of** *different* **planets, due to the Sun's attraction, vary inversely as the squares of their distances from the Sun.**

If, as we have every reason for believing, the planets are material bodies, Newton's laws of motion show that they cannot move as they do unless they are acted on by some force, otherwise they would either be at rest or move uniformly in a straight line. Kepler's Second Law then enables us to determine the direction of this force, his First Law enables us to compare the force at different parts of the same orbit, and his Third Law enables us to compare the forces on different planets.

407. We have seen that the orbits of most of the planets are nearly circular, the eccentricities being small, except in the case of Mercury. Before proceeding to the general discussion of the dynamical interpretation of Kepler's Laws, it will be convenient therefore to consider the case where the orbits are supposed circular, having the Sun for centre. Kepler's Second Law shows that under such circumstances the planets will describe their orbits uniformly, and it hence follows that the acceleration of a planet has no component in the direction of motion, but is directed exactly towards the centre of the Sun. The law of force can now be deduced very simply, as follows :—

KEPLER'S THIRD LAW FOR CIRCULAR ORBITS.

408. **To compare the Sun's attractions on different Planets, assuming that the orbits are circular and that the squares of the periodic times are proportional to the cubes of the radii.**

Suppose a planet of mass M is moving with velocity v in a circle of radius r. Let T be the periodic time, P the force to the centre.

Since the normal acceleration in a circular orbit is v^2/r, therefore
$$P = \frac{Mv^2}{r}.$$

In the period T the planet describes the circumference $2\pi a$;
$$\therefore \quad vT = 2\pi r.$$
Substituting for v, we have
$$P = \frac{4\pi^2 Mr}{T^2} = \frac{M}{r^2} \times \frac{4\pi^2 r^3}{T^2}.$$

Let M' be the mass of another planet revolving in a circular orbit of radius r', T' its periodic time, P' the force of the Sun's attraction; then we have in like manner

$$P' = \frac{M'}{r'^2} \times \frac{4\pi^2 r'^3}{T'^2}.$$

By Kepler's Third Law,

$$\frac{r^3}{T^2} = \frac{r'^3}{T'^2};$$

$$\therefore \quad P : P' = \frac{M}{r^2} : \frac{M'}{r'^2}.$$

Therefore the forces on different planets vary directly as their masses and inversely as the squares of their distances from the Sun.

COROLLARY 1.—Let $P = CM/r^2$; then C is called the **absolute intensity** of the Sun's attraction, and we see that

The absolute intensity of the Sun's attraction is the same for all planets.

For

$$C = \frac{4\pi^2 r^3}{T^2} = \frac{4\pi^2 r'^3}{T'^3}.$$

The constant C evidently represents the force with which the Sun would attract a unit mass at unit distance, or the acceleration which the Sun would produce at unit distance.

COROLLARY 2.—If another body be revolving in an orbit of radius r' in a period T', under a *different* central force, which produces an acceleration C'/r^2 at distance r', we have

$$C' = \frac{4\pi^2 r'^3}{T'^2} \quad \text{and} \quad C = \frac{4\pi^2 r^3}{T^2};$$

$$\therefore \quad C'T'^2 : CT^2 = r'^3 : r^3,$$

a formula which enables us to compare the absolute intensities of two *different* centres of force, which attract inversely as the squares of the distances, when the periodic times and distances of two bodies revolving about them are known.

409. To compare the velocities and angular velocities of two planets moving in circular orbits.—If v, v' are the velocities, n, n' the angular velocities (in radians per unit time), we have

$$n = 2\pi/T, \quad n' = 2\pi/T';$$
$$\therefore \quad n : n' = T^{-1} : T'^{-1} = r^{-\frac{3}{2}} : r'^{-\frac{3}{2}}.$$

Also

$$v = rn, \quad v' = r'n';$$
$$\therefore \quad v : v' = r^{-\frac{1}{2}} : r'^{-\frac{1}{2}}.$$

EXAMPLES.

1. If the Earth's period were doubled, to find what would be its new distance from the Sun.

If r, r' be the old and new distances, Kepler's Third Law gives

$$r'^3 : r^3 = 2^2 : 1^2;$$
$$\therefore \quad r' = r \times \sqrt[3]{4} = 92{,}000{,}000 \times 1{\cdot}587$$
$$= 146{,}000{,}000 \text{ miles.}$$

2. If the Earth's velocity were doubled, its orbit remaining circular, to find its new distance.

Here

$$r' : r = v^2 : v'^2 = 1 : 4;$$
$$\therefore \quad r' = \tfrac{1}{4}r = 23{,}000{,}000 \text{ miles.}$$

3. If the Earth's angular velocity were doubled, to find its new distance.

The new angular velocity being double the old, the new period would be half the old, and therefore

$$r'^3 : r^3 = (\tfrac{1}{2})^2 : 1^2;$$
$$\therefore \quad r' = r \times \sqrt[3]{\tfrac{1}{4}} = r/\sqrt[3]{4} = 92{,}000{,}000 + 1{\cdot}587$$
$$= 92{,}000{,}000 \times {\cdot}63 = 58{,}000{,}000 \text{ miles.}$$

4. To find what would be the coefficient of aberration to an observer situated on Venus.

The coefficient of aberration (in circular measure) is the ratio of the observer's velocity to the velocity of light; hence, if k, k' are the coefficients on the Earth and Venus,

$$\frac{k'}{k} = \frac{v'}{v} = \frac{r'^{-\frac{1}{2}}}{r^{-\frac{1}{2}}} = \sqrt{\frac{r}{r'}} = \sqrt{\frac{100}{72}};$$
$$\therefore \quad k' = 20{\cdot}493'' \times \sqrt{(1{\cdot}3\dot{8})} = 20{\cdot}493'' \times 1{\cdot}1785$$
$$= 24{\cdot}151''.$$

THE LAW OF UNIVERSAL GRAVITATION. 343

We shall now prove Newton's deductions from Kepler's Laws, for the general case of elliptic orbits, employing, however, different and simpler proofs to those used by Newton.

410. Areal Velocity. — Definition.—If a point P is moving in any path MPK about a centre S, the rate of increase of the area of the sector MSP, bounded by the fixed line SM and the radius vector SP, is called the **areal velocity** of P about the point S.

If the radius vector SP describes equal areas in equal times, in accordance with Kepler's Second Law, the areal velocity of P about S is of course constant, and is then measured by the area of the sector described in a unit of time.

If the rate of description of areas is not constant, we must, in measuring the areal velocity *at* any point, pursue a similar course to that adopted in measuring variable velocity *at* any instant, as follows:—

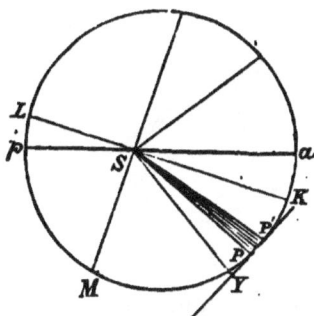

Fig. 136.

If the radius vector describes the sector PSP' in the interval of time t, then the **average areal velocity** over the arc PP' is measured by the ratio

$$\frac{\text{area } PSP'}{\text{time } t}.$$

(Thus the average areal velocity is the areal velocity with which a radius vector, sweeping out equal areas in equal times, would describe the sector PSP' in the same time t.)

The areal velocity *at* a point P is the limiting value of the average areal velocity over the arc PP' when this arc is *infinitesimally small*.

411. Relation between the Areal Velocity and the Actual (linear) Velocity.—Let PP' be the small arc described by a body in any small interval of time t. Let v be the actual or linear velocity of the body, h its areal velocity. Since the arc PP' is supposed small, we have

$$PP' = vt,$$
$$\text{area } PSP' = ht.$$

Draw SY perpendicular on the chord PP' produced. Then

$$\Delta PSP' = \tfrac{1}{2}(\text{base}) \times (\text{altitude})$$
$$= \tfrac{1}{2}PP' \times SY;$$
$$\therefore \quad ht = \tfrac{1}{2}vt \times SY,$$

or $\qquad\qquad h = \tfrac{1}{2}v \cdot SY.$

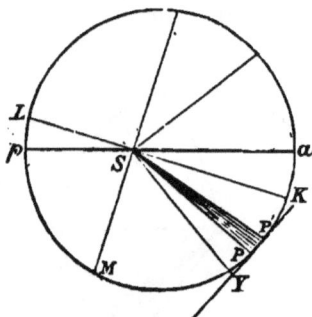

Fɪɢ. 137.

But when the arc PP' is infinitesimally small, PY is the tangent at P, and SY is therefore the perpendicular from S on the tangent at P. If this perpendicular be denoted by p, we have therefore

$$h = \tfrac{1}{2}vp \dots\dots\dots\dots\dots\dots\dots\dots\text{(i.)},$$

or (**areal vel. about S**)

$$= \tfrac{1}{2}(\text{velocity}) \times (\text{perp. from } S \text{ on tangent}).$$

Corollary.—**For planets moving in circular orbits of radii r, r',** $\quad h = \tfrac{1}{2}vr$, and $h' = \tfrac{1}{2}v'r'$.

But $\qquad\qquad v : v' = r^{-\frac{1}{2}} : r'^{-\frac{1}{2}};$

$\therefore \qquad\qquad h : h' = r^{\frac{1}{2}} : r'^{\frac{1}{2}};$

hence the areal velocity of a planet moving in a circular orbit is proportional to the square root of the radius.

412. PROPOSITION I. If a particle moves in such a manner that its areal velocity about a fixed point is constant, to prove that the resultant force on the particle is always directed towards the fixed point. [Newton's Deduction from Kepler's Second Law.]

Let a body be moving in the curve PQ in such a way that its areal velocity about S remains constant. Let v, v' be the velocities at P, Q, and let PY, QY', the corresponding directions of motion, intersect in R. Drop SY, SY' perpendicular on PY, QY'.

Since the areal velocities at P and Q are equal,

$$\therefore \ v \cdot SY = v' \cdot SY'.$$

But $$SY = SR \sin SRY,$$
$$SY' = SR \sin SRY'.$$

$$\therefore \ v \sin SRY = v' \sin SRY'.$$

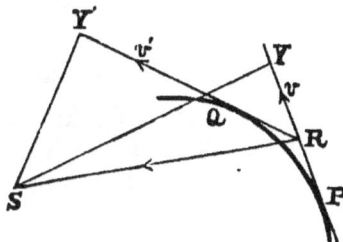

FIG. 138.

i.e., Component velocity at P perpendicular to SR
= component vel. at Q perp. to SR.

Therefore, as the particle moves from P to Q, its velocity perpendicular to RS is unaltered, and therefore the total change of velocity is parallel to RS.

This is true whether the arc PQ be large or small. But if the arc PQ be taken infinitesimally small, the average rate of change of velocity over PQ measures the acceleration at P, and R coincides with P.

Therefore the direction of the acceleration of the particle at any point of its path always passes through S, and therefore the force acting on the particle also always passes through S.

413. Conversely, *if the force on the particle always passes through S, the areal velocity about S remains constant.* For in passing from P to Q, the direction of motion is changed from PR to RQ, and the same change of velocity could therefore be produced by a suitable single blow or instantaneous impulse acting at R. And since the force at every point of PQ always passes through S, this equivalent impulse must evidently also pass through S; it must therefore act along RS. Hence the velocity perpendicular to RS is unaltered by the whole impulse, and is the same at P as at Q; therefore

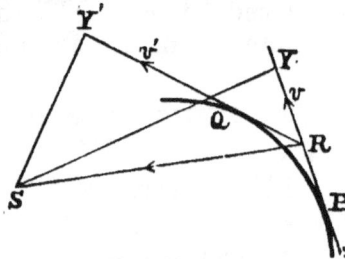

Fig. 139.

$$v \sin SRY = v' \sin SRY';$$

therefore

$$v \cdot SY = v' \cdot SY';$$

therefore

areal vel. at P = areal vel. at Q.

414. Proposition II. A particle describes an ellipse under a force directed towards the focus; to show that the force varies inversely as the square of the distance.

[Newton's Deduction from Kepler's First Law.]
If h is the constant areal velocity, we have, by (i.),

$$v = 2h/p.$$

We will now express the kinetic energy of the particle in terms of r, its distance from the focus. Let its mass be M.

In the Appendix (Ellipse 11) it is proved that for the ellipse whose major and minor axes are $2a$, $2b$,

$$\frac{1}{p^2} \equiv \frac{1}{ST^2} = \frac{a}{b^2}\left(\frac{2}{r} - \frac{1}{a}\right).$$

Therefore

$$v^2 = \frac{4h^2}{p^2} = \frac{4h^2 a}{b^2}\left(\frac{2}{r} - \frac{1}{a}\right),$$

and kinetic energy at distance r

$$= \tfrac{1}{2}Mv^2 = \tfrac{1}{2}M\frac{4h^2}{p^2} = M\frac{2h^2 a}{b^2}\left(\frac{2}{r} - \frac{1}{a}\right)\ldots\ldots\ldots\text{(ii.).}$$

If v' is the velocity at distance r', we have, similarly,

$$\tfrac{1}{2}Mv'^2 = \tfrac{1}{2}M\frac{4h^2}{p'^2} = M\frac{2h^2a}{b^2}\left(\frac{2}{r'} - \frac{1}{a}\right),$$

and therefore, for the increase of kinetic energy,

$$\tfrac{1}{2}Mv'^2 - \tfrac{1}{2}Mv^2 = \frac{4Mh^2a}{b^2}\left(\frac{1}{r'} - \frac{1}{r}\right) \quad \dots\dots\dots\text{(iii.)}.$$

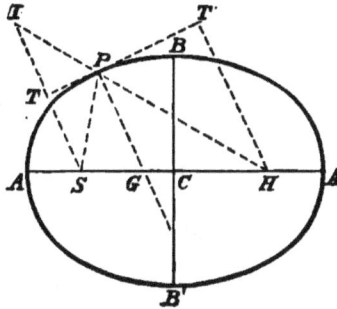

Fig. 140.

Now the increase of kinetic energy is equal to the work done by the impressed force in bringing the particle from distance r to distance r'. The resolved part of the displacement in the direction of the force is $r - r'$. Hence if P' denote the *average value* of the force between the distances r and r', we have

$$\text{Work done} = P'(r - r') = \tfrac{1}{2}Mv'^2 - \tfrac{1}{2}Mv^2 = \frac{4Mh^2a}{b^2}\left(\frac{1}{r'} - \frac{1}{r}\right)$$

$$= \frac{4Mh^2a}{b^2}\frac{r - r'}{rr'} \quad \dots\dots\dots\dots\dots\text{(iv.)};$$

$$\therefore \quad P' = \frac{4Mh^2a}{b^2rr'} \quad \dots\dots\dots\dots\dots\dots\text{(v.)}.$$

Put $r' = r$; then the average force P' becomes the actual force P at distance r. Therefore

$$P\,(\text{Force at distance } r) = \frac{4Mh^2a}{b^2r^2}\dots\dots\dots\text{(vi.)}.$$

This is proportional to $1/r^2$.

Therefore the force varies inversely as the square of the distance.

415. Proposition III. Having given that the squares of the periodic times of the planets are proportional to the cubes of the semi-axes major of their orbits, to compare the forces acting on different planets. [Newton's Deduction from Kepler's Third Law.]

Let T be the periodic time of any planet; then, by hypothesis, the ratio

$$\frac{T^2}{a^3}.$$

is the same for all planets.

In the last proposition (vi.) we showed that the force at distance r is given by

$$P = \frac{4Mh^2a}{b^2r^2}.$$

Let this be put $= MC/r^2$, where C is some constant; then

$$C = \frac{4h^2a}{b^2} \quad \dots\dots\dots\dots\dots\dots\dots(\text{vii.}).$$

Now in the period T the radius vector sweeps out the area of the ellipse, and this area is πab (Appendix, Ellipse 13). Hence, since the areal velocity is h, we have

$$hT = \pi ab.$$

Substituting the value of h from this equation in (vii.), we have

$$C = \frac{4\pi^2a^3b^2}{T^2b^2} = \frac{4\pi^2a^3}{T^2} \quad \dots\dots\dots\dots\dots(\text{viii.}).$$

But a^3/T^2 is the same for all the planets; therefore C is constant for all the planets, and since the force

$$P = \frac{MC}{r^2} \quad \dots\dots\dots\dots\dots\dots\dots(\text{ix.}),$$

it follows that

The forces on different planets are proportional to their masses divided by the squares of their distances from the Sun.

Or, as in § 408, Cor. 1,

The absolute intensity of the Sun's attraction (C) is the same for all the planets.

Corollary.—Let accented letters refer to the orbit of another particle revolving round a *different* centre of force of intensity C'. Then, by (viii.),

$$T^2C : T'^2C' = a^3 : a'^3.$$

416. Other Consequences of Kepler's Laws.

(i.) In § 150 we showed that, in consequence of Kepler's Second Law being satisfied by the Earth in its annual orbit, the Sun's apparent motion in longitude is inversely proportional to the square of the Earth's distance from it. Since the areal velocity of any planet about the Sun always remains constant, it may be shown in like manner that its angular velocity is inversely proportional to the square of its distance from the Sun.

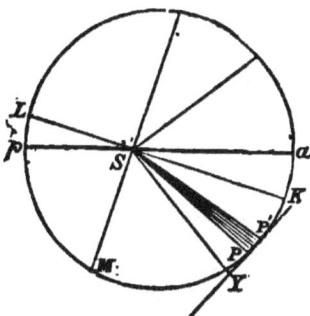

FIG. 141.

For, if the planet's radius vector revolves from SP to SP in the time t, and if the arc PP' is very small, we have

area $SPP' = \frac{1}{2}SP^2 \times \angle PSP'$ (§ 150),

the angle being measured in radians;

$$\therefore \quad \frac{\text{area } SPP'}{t} = \frac{1}{2}SP^2 \times \frac{\angle PSP'}{t},$$

i.e., (areal velocity of P) = $\frac{1}{2}SP^2 \times$ (angular velocity of P), provided that the angular velocity is measured in *radians* per unit of time.

If n denote the angular velocity, h the areal velocity, and r the distance SP, we have therefore

$$h = \frac{1}{2}r^2 n.$$

And since h is constant, n is inversely proportional to r^2.

* (ii.) If the mass of the planet is M, its momentum is Mv along PY, and the moment of this momentum about S is

$$= Mv \times SY = Mvp = 2hM. \quad (\text{§ 411.})$$

This is the planet's *angular momentum*, and is constant, since h is constant.

*417. Having given, in magnitude and direction, the velocity of a planet at any point of its orbit, to construct the ellipse described under the Sun's attraction.

Let the attraction at distance r be C/r^2 per unit mass, where C is given. Suppose that at the point P of the orbit the planet is moving with velocity v in the direction PT. We have

$$v \times ST = 2h, \text{ which determines } h.$$

Also, from (vii.),

$$C = 4h^2 a/b^2.$$

Substituting in (ii.),

$$v^2 = C \left(\frac{2}{r} - \frac{1}{a} \right) \dots \text{(x.)}.$$

Hence, by considering the planet at P, we have

$$v^2 = C \left(\frac{2}{SP} - \frac{1}{a} \right) \dots \text{(x.a)}.$$

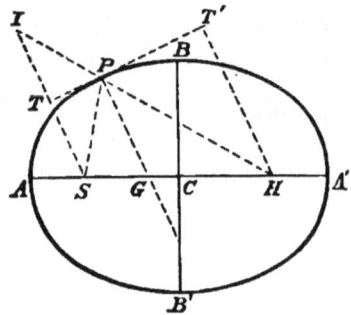

FIG. 142.

Now v, C, and SP are known; hence the last equation determines the semi-axis major a. If $r = SP$, we have

$$2a = \frac{2Cr}{2C - rv^2}.$$

Let H be the other focus of the ellipse. Then it is known (Ellipse 8) that HP, SP make equal angles with PT. Also $SP + HP = 2a$. Hence, we can construct the position of H by making $\angle TPI = \angle TPS$, and producing IP to a point H such that

$$PH = 2a - SP.$$

The ellipse can now be constructed as in Appendix (Ellipse 2).

COROLLARY 1.—Since $SP + HP = 2a$, equation (x.) gives

$$v^2 = \frac{C \cdot HP}{SP \cdot a}.$$

COROLLARY 2.—Substituting for h in terms of C, we see from equation (iv.) that the work done when the body moves from distance r to distance r' is

$$= MC \left(\frac{1}{r'} - \frac{1}{r} \right). \text{ *}$$

---

* This result is also proved independently in many treatises on dynamics, but a fuller investigation would be out of place here.

Hence the work done by a mass M in falling from distance $2a$ to distance r is

$$= MC\left(\frac{1}{r} - \frac{1}{2a}\right) = \tfrac{1}{2}Mv^2 \dots \dots \dots \dots \dots \text{ by } (x).$$

= kinetic energy of the planet when at distance r.

Therefore, if a circle be described about the centre of force S, with radius equal to the major axis $2a$, the velocity at any point of the orbit is that which the planet would acquire by falling freely from the circle to that point under the action of the attracting force.

COROLLARY 3.—If the planet be revolving in a circle, $r = a$, and therefore $v^2 = C/r = C/a$, as in § 408.

COROLLARY 4.—If $v^2 = 2C/r$, (x.) gives $1/a = 0$; ∴ $a = \infty$.

Hence the velocity is that acquired by falling from an infinite distance. In this case, the orbit is not an ellipse, but a *parabola*, a conic section satisfying the "focus and directrix" definition of Appendix (1), but having its eccentricity equal to unity.

If $v^2 > 2C/r$, the velocity is greater than that due to falling from infinity, a comes out negative, and the orbit is a *hyperbola*, a conic section satisfying the focus and directrix definition, but having its eccentricity e greater than unity.

A few *comets* have been observed to describe parabolas and hyperbolas about the Sun. In such a case the motion is *not periodic;* the comet gradually moves away to an infinite distance, and is lost for ever, unless the attraction of some other heavenly body should happen to divert its course, and send it back into the solar system.

EXAMPLE.—To find how long the Earth would take to fall into the Sun if its velocity were suddenly destroyed.

If the Earth's velocity were *very nearly*, but not quite destroyed, it would describe a very narrow ellipse, very nearly coinciding with the straight line joining the point of projection to the Sun. The major axis of this ellipse would be very nearly equal to the Earth's initial distance from the Sun, and therefore the Earth would have very nearly gone half round the narrow ellipse when it would collide with the surface of the Sun.

Hence, if r denote the Earth's distance from the Sun, the semi-major axis of the narrow ellipse is $\tfrac{1}{2}r$, and the periodic time in this eilipse would be $(\tfrac{1}{2})^{\frac{3}{2}}$ years. The Earth would therefore collide with the Sun in

$$\tfrac{1}{2} \times (\tfrac{1}{2})^{\frac{3}{2}} \text{ years} = \frac{1}{4\sqrt{2}} \text{ years} = \frac{\sqrt{2}}{8} \text{ years}$$

$$= \frac{365}{8} \times 1{\cdot}4142 \text{ days} = 64\tfrac{1}{2} \text{ days nearly.}$$

SECTION II.—*Newton's Law of Gravitation—Comparison of the Masses of the Sun and Planets.*

418. In the last section we showed that the Sun attracts any planet of mass M at distance r with a force CM/r^2, where C is a constant. If we assume the truth of Newton's Third Law of Motion (*i.e.*, that action and reaction are equal and opposite), the planet must also attract the Sun with an equal and opposite force CM/r^2. Since in the former case the force is proportional to the mass of the *attracted* body, and in the latter to the mass of the *attracting* body, it is reasonable to suppose that the attraction between two bodies is proportional to the mass of each.

Moreover, the motions of the various satellites, such as the Moon, confirm the theory that they revolve in their orbits under the attraction of their respective primary planets. From evidence of this character Newton, after many years of careful investigation, enunciated his **Law of Universal Gravitation**, which he stated thus:—

Every particle in the universe attracts every other particle with a force proportional to the quantities of matter in each, and inversely proportional to the square of the distance between them.

By *quantity of matter* is, of course, meant *mass*, and the word *attracts* implies that the force between two particles acts in the straight line joining them and tends to bring them together.

If M, M' be the masses of two particles, and r the distance between them, the law asserts that either particle is acted on by a force, directed towards the other, of magnitude

$$k\,\frac{MM'}{r^2},$$

where k has the same value for all bodies in the universe. The constant k is called the *constant of gravitation*.

*419. Astronomical Unit of Mass.—Taking any fundamental units of length and time, it is possible to choose the unit of mass such that $k = 1$. This unit is called the *astronomical unit of mass*. Hence, if M, M' are expressed in astronomical units, the force between the particles is *equal to* MM'/r^2. It is, however, usually more convenient to keep the unit of mass arbitrary, and to retain the constant k.

420. Remarks on the Law of Gravitation.—Newton's Law states that not only do the Sun, the planets and their satellites, and the stars, mutually attract one another, but every pound of matter on one celestial body attracts every other pound of matter, on either the same or another body. But it is well-known that two spheres attract one another in just the same way as if the whole of the mass of either were concentrated at its centre, provided that the spheres are either homogeneous or made up of concentric spherical layers, each of uniform density. Since the Sun and planets are very nearly spherical, and their dimensions are very small compared with their distances, we see that their attractions may be very approximately found by regarding them as mere particles, instead of taking separate account of the individual particles forming them.

Moreover, every planet is attracted by every other planet, as well as by the Sun. But the mass of the Sun, and consequently its attraction, is so much greater than that of any other member of the solar system, that the planetary motions are only very slightly influenced by the mutual attractions. Kepler's Laws, therefore, still hold *approximately*, but the orbits are subject to small and slow changes or perturbations.

The Moon, on the other hand, is far nearer to the Earth than to the Sun; hence the Moon's orbital motion is mainly due to the Earth's attraction. The chief effect of the Sun's attraction on the Earth and Moon is to cause them together to describe the annual orbit; but it also produces perturbations or disturbances in the Moon's relative orbit (§ 272) with which we are not here concerned.

The fixed stars also attract one another and attract the solar system, which in its turn attracts the stars. The proper motions of stars are probably due to this cause; but when we consider the vast distances of the stars, and remember that the attraction varies inversely as the square of the distance, it is evident that the relative accelerations are mostly too feeble to have produced any sensible changes of motion within historic times, and that countless ages must elapse before such changes can be discerned.

421. Correction of Kepler's Third Law.—From the fact that a planet attracts the Sun with a force equal to that with the Sun attracts the planets, it may be shown that Kepler's Third Law cannot be *strictly* true, as a consequence of the law of gravitation. Not only will the planet move under the Sun's attraction, but the Sun will also move under the planet's attraction. But since the forces on the two bodies are equal, while the mass of the Sun is very great compared with the mass of any planet, it follows that the *acceleration* of the Sun is very small compared with that of the planet, and hence the Sun remains very nearly at rest.

We may, however, obtain a modification of Kepler's Third Law, in which the planet's reciprocal attraction is allowed for as follows :—

Let S, M be the masses of the Sun and planet; then the attraction between them is

$$k\,\frac{SM}{r^2}.$$

This attraction, acting on the mass M of the planet, produces an acceleration of the planet towards the Sun equal to

$$k\,\frac{S}{r^2}.$$

The corresponding attraction on the mass S of the Sun produces an acceleration, in the reverse direction, of

$$k\,\frac{M}{r^2}.$$

Hence the whole acceleration of the planet *relative* to the Sun is

$$k\,\frac{S+M}{r^2},$$

instead of kS/r^2, as it would be if the Sun were at rest. Hence the *absolute intensity* of the planet's acceleration towards the Sun is $k\,(S+M)$, and this depends on the values of both M and S. Let now T be the periodic time, r the planet's mean distance from the Sun, or the semi-axis major of the relative orbit ; then, by § 408 (for a circular orbit), or § 415 (for an elliptical orbit),

$$k\,(S+M)\,T^2 = 4\pi^2 r^3$$

If M' be the mass of another planet, we have in like manner for its orbit $k(S+M') T'^2 = 4\pi^2 r'^3$.

Therefore $T^2(S+M) : T'^2(S+M') = r^3 : r'^3$, the correct relation between the periods and mean distances.

It is known that different planets have different masses. Hence, the fact that Kepler's Third Law is approximately true shows that the masses of the planets are small compared with that of the Sun.

422. Motion relative to Centre of Mass.—The mutual attractions of the Sun and planet have no influence on the position of the centre of mass (commonly called the " centre of gravity ") of the solar system ; hence, in considering the relative motions, that point may be treated as fixed. It is known from general dynamical principles that when a system of bodies are under the influence of their mutual reactions or attractions alone, the centre of mass of the whole system is not accelerated. But it may be interesting to prove independently that when two bodies, such as the Sun and a planet, attract one another, they both revolve about their centre of mass.

Let us suppose (to take a simple case) the relative orbit circular and of radius $(SP =)$ r, the angular velocity being n. Then, if G be the point about which the planet (P) and Sun (S) revolve, individually, we have

$$n^2 \times GP = \text{accel. of planet} = kS/r^2 ;$$
$$n^2 \times GS = \text{accel. of Sun} = kM/r^2.$$

Hence $M \times GP = S \times GS ;$ which relation shows that G is the common centre of mass, as was to be proved.

In the case of three or more bodies, such as the Sun and planets, the centre of mass is still the common centre about which they revolve, but the corresponding investigation is more difficult, owing to the effect of the mutual attractions of the planets in producing perturbations.

It may be mentioned that the mass of the Sun is so large, compared with those of the planets, that, although the further planets are so very distant, the centre of mass of the whole solar system always lies very near the Sun.

423. Verification of the Theory of Gravitation for the Earth and Moon.—Before considering the motions of the planets about the Sun, Newton investigated the orbital motion of the Moon about the Earth, with the view of discovering whether the Earth's attractive force, which retains the Moon in its orbit, is the same force as that which produces the phenomenon of gravity at the Earth's surface.

If we assume that the force varies inversely at the square of the distance, and that the Moon's distance is 60 times the Earth's radius, the acceleration of gravity at the Moon should be $(\frac{1}{60})^2 g$, where g is the acceleration of gravity on the Earth's surface.

But the acceleration g = 32·2 feet per sec. per sec. ;
∴ accel. at Moon's distance = 32·2/3600 feet per sec. per sec.
= 32·2 feet per min. per min.

From the length of the lunar month and the Moon's distance in miles, Newton calculated what must be the normal acceleration of the Moon in its orbit. At the time of his first investigation (1666) the Earth's radius and the Moon's distance were but imperfectly known, and the Moon's normal acceleration, as thus computed, came out only about 27 feet per minute per minute. Some fifteen years later, the Earth's radius, and consequently the Moon's distance, had been measured with much greater accuracy, and, working with the new values, Newton found that the Moon's normal acceleration to the Earth agreed with that given by his theory.

Taking the lunar sidereal month as 27·3 days, the Earth's radius as 3960 miles, and the radius of the Moon's orbit as 60 times the Earth's radius, the angular velocity (n) of the Moon, in radians, per minute is

$$\frac{2\pi}{27\cdot3 \times 24 \times 60}.$$

The Moon's distance in feet (d) = 3960 × 60 × 5280.

Hence the Moon's normal acceleration (n^2d) in feet per minute per minute

$$= \frac{3960 \times 60 \times 5280 \times 4\pi^2}{(27\cdot3)^2 \times 24^2 \times 60^2} = \frac{2 \times 110^2 \times \pi^2}{(27\cdot3)^2 \times 10}$$

$$= 32 \text{ approximately,}$$

thus agreeing with that given by the law of gravitation.

EXAMPLE.—Having given that a body at the Earth's equator loses $1/289$ of its weight in consequence of centrifugal force,

(i.) To calculate the period in which a projectile could revolve in a circular orbit, close to, but without touching the Earth, and

(ii.) To deduce the Moon's distance.

(i.) The centrifugal force on the body would have to be equal to its weight, and would therefore have to be 289 times as great as that at the Earth's equator.

Hence the projectile would have to move $\sqrt{289}$, or 17 times as fast as a point on the Earth's equator, and would therefore have to perform 17 revolutions per day.[*]

Therefore the period of revolution $= \frac{1}{17}$ of a day.

(ii.) Assuming the law of gravitation, the periodic times and distances of the projectile and Moon must be connected by Kepler's Third Law. Hence, taking the Moon's sidereal period as $27\frac{1}{3}$ days, we have, if $a = $ Earth's rad., $d = $ Moon's dist.,

$$d^3 : a^3 = (27\tfrac{1}{3})^2 : (\tfrac{1}{17})^2 ;$$

$$\therefore d^3 = a^3 \times (17 \times 27\tfrac{1}{3})^2 = a^3 \{ \tfrac{1394}{3} \}^2 = a^3 . 215915\dot{\cdot}1 ;$$

$$\therefore d = a \times \sqrt[3]{215915\dot{\cdot}1} = 59\dot{\cdot}99a ;$$

$\therefore$ distance of Moon $= 60 \times$ Earth's radius almost exactly.

424. Effect of Moon's Attraction.—Moon's Mass.—

If we take account of the Moon's attraction on the Earth we must introduce a correction analogous to that made in Kepler's Third Law (§ 421). If M, m are the masses of the Earth and Moon, the whole relative acceleration is $k(M+m)/d^2$, instead of kM/d^2. But, if g_0 is the acceleration of gravity on the Earth's surface, $g_0 = kM/a^2$;

$$\therefore k = g_0 a^2/M,$$

and, if T is the length of the sidereal month, then, by § 421,

$$4\pi^2 d^3 = k(M+m)T^2 = g_0 a^2 \frac{M+m}{M} T^2.$$

$$\therefore 1 + \frac{m}{M} = \frac{4\pi^2 d^3}{g_0 a^2 T^2}.$$

This formula might be used (and has been used by Airy) to find m/M, the ratio of the Moon's to the Earth's mass, in terms of the observed values of a, d, g_0, T. It is not, however, a very accurate method, owing to the smallness of m/M.

---

[*] *Relative to the Earth* it would perform 16 or 18 revolutions per day, according to whether it was revolving in the same or the opposite direction to the Earth.

425. To find the ratio of the Sun's Mass to that of the Earth.

Let S, M, m be the masses of the Sun, Earth, and Moon, d, r the distances of the Moon and Sun from the Earth, T, Y the lengths of the sidereal lunar month and year respectively. Then, if k be the gravitation constant, the Earth's attraction on the Moon is $= kMm/d^2$, and its intensity is kM.

The Sun's attraction on the Earth is $= kSM/r^2$, and its intensity is kS.

Therefore, by § 415, Corollary,

$$kM \cdot T^2 = 4\pi^2 d^3, \quad kS \cdot Y^2 = 4\pi^2 r^3;$$

$$\therefore \quad S : M = \frac{r^3}{Y^2} : \frac{d^3}{T^2}, \quad \text{or} \quad \frac{S}{M} = \frac{r^3}{d^3}\frac{T^2}{Y^2};$$

whence the ratio of the Sun's to the Earth's mass may be found.

If we take account of the attraction of the smaller body on the larger, the whole acceleration of the Earth, relative to the Sun, is $k(S + M + m)/r^2$ (since the Sun is attracted by the Moon as well as the Earth), and that of the Moon, relative to the Earth, is $k(M+m)/d^2$. Hence the corrected or more exact formula is

$$S + M + m : M + m = \frac{r^3}{Y^2} : \frac{d^3}{T^2}.$$

Since the Moon's mass is about $\frac{1}{81}$ of that of the Earth, the first or approximate formula can only be used if the calculations are not carried beyond two significant figures.

In this manner it is found that the Sun's mass is about **331,100** times that of the Earth.

CENTER**EXAMPLES.**

1. To compare, roughly, the masses of the Earth and Sun, taking the Sun's distance to be 390 times the Moon's, and the number of sidereal months in the year to be 13.

We have
$$S : M = \frac{390^3}{13^2} : 1^2;$$

$$\therefore \quad \frac{\text{mass of Sun}}{\text{mass of Earth}} = \frac{390^3}{13^2} = 30^2 \times 390 = 351,000.$$

To the degree of accuracy possible by this method, the Sun's mass is therefore 350,000 times that of the Earth.

2. To find the ratio of the masses, taking the Moon's mass as $\frac{1}{81}$ of the Earth's, and the number of sidereal months in the year as $13\frac{1}{3}$.

$$\frac{S+M+m}{M+m} = \frac{390^3}{(13\frac{1}{3})^2} = \frac{390^3 \times 3^2}{40^2} = \frac{5338710}{16} = 333669;$$

$$\therefore \quad S = 333668\,(M+m) = 333668\,(1 + \tfrac{1}{81})\,M = 337{,}787\,M.$$

426. To determine the mass of a planet which has one or more satellites.

The method of the last paragraph is obviously applicable to the case of any planet which has a satellite. We require to know the mean distance and the periodic time of the satellite. The former may be easily found by observing the maximum angular distance of the satellite from its primary, the distance of the planet from the Earth at the time of observation having been previously computed. The periodic time of the satellite may also be easily observed.

Let M', m' be the masses of the planet and satellite, d' their distance apart, r' their distance from the Sun, T' the period of revolution of the satellite, Y' the planet's period of revolution round the Sun. Using unaccented letters to represent the corresponding quantities for the Earth and Moon we have, roughly,

$$\frac{4\pi^2}{k} = \frac{M'T'^2}{d'^3} = \frac{SY'^2}{r'^3} = \frac{SY^2}{r^3} = \frac{MT^3}{d^3},$$

or, more accurately,

$$\frac{4\pi^2}{k} = \frac{(M'+m')\,T'^2}{d'^3} = \frac{(S+M'+m')\,Y'^2}{r'^3}$$

$$= \frac{(S+M+m)\,Y^2}{r^3} = \frac{(M+m)\,T^2}{d^3};$$

whence the mass of the planet, or, more correctly, the sum of the masses of the planet and satellite, may be determined in terms of the mass of the Sun, or the sum of the masses of the. Earth and Moon. We do not require to know the periodic time and mean distance of the planet from the Sun, since the above expressions enable us to express the required mass, $M'+m'$, in terms of the year and mean distance of the Earth, or in terms of the lunar month and the mean distance of the Moon.

EXAMPLE.—To find the mass of Uranus in terms of that of the Sun, having given that its satellite *Titania* revolves in a period of 8 days 17 hours at a distance from the planet = ·003 times the distance of the Earth from the Sun.

Let M be the mass of Uranus, then we have

$$M : S = \frac{d^3}{T^2} : \frac{r^3}{Y^2},$$

and, by Kepler's Third Law, r^3/Y^2 is the same for Uranus as for the Earth. Hence

$$M : S = \frac{(\cdot003)^3}{(8d.\ 17h.)^2} : \frac{1^3}{(365d.\ 6h.)^2};$$

$$\therefore \quad \frac{M}{S} = \left(\frac{3}{1000}\right)^3 \times \left(\frac{365d.\ 6h.}{8d.\ 17h.}\right)^2$$

$$= \frac{27}{10^9} \times \left(\frac{8766}{209}\right)^2$$

$$= \frac{1}{21053} \text{ nearly.}$$

Thus, the mass of Uranus is to that of the Sun in the ratio of 1 to 21,053.

*427. The Masses of Mercury and Venus (which have no satellites) could *theoretically* be found by determining their mean distances from the Sun by direct observation, and comparing them with those calculated from their periodic times by Kepler's Third Law. For, if M' is the mass of such a planet, we have

$$\frac{(S + M')\ Y'^2}{r'^3} = \frac{(S + M + m)\ Y^2}{r^3}.$$

This enables us to find the sum of the masses of the Sun and planet, and, the Sun's mass being known, the planet's mass could be found.

This method is, however, worthless, because the masses of Mercury and Venus are only about $\frac{1}{5000000}$ and $\frac{1}{400000}$ of that of the Sun, and in order to calculate one significant figure of the fraction M'/S it would be necessary to know all the data correct to about seven significant figures, a degree of accuracy unattainable in practice.

For this reason it is necessary to calculate the masses of these planets by means of the perturbations they produce on one another and on the Earth; these perturbations will be discussed in the next chapter.

428. Centre of Mass of the Solar System.—When the masses of the various planets have been found in terms of the Sun's mass, the position of the centre of mass of the system can be found for any given configuration, and can thus be shown to always lie very near the Sun.

EXAMPLES.

1. To find the distance of the centre of mass of the Earth and Sun from the centre of the Sun.

Here the mass of the Sun is 331,100 times the Earth's mass, and the distance between their centres is about 92,000,000 miles. Hence, the centre of mass of the two is at a distance from the Sun's centre of about

$$\frac{92,000,000}{331,100+1} = 278 \text{ miles.}$$

2. To find the centre of mass of Uranus and the Sun, and to show that it lies within the Sun.

The distance of Uranus from the Sun is 19·2 times the Earth's distance, and its mass is 1/21053 of the Sun's. Hence the C.M. is at a distance from the Sun's centre of

$$\frac{92,000,000 \times 19·2}{21053+1} \text{ miles} = 83,900 \text{ miles.}$$

The Sun's semi-diameter is 433,200 miles; hence the centre of mass of the Sun and Uranus is at a distance from the Sun's centre of rather less than ⅕ the radius.

3. In the case of Jupiter, the mean distance is 5·2 times that of the Earth, and the mass is 1/1050 of that of the Sun; hence the C.M. is at a distance

$$\frac{5·2 \times 92,000,000}{1050+1} = 455,000 \text{ miles.}$$

This is just greater than the Sun's radius (433,200), showing that the centre of mass lies just *without* the Sun's surface.

Section III.—*The Earth's Mass and Density.*

429. The so-called " Weight of the Earth " really
means the Earth's **mass,** and the operation called "weighing
the Earth," in some of the older text-books, means finding the
mass of the Earth.   In the last section we explained how to
compare the masses of the Sun and certain planets with that
of the Earth, and in the next chapter we shall give methods
applicable to a planet having no satellites.   But before the
masses can be expressed in pounds or tons it is necessary to
determine the Earth's mass in these units.   The methods of
doing this all depend on comparing the Earth's attraction
with that of a body of known mass and distance ; and the only
difficulty lies in determining the latter attraction, since the
force between two bodies of ordinary dimensions is always
extremely small.   The following methods have been used.
The first two are by far the best.

(1) By the "Cavendish Experiment," or the balance.
(2) By observations of the influence of tides in estuaries.
(3) By the " Mountain " method.
(4) By pendulum experiments in mines.

430. The " Cavendish Experiment " owes its name to
its having been first used to determine the Earth's mass by
Cavendish, about the year 1798.   The essential principle of
the method consists in comparing the attractions of two heavy
balls of known size and weight with the Earth's attraction.
Since the attraction of a sphere at any point is proportional
directly to the mass of the sphere and inversely to the square
of the distance from its centre, it is evident that by comparing
the attractions of different spheres—such as the Earth and the
experimental ball of metal—we can find the ratio of their
masses.

The comparison is effected by means of a **torsion balance.**
Two equal small balls *A*, *B* are fixed to the ends of a light
beam suspended from its middle point *O* by means of a slender
vertical thread or "torsion fibre" (in his recent experiments,
Professor C. V. Boys has used a fine fibre of spun quartz), so
as to be capable of twisting about *O* in a horizontal plane
(the plane of the paper in Fig. 143).   Two heavy metal balls
C, *D*, are brought near the small balls *A*, *B* (as shown in the

figure), and their attraction causes the beam to turn about O, say from its original position of rest XX' to the position AB. As the beam turns the fibre twists; this twisting is resisted by the elasticity of the fibre, which produces a couple, proportional to the angle of twist XOA, tending to untwist it again. Let us call this couple $f \times \angle XOA$, where f is a constant depending on the fibre, called its "*torsional rigidity.*"

The beam AB assumes a position of equilibrium when the moments about O of the attractions of the large spheres C, D on the balls A, B, just balance the "untwisting couple" $f \times \angle XOA$. The angle XOA being measured, and the dimensions of the apparatus being supposed known, the attractions of the spheres can now be determined in terms of the torsional rigidity.

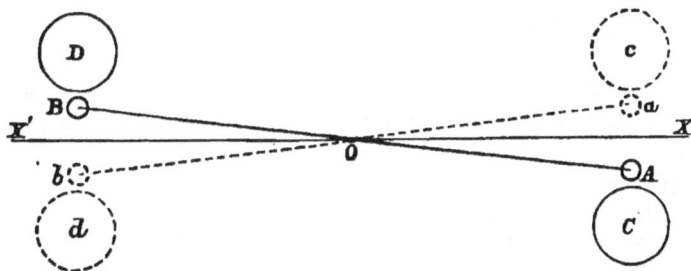

FIG. 143.

The value of f is found in terms of absolute units of couple by observing the time of a small oscillation of the beam when the balls A, B have been removed. [The beam will then swing backwards and forwards like the balance wheel of a chronometer (§ 204). The greater the torsional rigidity, the more frequently will it reverse the motion of the beam, and the more frequent will be the oscillations.*]

Hence finally the attractions between the known masses C, D and A, B are found in terms of known units of force, and by comparing these attractions with that of gravity the Earth's mass is found.

---

* The student who has read Rigid Dynamics should work out the formula.

In practice, instead of measuring the angle XOA, the masses C, D are subsequently placed on the reverse side of the beam, say with their centres at c, d, and they now deflect the beam in the reverse direction, say to ab. The angle measured is the whole angle aOA, and this angle is *twice* the angle XOA, if the positions CD and cd are symmetrically arranged with respect to the line XOX'.

In the earlier experiments the beam AB was six feet long, and the masses C, D were balls of lead a foot in diameter. Quite recently, however, Professor C V. Boys, by the use of a quartz fibre for the suspending thread, has performed the experiment on a much smaller scale, the whole apparatus being only a few inches in size and being highly sensitive. He uses cylinders instead of spheres for the attracting bodies, and this introduces extra complications in the calculations.

Although the above description shows the general principle of the method, many further precautions are required to ensure accuracy. A full description of these would be out of place here.

431. The common balance has also been used to determine the Earth's mass. In this case the differences of weight of a body are observed when a large attracting mass is placed successively above and below the scale-pan containing it.

EXAMPLE.—To find the Earth's mass in tons, having given that the attraction of a leaden ball, weighing 3 cwt., on a body placed at a distance of 6 inches from its centre is ·0000000432 of the weight of the body.

Let M be the mass of the Earth in tons.

The mass of the ball in tons is $= \frac{3}{20}$.

The Earth's radius in feet $= 3960 \times 5280 = 20,900,000$ roughly; and the distance of the body from the ball in feet $= \frac{1}{2}$.

Hence, since the attractions of the Earth and ball are proportional directly to the masses and inversely to the squares of the distances from their centres,

$$\therefore \quad ·0000000432 : 1 = \frac{\frac{3}{20}}{(\frac{1}{2})^2} : \frac{M}{(20,900,000)^2};$$

$$\therefore \quad M = \frac{(20,900,000)^2 \times \frac{3}{20}}{\times ·0000000432} = \frac{3 \times 209^2 \times 10^{10}}{5 \times 432 \times 10^{-10}}$$

$$= \frac{3}{5} \times \frac{209^2}{432} \times 10^{20} = \frac{43681 \times 3}{2160} \times 10^{20}$$

$$= 6067 \times 10^{18}.$$

Hence the mass of the Earth is (roughly) **6067 million billion tons.**

432. To determine the Earth's Mass by observa·tions of the Attraction of Tides in Estuaries.—A method which admits of very great accuracy is that in which the mass of the Earth is found by comparing it with that of the water brought by the tide into an estuary. Consider an observatory situated (like Edinburgh Observatory) due south of an arm of the sea, whose general direction is east and west. The direction of its zenith, as shown either by a plummet or by the normal to the surface of a bowl of mercury, is not the same at high tide as at low, because the additional mass of water at high tide produces an attraction which deflects the plummet and the nadir point northward, and hence displaces the zenith towards the south. Hence the latitude of the observatory is less at high tide than at low ; and the difference is a measurable quantity. The great advantage of this method is that the mass which deflects the plumb-line can be measured with great certainty ; for the density of the sea-water is exactly known (and, unlike that of the rocks in the next methods, is uniform throughout) and the shape and height of the layer of water brought in are known from the ordnance maps, and the tide measurements at the port.

*433. In the Pendulum Method the values of g, the acceleration of gravity, are compared by comparing the oscillations of two pendulums at the top and bottom of a deep mine. The difference of the two values is due to the attraction of that portion of the Earth which is above the bottom of the mine ; this exerts a downward pull on the upper pendulum, and an upward pull on the lower one. If the Earth were homogeneous throughout, the values of g at the top and bottom would be directly proportional to the corresponding distances from the Earth's centre. If this is not observed to be the case, the discrepancy enables us to find the ratio of the density of the Earth to that of the rocks in the neighbourhood of the mine. If the latter density is known, the Earth's density can be found, and knowing its volume, its mass can be computed. But this method is very liable to considerable errors, arising from imperfect knowledge of the density of the rocks overlying the mine.

*434. **In the Mountain Method** the Earth's attraction is com‑
pared with that of a mountain projecting above its surface.　Suppose
a mountain range, such as Schiehallien in Scotland, running due E.
and W.; then at a place at its foot on the S. side the attraction of
the mountain will pull the plummet of a plumb line towards the N.,
and at a place on the N. side the mountain will pull the plummet to
the S.　Hence the Z.D. of a star, as observed by means of zenith
sectors, will be different at the two sides, and from this difference
the ratio of the Earth's to the mountain's attraction may be found.

In order to deduce the Earth's density it is then necessary to
determine accurately the dimensions and density of the mountain.
This renders the method very inexact, for it is impossible to find
with certainty the density of the rocks throughout every part of the
mountain.

**435. Determination of Densities.—Gravity on the
Surface of the Sun and Planets.—**When the mass and
volume of a celestial body have been computed, its average
density can, of course, be readily found.　By dividing the
mass in pounds by the volume in cubic feet, we find the
average mass per cubic foot, and since we know that the
mass of a cubic foot of water is about $62\frac{1}{2}$ lbs., it is easy to
compare the average density with that of water.　The determination of densities is particularly interesting, on account of
the evidence it furnishes regarding the physical condition of
the members of the solar system.　The Earth's density is
about **5·58.**

From knowing the ratios of the mass and diameter of the
Sun or a planet to that of the Earth, we can compare the
intensity of its attraction at a point on its surface with the
intensity of gravity on the Earth.

It may be noticed that attraction of a sphere at its surface is proportional to the product of the density and the radius.

For the attraction is proportional to mass $\div$ (radius)^2, and the
mass is proportional to the density $\times$ (radius)^3; $\therefore$ the attraction
at the surface is proportional to the density $\times$ radius.

EXAMPLES.

1. To find the Earth's average density and mass, having given
that the attraction of a ball of lead a foot in diameter, on a particle
placed close to its surface, is less than the Earth's attraction in the
proportion of 1 : 20,500,000, and that the density of lead is 11·4 times
that of water.

Let D be the average density of the Earth. Then, since the radii of the Earth and the leaden ball are $\frac{1}{2}$ and 20,900,000 feet respectively, and the attractions at their surfaces are proportional to their densities multiplied by their radii,

$$\therefore \quad 1 : 20,500,000 = 11\cdot4 \times \tfrac{1}{2} : D \times 20,900,000 \, ;$$

$\therefore$ average density of Earth $D = 5\cdot7 \times \frac{205}{209} = 5\cdot6$.

Hence the average mass of a cubic foot of the material forming the Earth is $5\cdot6 \times 62\cdot5$ pounds. But the Earth is a sphere of volume

$$\tfrac{4}{3}\pi \, (20,900,000)^3 \text{ cubic feet.}$$

Hence the mass of the Earth, with these data,

$$= \tfrac{4}{3}\pi \times 209^3 \times 10^{15} \times 5\cdot6 \times 62\cdot5 \text{ pounds}$$
$$= 1338 \times 10^{22} \text{ pounds} = 597 \times 10^{19} \text{ tons.}$$

2. To calculate the mean density of the Sun from the following data:—

Mass of $\odot$ = 330,000 . (mass of $\oplus$) ;
Density of $\oplus$ = 5·58 ;
$\odot$'s parallax = 8·8''; $\quad$ $\odot$'s angular semi-diameter = 16'.

The radii of the Sun and Earth being in the ratio of the Sun's angular semi-diameter to its parallax (§ 258), we have

$$\frac{\odot\text{'s radius}}{\oplus\text{'s radius}} = \frac{16'}{8\cdot8''} = \frac{960}{8\cdot8} = 109\cdot1 \, ;$$

$\therefore$ volume of Sun $= (109\cdot1)^3$. (vol. of Earth)
$$= 1,298,000 . (\text{vol. of Earth}) \text{ roughly.}$$

But mass of Sun $= 330,000 . (\text{mass of Earth})$;

$$\therefore \quad \frac{\text{density of Sun}}{\text{density of Earth}} = \frac{330}{1298} = \frac{1}{3\cdot9} \text{ very nearly ;}$$

$$\therefore \quad \text{density of Sun} = 1\cdot4.$$

3. To find the number of poundals in the weight of a pound at the surface of Jupiter, taking the planet's radius as 43,200 miles and density $1\frac{1}{3}$ times that of water.

Taking the Earth's radius as 3960 miles and density as 5·58, we have

$$(\text{gravity at surface of Jupiter}) : (\text{gravity on Earth})$$
$$= 1\cdot33 \times 43,200 : 5\cdot58 \times 3960.$$

But at the Earth's surface the weight of a pound

$$= 32\cdot2 \text{ poundals} \, ;$$

therefore on the surface of Jupiter the weight of a pound

$$= 32\cdot2 \times \frac{1\cdot33 \times 43200}{5\cdot58 \times 3960} \text{ poundals}$$

$$= 83\cdot7 \text{ poundals.}$$

EXAMPLES.—XIII.

1. Taking Neptune's period as 80 years, and the Earth's velocity as $9\frac{1}{4}$ miles per second, find the orbital velocity of Neptune.

2. If we suppose the Moon to be 61 times as far from the Earth's centre as we are, find how far the Earth's attraction can pull the Moon from rest in a minute.

3. If the Earth possessed a satellite revolving at a distance of only 6,000 miles from the Earth's surface, what would be approximately its periodic time, assuming the Earth to be a sphere of 4,000 miles radius?

4. Assuming the distance between the Earth's centre and the Moon's to be 240,000 miles, and the period of the Moon's revolution 28 days, find how long the month would be if the distance were only 80,000 miles.

5. Calculate the mass of the Sun in terms of that of Mars, given that the Earth's mean distance and period are 92×10^6 miles and $365\frac{1}{4}$ days, and the mean distance and period of the outer satellite of Mars are 14,650 miles and 1d. 6h. 18m.

6. Show that the periodic time of an asteroid is $3\frac{1}{2}$ years, having given that its mean distance is 2·305 times that of the Earth.

7. Show that we could find the Sun's mass in terms of the Earth's, from exact observation of the periods and mean distances of the Earth and an asteroid, by the error produced in Kepler's Third Law in consequence of the Earth's mass.

8. Show that an increase of 10 per cent. in the Earth's distance from the Sun would increase the length of the year by 56·14 days.

9. The masses of the Earth and Jupiter are approximately $\frac{1}{300000}$ and $\frac{1}{1000}$ respectively of the Sun's mass, and their distances from the Sun are as 1 : 5. Show that Kepler's Laws would give the periodic time of Jupiter too great by more than 2 days.

10. Prove that the mass of the Sun is 2×10^{27} tons, given that the mean acceleration of gravity on the Earth's surface is 9·81 metres per second per second, the mean density of the Earth is 5·67, the Sun's mean distance $1·5 \times 10^8$ kilometres, a quadrant of the Earth's circumference 10,000 kilometres, and taking a metre cube of water to be a ton.

11. Having given that the constant of aberration for the Earth is 20·49″, and that the distance of Jupiter from the Sun is 5·2 times the distance of the Earth from the Sun, calculate the constant of aberration for Jupiter.

12. If the mass of Jupiter is $\frac{1}{1050}$ of the mass of the Sun, show that the change in the constant of aberration caused by taking into account the mass of Jupiter is 0·004″ nearly (see Question 11).

13. Find the centre of mass of Jupiter and the Sun. Hence find the centre of mass of Jupiter, the Sun, and Earth, (1) when Jupiter is in conjunction, (2) when in opposition. (Sun's mass = 1,048 times Jupiter's = 332,000 times Earth's. Jupiter's mean distance = 480,000,000 miles; Earth's = 93,000,000 miles.)

14. If the intensity of gravity at the Earth's surface be 32·185 feet per second per second, what will be its value when we ascend in a balloon to a height of 10,000 feet? (Take Earth's radius = 4,000 miles and neglect centrifugal force.) Would the intensity be the same on the top of a mountain 10,000 feet high? If not, why not?

15. Show how by comparing the number of oscillations of a pendulum at the top and bottom of a mountain of known density, the Earth's mass could be found.

16. How would the tides in the Thames affect the determination of meridian altitudes at Greenwich observatory theoretically?

17. If the mean diameter of Jupiter be 86,000 miles, and his mass 315 times that of the Earth, find the average density of Jupiter.

18. If the Sun's diameter be 109 times that of the Earth, his mass 330,000 times greater, and if an article weighing one pound on the Earth were removed to the Sun's surface, find in poundals what its weight would be there.

19. Taking the Moon's mass as $\frac{1}{81}$ that of the Earth, show that the attraction which the Moon exerts upon bodies at its surface is only about 1-5th that of gravity at the Earth's surface.

20. If the Earth were suddenly arrested in its course at an eclipse of the Sun, what kind of orbit would the Moon begin to describe?

EXAMINATION PAPER.—XIII.

1. State reasons for supposing that the Earth moves round the Sun, and not the Sun round the Earth.

2. State Kepler's Laws, and give Newton's deductions therefrom.

3. If the Sun attracts the Earth, why does not the Earth fall into the Sun?

4. Show that the angular velocities of two planets are as the cubes of their linear velocities.

5. State Newton's Law of Gravitation, and prove Kepler's Third Law from it for the case of circular orbits, taking the planets small.

6. Explain clearly (and illustrate by figures or otherwise) what is meant by a force varying inversely as the square of the distance.

7. Are Kepler's Laws perfectly correct? Give the reason for your answer. What is the correct form of the Third Law if the masses of the planets are supposed appreciable as compared with the mass of the Sun?

8. How can the mass of Jupiter be found?

9. Show that if a body describes equal areas in equal times about a point, it must be acted on by a force to that point.

10. Find the law of force to the focus under which a body will describe an ellipse; and if C be the acceleration produced by the force at unit distance, T the periodic time, and $2a$ the major axis of the ellipse, find the relation between C, a, T.

CHAPTER XIV.

FURTHER APPLICATIONS OF THE LAW OF GRAVITATION.

SECTION 1.—*The Moon's Mass—Concavity of Lunar Orbit.*

436. The Earth's Displacement due to the Moon.— In Section II. of the last chapter we saw that when two bodies are under their mutual attraction they revolve about their common centre of mass. Thus, instead of the Moon revolving about the Earth in a period of $27\frac{1}{3}$ days, both bodies revolve about their centre of mass in this period, although from the Moon's smaller size its motion is more marked.

In this case both the Earth and Moon are under the attraction of a third body—the Sun—which causes them together to describe the annual orbit. But the Sun's distance is so great compared with the distance apart of the Earth and Moon, that its attraction is very nearly the same, both in intensity and direction, on both bodies. To a first approximation, therefore, the resultant attraction of the Sun is the same as if the masses of both the Earth and Moon were collected at their common centre of mass. Hence it is strictly the centre of mass of the Earth and Moon, and not the centre of the Earth, which revolves in an ellipse about the Sun with uniform areal velocity, in accordance with the laws stated in § 155. And, owing to the revolution of the Moon, the Earth's centre revolves round this point once in a sidereal month, threading its way alternately in and out of the ellipse described, and being alternately before and behind its mean position.

This displacement of the Earth has been used for finding the Moon's mass in terms of the Earth's, by determining the common centre of mass of the Earth and Moon, as follows.

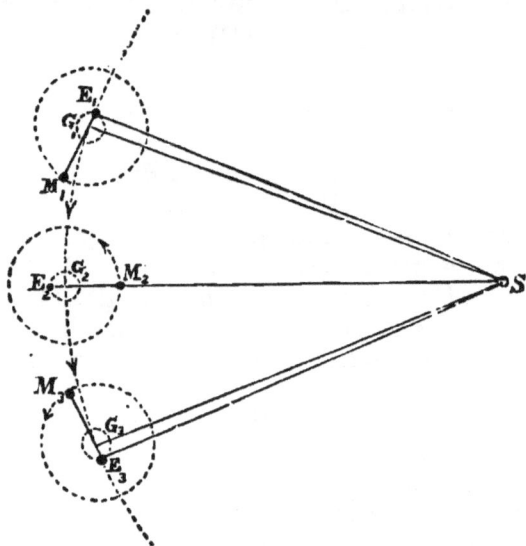

FIG. 144.

Let E_1, M_1, G_1 (Fig. 144) be the positions of the centres of the Earth and Moon, and their centre of mass, at the Moon's last quarter, E_2, M_2, G_2 and E_3, M_3, G_3 their positions at new Moon and at first quarter respectively, S the Sun's centre.

Then, at last quarter, E_1 is behind G_1, and the Sun's longitude, as seen from E_1, is less than it would be as seen from G_1 by the angle $E_1 S G_1$. At first quarter, E_3 is in front of G_3, and therefore the Sun's longitude is greater at E_3 than at G_3 by the angle $G_3 S E_3$. If, then, the observed coordinates of the Sun be compared with those calculated on the supposition that the Earth moves uniformly (i.e., with uniform areal velocity), its longitude will be found to be decreased at last quarter and increased at first quarter.

From observing these displacements the Moon's mass may be found. For, knowing the angle of displacement $E_1 S G_1$ and the Sun's distance, the length $E_1 G_1$ may be found. Also the Moon's distance $E_1 M_1$ is known. And, since G_1 is the centre of mass of the Earth and Moon,

mass of Moon : mass of Earth $= E_1 G_1 : G_1 M_1$;

whence the mass of the Moon can be found.

The Sun's displacement at the quarters could be found by meridian observations of the Sun's R.A. with a transit circle. The displacement of the Earth will also give rise to an apparent displacement, having a period of about one month, in the position of any near planet; this could be detected by observations on Mars, when in opposition, similar to those used in finding solar parallax (§ 339).

From this and other methods it is found that the mass of the Moon is about **1/81** of that of the Earth. The Moon's *density*, as thus deduced, is about 3·44, or $\frac{3}{5}$ of that of the Earth.

EXAMPLE.—To compare the masses of the Moon and Earth, having given that the Sun's displacement in longitude at the Moon's quadratures is equal to $\frac{3}{4}$ of the Sun's parallax.

Since $\angle E_1 S G_1 = \frac{3}{4}$ the angle subtended by Earth's radius at S,

therefore $E_1 G_1 = \frac{3}{4}$ (Earth's radius).

But $E_1 M_1 = 60$ (Earth's radius);

∴ $E_1 M_1 = 80 . E_1 G_1$;

∴ $G_1 M_1 = 79 . E_1 G_1$,

and   mass of Moon : mass of Earth $= E_1 G_1 : G_1 M_1 = 1 : 79$;

∴   the Moon's mass $= 1/79$ of the Earth's mass.

437. Application to Determination of Solar Parallax.

—If the Moon's mass be found by any other method, the above phenomena give us a means of finding the Sun's parallax and distance. For we then know $E_1 G_1 : G_1 M_1$, and therefore $E_1 G_1$ and the angle $E_1 S G_1$ is found by observation. But the exact ratio of $E_1 S G_1$ to the parallax is known, for it is equal to that of $E_1 G_1$ to the Earth's radius; hence the Sun's parallax and distance can be found. Since the Moon's mass can be found with extreme accuracy by many different methods, this method is quite as accurate as many that have been used for finding the solar parallax.

*438. Concavity of the Moon's Path about the Sun. - The Moon, by its monthly orbital motion about the Earth, threads its way alternately inside and outside of the ellipse which the centre of mass of the Earth and Moon describes in its annual orbit about the Sun.

Hence the path described by the Moon in the course of the year is
a wavy curve, forming a series of about thirteen undulations about
the ellipse. It might be thought that these undulations turned
alternately their concave and convex side towards the Sun, but the
Moon's path is really always concave ; that is, it always bends
towards the Sun, as shown in Fig. 145, which shows how the path
passes to the inside of the ellipse without becoming convex.

To show this it is necessary to prove that the Moon is always
being accelerated towards the Sun. Let n, n' be the angular velocities of the Moon about the Earth and the Earth about the Sun
respectively. Then, when the Moon is new, as at M_2 (Fig. 145), its
acceleration towards G_2, relative to G_2, is $n^2 . M_2 G_2$. But G_2 has a
normal acceleration $n'^2 G_2 S$ towards S. Hence the resultant acceleration of the Moon M_2 towards S is $n'^2 G_2 S - n^2 M_2 G_2$.

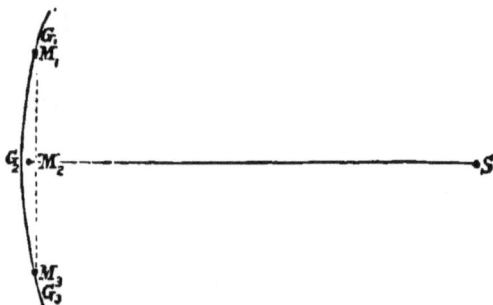

FIG. 145.

Now, there are about $13\frac{1}{4}$ sidereal months in the year; therefore
$n = 13\frac{1}{4} n'$. Also $E_2 S$ is nearly 400 times $E_2 M_2$, and therefore $G_2 S$ is
slightly over 400 times $G_2 M_2$. Therefore roughly

$$n'^2 G_2 S : n^2 M_2 G_2 = 400 : 182 ; \quad \therefore \ n'^2 G_2 S > n^2 G_2 M_2.$$

Thus, the resultant acceleration of M_2 is directed *towards*, not *away
from* S, even at M_2, where the acceleration, relative to G_2, is directly
opposed to that of G_2. Therefore the Moon's path is constantly
being bent (or deflected from the tangent at M_2) in the direction of
the Sun, and is concave towards the Sun.

*439. Alternate Concavity and Convexity of the Path of a
Point on the Earth.—In consequence of the Earth's diurnal rotation, combined with its annual motion, a point on the Earth's equator
describes a wavy curve forming 365 undulations about the path
described by the Earth's centre. In this case, however, it may be
easily shown in the same way that the acceleration of the point
towards the Earth's centre is *greater* than the acceleration of the
Earth's centre towards the Sun. The path is, therefore, not
always concave to the Sun, being bent away from the Sun in
the neighbourhood of the points where the two component accelerations act in opposite directions.

SECTION II.— *The Tides.*

In the last section we investigated the displacements due to the ·Moon's attraction on the Earth as a whole. We shall now consider the effects arising from the fact that the Moon's attractive force is not quite the same either in magnitude or direction at different parts of the Earth, and shall show how the small differences in the attraction give rise to the tides.

440. The Moon's or Sun's Disturbing Force.—Let C, M be the centres of the Earth and Moon; $A C A'$ the Earth's diameter through M; B, B' points on the Earth such that $MC = MB = MB'$. Let M, m denote the masses of the Earth and Moon, a the Earth's radius, d the Moon's distance.

The resultant attraction of the Moon on the Earth as a whole is kMm/CM^2, and the Earth is therefore moving with acceleration km/CM^2 towards the common centre of mass of the Earth and Moon, as shown in §§ 422, 424.

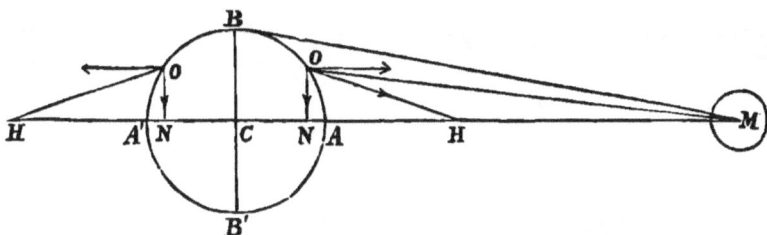

FIG. 146.

(i.) Now at the sublunar point A the Moon's attraction on unit mass is km/AM^2 and is greater than that at C (since $AM < CM$). Hence the Moon tends to accelerate A more than C and thus to draw a body at A *away from* the Earth, with relative acceleration F, where

$$F = km \left(\frac{1}{AM^2} - \frac{1}{CM^2} \right) = km \frac{CA\,(CM+AM)}{CM^2 . AM^2}$$

$$= km \frac{a\,(2d-a)}{d^2\,(d-a)^2} = km \frac{2a}{d^3} \frac{1-a/2d}{(1-a/d)^2}.$$

Since a/d is a small fraction, we have, to a first approximation,

$$F = km \frac{2a}{d^3} = 2k \frac{m}{d^3} CA.$$

(ii.) At A' the Moon's attraction per unit mass is $km/A'M^2$, and is less than that at C, since $A'M > CM$. Hence the Moon tends to accelerate C more than A', and thus to draw the Earth *away from* A' with relative acceleration F', where

$$F' = km \left(\frac{1}{CM^2} - \frac{1}{A'M^2} \right) = km \frac{CA'(CM+A'M)}{CM^2 \cdot AM'^2}$$

$$= km \frac{a(2d+a)}{d^2(d+a)^2} = km \frac{2a}{d^3} \frac{1+a/2d}{(1+a/d)^2}.$$

To a first approximation, therefore,

$$F' = km \frac{2a}{d^3} = 2k \frac{m}{d^3} CA'.$$

Thus a body *either* at A or A' tends to separate from the Earth, as if acted on by a force *away from* C, of magnitude approximately $= 2km.a/d^3$ per unit mass.

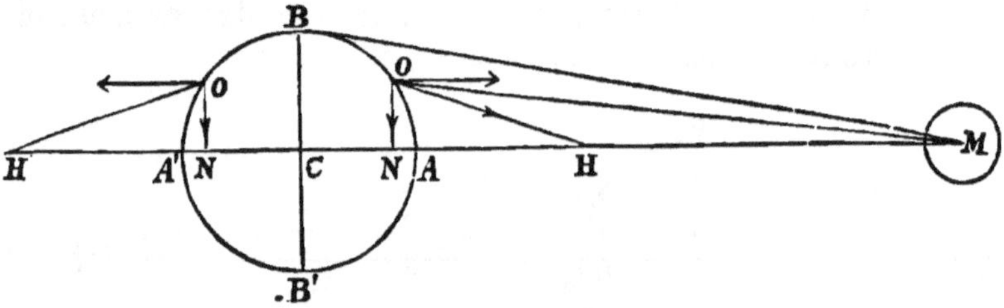

FIG. 147.

(iii.) Consider now the effect of the Moon's attraction on a body at B. This produces a force per unit mass of km/BM^2, which may be resolved into components

$$k \frac{m}{BM^2} \times \frac{CM}{BM} \text{ parallel to } CM,$$

and

$$k \frac{m}{BM^2} \times \frac{BC}{BM} \text{ along } BC.$$

Since we have taken $BM = CM$, the first component is equal to km/CM^2; that is, to the force at C. This component therefore tends to make a body at B move with the rest of the Earth, and produces no relative acceleration. Therefore the Moon tends to draw a body at B *towards* the Earth with relative acceleration f, represented by the second component;

thus

$$f = km \frac{BC}{BM^3}.$$

The point B is approximately the end of the diameter BCB' perpendicular to AC (since BM, CM, $B'M$ are nearly parallel in the neighbourhood of the Earth).

Hence the relative acceleration at B is approximately perpendicular to CM, and its magnitude

$$f = km\,\frac{a}{d^3} = km\,\frac{BC}{d^3}.$$

Similarly at B' the Moon tends to draw a body *towards* C, with relative acceleration $f = kma/d^3$.

At either of these points, B, B', therefore, a body tends to approach the Earth, as if acted on by a force *towards* the Earth's centre, of magnitude kma/d^3 per unit mass. Generally, the Moon's attraction at any point O tends to accelerate a body, relatively to the Earth, as if it were acted on by a force depending on the difference in magnitude and direction between the Moon's attractions at that point and at the Earth's centre.

This apparent force is called the **Moon's disturbing force or tide-generating force.** We see that the disturbing force produces a pull along AA' and a squeeze along BB'.

A similar consequence arises from the attraction of the Sun. The Sun's *actual* attraction on the Earth as a whole keeps the Earth in its annual orbit, but the variations in the attraction at different points give rise to an apparent distribution of force on the Earth which is the **Sun's disturbing force or tide-generating force.**

441. To find approximately the Moon's or Sun's Disturbing Force at any point.

Let O be any point of the Earth. Draw ON perpendicular on CM. Then the difference of the Moon's attractions at O and N tends to accelerate O towards N, with a relative acceleration $km \,.\, NO/d^3$ [by § 440 (iii.)]. Also, the difference of the attractions at N, C tends to accelerate N away from C with a relative acceleration $2km \,.\, CN/d^3$ [by § 440 (i.)].

The whole acceleration of O, relative to C, is compounded of these two relative accelerations. Therefore, if X, Y be the components of the disturbing force at O in the directions CN, NO,

$$X = 2km\,.\,\frac{CN}{d^3}, \quad Y = -km\,.\,\frac{NO}{d^3}.$$

442. Hence the following geometrical construction :—
On CN produced take a point II such that

$$NII = 2CN.$$

Then the line OII represents the disturbing force at O in direction, and its magnitude is

$$F = km \cdot \frac{OII}{d^3}.$$

The Sun's tide-raising force may be found exactly in the same way. The force is everywhere directed towards a point on the diameter of the Earth through the Sun, found by a similar construction to the above. And if r, S denote the Sun's distance and mass, the force is proportional to S/r^3 instead of m/d^3.

In all these investigations we see that the *tide-raising force due to an attracting body is proportional directly to its mass and inversely to the* **cube** (not the square) *of its distance.*

From this it is easy to compare the tide-raising forces due to different bodies acting at different distances.

EXAMPLES.

1. To compare the tide-raising forces due to the Sun and Moon.

The masses of the Sun and Moon are respectively 331,000 and $\frac{1}{81}$ times the Earth's mass. Also, the Sun's distance is about 390 times the Moon's.

∴ Sun's tide-raising force : Moon's tide-raising force

$$= \frac{331,000}{(390)^3} : \frac{1}{81} = 331 : \frac{(39)^3}{3 \times 3^3} = 331 \times 3 : 13^3 = 993 : 2197$$

$$= 33 : 73 \text{ nearly} = 3 : 7 \text{ nearly.}$$

Thus the Sun's tide-raising force is about three-sevenths of that of the Moon.

2. To find what would be the change in the Moon's tide-raising force if the Moon's distance were doubled and its mass were increased sixfold.

If f, f' be the old and new tide-raising forces at corresponding points,

$$f' : f = \frac{6}{2^3} : \frac{1}{1^3}; \quad \therefore f' = \tfrac{3}{4}f.$$

Therefore the tide-raising force would have **three-quarters** of its present value.

3. To compare the Moon's tide-raising forces at perigee and apogee.

The greatest and least distances of the Moon being in the ratio of

$1 + \frac{1}{18}$ to $1 - \frac{1}{18}$, or 19 to 17 (§ 270), the tide-raising power at perigee is greater than at apogee in the ratio of $19^3 : 17^3$ or $6859 : 4913$, or roughly 7 : 5.

4. To compare the maximum and minimum values of the Sun's tide-raising force.

The eccentricity of the Earth's orbit being $\frac{1}{60}$, these are in the ratio of $(1 + \frac{1}{60})^3 : (1 - \frac{1}{60})^3$, or approximately $1 + \frac{3}{60} : 1 - \frac{3}{60}$, or 21 : 19. As before, the force is greatest at perigee and least at apogee.

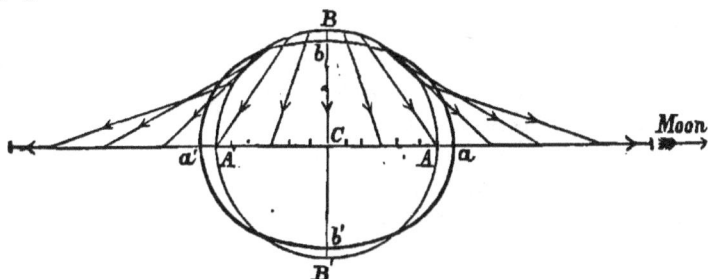

FIG. 148.

443. The Equilibrium Theory of the Tides. — Let us imagine the Earth to be a solid sphere covered with an ocean of uniform depth. If we plot out the disturbing forces at different points of the Earth by the construction of § 442, we shall find the distribution represented in Fig. 148, the lines representing the forces both in magnitude and direction. Here the disturbing force tends to raise the ocean at the sub-lunar point A and at the opposite point A', and to depress it at the points B, B'. At intermediate points it tends to draw the water away from B and B', towards A and A'.

Hence the surface of the ocean will assume an oval form, as represented by the thick line in Fig. 148, and there will be high water at the sublunar point A and the opposite point A', low water along the circle of the Earth BB', distant $90°$ from the sublunar point. Thus we have the same tides occurring simultaneously at opposite sides of the Earth.

It may be shown that the oval curve $aba'b'$ is an ellipse whose major axis is aa'. The surface of the ocean, therefore, assumes the form of the figure produced by revolving this ellipse about its *major* axis. This figure is called a **prolate spheroid**, and is thus distinguished from an oblate spheroid, which is formed by revolution about the minor axis.

ASTRON.   2 c

But though this is the form which the ocean would assume if it were at rest, a stricter mathematical investigation shows that the Earth's rotation would cause the surface of the sea to assume a very different form.

In fact, if the Earth were covered over with a *sufficiently shallow* ocean of uniform depth, and rotating, we should really have *low tide* very near the sublunar point A and its antipodal point A', and high tide at the two points on the Earth's equator distant 90° from the Moon (Fig. 149).

If the Moon were to move in the equator, the equilibrium theory would always give low water at the poles. This phenomenon is uninfluenced by the Earth's rotation, and since the Moon is never more than about 28° from the equator, we see that the Moon's tide-raising force has the general effect of drawing some of the ocean from the poles towards the equator.

*444. A few other consequences of the equilibrium theory may also be enumerated. (1) According to it the height of the tides, or the difference of height between high and low water at any place, is directly proportional to the tide-generating force, and consequently, with the results of Example 1 of § 442, the heights of the solar and lunar tides are in the proportion of 3 to 7. (2) Since the distortion of the mass of liquid is resisted by gravity, the height of the tide depends on the *ratio* of the tide-producing force to gravity, and therefore is inversely proportional to the intensity of gravity, and therefore to the density of the Earth; if the density were halved, the height of the tides would be doubled. (3) If the diameter of the Earth were doubled, its density remaining the same, the intensity of gravity and the tide-producing force would both be doubled, since both are proportional to the Earth's radius. This would cause the ocean to assume the *same shape* as before, only all its dimensions would be doubled.† Consequently the height of the tide would also be doubled, and it thus appears that the height of the tide is proportional to the Earth's radius.

We thus have the means of comparing the tides which would be produced on different celestial bodies, for the above properties show that the height of tide is proportional to ma/Dd^3, where a and D are the radius and density of the body under consideration, m, d the mass and distance of the disturbing body.

*445. **Canal Theory of the Tides.**—As an illustration, let us consider what would happen in a circular canal, not extremely deep, supposed to extend round the equator of a re-

---

† Of course this is not a very strict proof.

volving globe. Then, in Fig. 149, it is clear that the direction of the disturbing force would, if it acted alone, cause the water in the quadrants AB and AB' to flow towards A; and, in the quadrants $A'B$ and $A'B'$, towards A'. Hence this force acts in the *same* direction as the Earth's rotation in the quadrants $B'A$ and BA', and in the *opposite* direction in AB and $A'B'$. Hence, as the water is carried from A to B, it is constantly being retarded, from B to A' it is accelerated, from A' to B' it is retarded, and from B' to A it is again accelerated, the average velocity being, of course, that of the Earth's rotation. Hence the velocity is *least* at B and B', and *greatest* at A and A'.

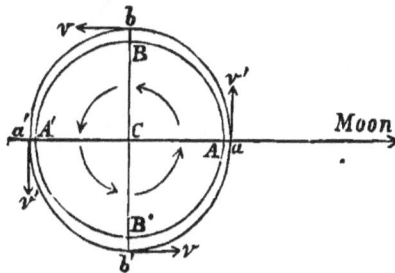

Fig. 149.

Now, it is easy to see that when water moves steadily in a uniform canal it must be shallow where it is swift and deep where it is slow. For, if we consider any portion of the canal, say AB, the quantity that flows in at one end A is equal to the quantity that flows out at the other end B. But it is evident that if the depth of the canal were doubled at any point without altering the velocity of the liquid, twice as much liquid would flow through the canal; consequently, in order that the amount which flows through might be the same as before, we should have to *halve* the velocity of the liquid. This shows that where the canal is deepest the water must be travelling most slowly. Conversely, where the velocity is least the depth must be greatest, and where the velocity is greatest the depth must be least. Hence the depth is *least* at A and A', and *greatest* at B and B', just the opposite to what we should have expected from the equilibrium theory.

In a canal constructed round any parallel of latitude the same would be the case; and hence, if we could imagine a uniform ocean replaced by a series of such parallel canals, low tide would occur at every place when the Moon was in the meridian.

This theory (due to Newton), though sounder than Laplace's equilibrium theory, is still not quite mathematically correct. The true explanation of the tides, even in an ocean of uniform depth, is far more complicated, and quite beyond the scope of this book.

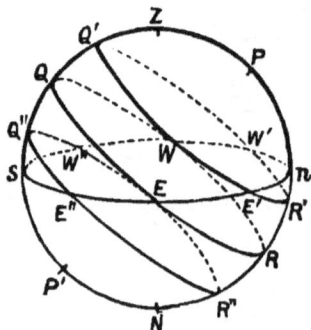

Fig. 150.

446. Lunar Day and Lunar Time. — According to either hypothesis, the recurrence of high and low water depends on the Moon's motion relative to the meridian; hence, in investigating this, it is convenient to introduce another kind of time, depending on the Moon's diurnal motion.

The **lunar day** is the interval between two consecutive upper transits of the Moon across the meridian.

In a lunation, or $29\frac{1}{2}$ mean solar days, the Moon performs one direct revolution relative to the Sun, and therefore performs *one* retrograde revolution *less* relative to the meridian. Thus $29\frac{1}{2}$ mean days $= 28\frac{1}{2}$ lunar days; whence the mean length of a lunar day

$$= (1 + \tfrac{2}{57}) \text{ mean solar days} = 24\text{h. } 50\text{m. } 32\text{s. nearly.}$$

The **lunar time** is measured by the Moon's hour angle, converted into hours, minutes, and seconds, at the rate of $15°$ to the hour.

*447. Semi-diurnal, Diurnal and Fortnightly Tides.

—It has been found convenient to regard the tides produced by the Moon's disturbing force as divided into three parts, whose periods are half a day, a day and a fortnight, the "day" being the lunar day of the last paragraph.

If we adopt the equilibrium theory as a working hypothesis, the lunar tide must be highest when the Moon is nearest to the zenith or nadir. Hence high tide takes place at the Moon's upper and lower transits, when its zenith distance and nadir distance are least respectively. But, for a place in N. lat. (Fig. 150) when the Moon's declination is N., it describes a small circle $Q'R'$, and its least zenith distance ZQ' is less than its least nadir distance NR'; hence the two tides are unequal in height. This phenomenon can be represented by supposing a **diurnal tide,** high only once in a lunar day, combined with a **semi-diurnal tide,** high twice in this period.

Again, the Moon's meridian Z.D. and N.D. go through a complete cycle of changes, owing to the change of the Moon's declination, whose period is a month. But after *half* a month, the Moon's declination will have the same value but opposite sign, and hence the diurnal circles $Q'R'$, $Q''R''$, equidistant from the equator QR, are described at intervals of a fortnight. But $NR'' = ZQ'$, $ZQ'' = NR'$; hence the two tides have the same heights. This can be represented by supposing a **fortnightly tide** of the proper height combined with the diurnal and semi-diurnal ones.

In just the same way the smaller tides caused by the Sun may be artificially represented by combining a **diurnal** and **semi-diurnal tide** (the solar day being used) and a **six-monthly tide.**

448. Spring and Neap Tides.—Priming and Lagging

—We have hitherto considered chiefly the tides due to the action of the Moon. In reality, however, the tides are due to the combined action of the Sun and Moon, the tide-raising forces due to these bodies being in the proportion of about 3 to 7 (Ex. 1, § 442). We shall make the assumption that the height of the tide at any place is the *algebraic sum* of the heights of the tides which would be produced at that place by the Sun and Moon separately.

At new or full Moon the Sun is nearly in the line AA', and the tide-raising powers of the Sun and Moon both act in the same direction, and tend to draw the water from B, B' to A, A'; hence the whole tide is that due to the *sum* of the separate disturbing forces of the Sun and Moon. The tides are then most marked, the height of high water and depth of low water being at their maximum. Such tides are called **Spring Tides.** We notice that the height of the spring tide $= 1 + \frac{3}{7}$ or $\frac{10}{7}$ of that of the lunar tide alone.

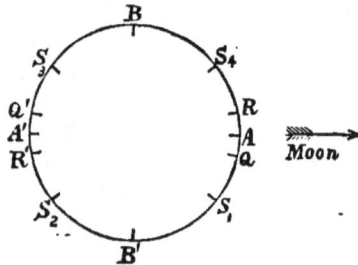

FIG. 151.

At the Moon's first or last quarter the Sun is in a line BB' perpendicular to AA'. Hence the Sun tends to draw the water away from A, A' to B, B', while the Moon tends to draw the water in the opposite direction. The Moon's action being the greater, preponderates, but the Sun's action diminishes the tides as much as possible. The variations are therefore at their minimum, although high water still occurs at the same time as it would if the Sun were absent. These tides are called **Neap Tides.** The height of the neap tide is the *difference* of the heights of the lunar and solar tides, and is therefore $\frac{4}{7}$ of that of the lunar tide.

Hence spring tides and neap tides are in the ratio of (roughly) **10 to 4.**

For any intermediate phase of the Moon, the Sun's action is somewhat different.

Between new Moon and first quarter, the Sun is over a point S_1 behind A. Here the Moon tends to draw the water towards A, A', and the Sun tends to draw the water towards S_1 and the antipodal point S_3. Therefore the combined action tends to draw the water towards two points Q, Q'

between A and S_1 and between A and S_3 respectively, whose longitudes are rather less than those of A and A' respectively. The resulting position of high water is therefore displaced to the west, and the high water occurs *earlier* than it would if due to the Moon's influence alone. The tides are then said to **prime**.

Between first quarter and full Moon the Sun is over a point S_2 between B' and A', and the combined action of the Sun and Moon tends to draw the water towards two points R, R', whose longitudes are slightly greater than those of A, A'. The resulting high tides are therefore displaced eastwards, and occur *later* than they would if the Sun were absent. The tides are then said to **lag**.

Between full Moon and last quarter the Sun is over some point S_3 between B and A', but the antipodal point S_1 is between A and B'; hence the tide **primes**.

Between last quarter and new Moon, when the Sun is at a point S_4 between B and A, it is evident in like manner that the tide **lags**.

Hence **Spring Tides** *occur at the* **syzygies** (conjunction and opposition).

Neap Tides *occur at the* **quadratures**.

From **syzygy** *to* **quadrature**, *the tide* **primes**.

From **quadrature** *to* **syzygy**, *the tide* **lags**.

The heights of the spring and neap tides vary with the varying distances of the Sun and Moon from the Earth. Spring tides are the highest possible when both the Sun and Moon are in perigee, while neap tides are the most marked when the Moon is in apogee but the Sun is in *perigee* (because the Sun then pulls *against* the Moon with the greatest power, as far as the Sun's action is concerned). Both the spring and neap tides, and also the priming and lagging, are on the whole most marked when the Sun is near perigee, *i.e.*, about January.

It may be here stated, without proof, that, taking the Sun's and Moon's tide-raising forces to be in the proportion of 3 to 7, the maximum interval of priming or lagging is found to be about 51 minutes.

449. Establishment of the Port.—Both the equilibrium and canal theories completely fail to represent the actual tides on the sea, owing to the irregular distribution of land and water on the Earth, combined with the varying depth of the ocean. These circumstances render the prediction of tides by calculation one of the most complicated problems of practical astronomy, and the computations have to be based largely on previous observations. In consequence of the barriers offered to the passage of tidal waves by large continents, lunar high tide does not occur either when the Moon crosses the meridian, as it would on the equilibrium theory, or when the Moon's hour angle is 90°, as it would on the canal theory. But this continental retardation causes the high tide to occur later than it would on the equilibrium theory, by an interval which is constant for any given place. This interval, reckoned in lunar hours, is called the **Establishment of the Port** for the place considered. Thus the establishment of the port at London Bridge is 1h. 58m., so that lunar high water occurs 1h. 58m. after the Moon's transit, *i.e.*, when the Moon's hour angle, reckoned in time, is 1h. 58m.

The same causes affect the solar tide as the lunar, hence the *Sun's* hour angle (or the local apparent time) at the *solar* high tide is also equal to the establishment of the port.

The actual high tide, being due to the Sun and Moon conjointly, is earlier or later than the lunar tide by the amount of priming or lagging. By adding a correction for this to the establishment of the port, the lunar time of high water may be found for any phase of the Moon; and we notice in particular that at the Moon's four quarters (syzygies and quadratures), the lunar time of high water is equal to the establishment of the port. And, knowing the lunar time of high water, the corresponding mean time can be found, for

(mean solar time) $-$ (lunar time)

$$= \text{(mean} \odot \text{'s hour angle)} - (\mathbb{C} \text{'s hour angle)}$$

$$= (\mathbb{C} \text{'s R.A.)} - \text{(mean} \odot \text{'s R.A.)}$$

[since R.A. and hour angle are measured in opposite directions].

Now the Moon's R.A. is given in the Nautical Almanack for every hour of every day in the year. Also the mean Sun's R.A. at noon is the sidereal time of mean noon, and is given in the Nautical Almanack. Hence the mean Sun's R.A. [which = (sidereal time) − (mean time)] is easily found for any intermediate time.

Hence the mean time of high water can be readily found. The establishments of different ports, and the times of high water at London Bridge, are given in the Nautical Almanack.

*450. If only a *very rough* calculation is required, we may proceed as in §§ 35, 40. We assume the Moon's R.A. to increase uniformly; we shall then have

$$(\text{☾'s R.A.}) − (\odot\text{'s R.A.}) = (\text{☾'s elongation}) ;$$
$$\therefore (\text{solar time}) = (\text{lunar time}) + (\text{☾'s elongation}).$$

Knowing the Moon's age, its elongation may be found, as in § 40, and this must be converted into time, at the rate of 1h. to 15°. We then have (time of high water)

$$= (\text{establishment}) + (\text{amount of lag.}) + (\text{☾'s elongation in time})_.$$

EXAMPLE.—To find, roughly, the time of high water at the Moon's first quarter, at London Bridge.

Here there is no priming or lagging. Hence the lunar time, or ☾'s hour angle at high water, is equal to the establishment, or 1h. 58m. Also the Moon's elongation is 90°. Hence the Sun's hour angle, in time, = 1h. 58m. + 6h.. and high water occurs about 7h. 58m.

*451. Tidal Constants.—The excess of the establishment of the port at any place, over that at London Bridge, expressed in mean time, is sometimes called the **Tidal Constant** of that place.

If we assume the amount of priming or lagging to be the same at both places, the tidal constant is the difference between the times of high water at London Bridge and the given place. Hence, knowing the tidal constant and the time of high water at London Bridge, the time at any other place can be found.

Tables of tidal constants, and of the heights of the spring and neap tides at different places, are given in *Whitaker's Almanack*.

EXAMPLE.—To find the times of high water at Cardiff and Portsmouth on January 25, 1892, the tide intervals from London Bridge being +4h. 58m. and −2h. 17m. From the Almanack we find times of high water at London Bridge are

	Jan. 24.	Jan. 25.	
	9h. 15m. aft.	9h. 53m. morn.,	10h. 31m. aft.
Add	4h. 58m.	4h. 58m.	4h. 58m.

∴ Times at Cardiff are
(Jan. 25) 2h.13m. morn. 2h. 51m. aft.

Again, subtract from first line   2h. 17m.   2h. 17m.

∴ times at Portsmouth are   (Jan. 25) 7h. 36m. morn.,   8h. 14m. aft.

452. **The Masses of the Sun and Moon** can be compared by observing the relative heights of the solar and lunar tide, the relative distances of the Sun and Moon being known. Or, if the ratio of the masses be supposed known, the distances could be compared by this method. In this manner Newton (A.D. 1687) found the masses of the Moon and Earth to be in the proportion of 1 : 40 nearly. D. Bernouilli (1738) found 1 : 70, and Lubbock (1862) found 1 : 67·3. The two last make the Moon's mass a little too great. Newton makes it double what it ought to be.

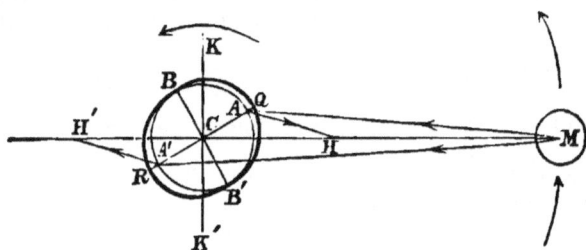

Fig. 152.

453. **Effects of Tidal Friction. — Retardation of Earth's Rotation.—Acceleration of Moon's Orbital Motion.** — All liquids possess a certain kind of friction, known as "viscosity," which tends to resist their motion when they are changing their form, and to convert part of their kinetic energy into heat. Owing to this friction between the Earth and the oceans, the Earth, in its diurnal rotation, tends to carry the tidal wave round slightly in front of the point underneath the Moon, taking the positions of high water forward from the line $H'CM$ to $A'CA$. The Moon, on the contrary, tends to draw the water back from A, A', the disturbing forces AH, $A'H'$ forming a couple, which is resisted only by the Earth's friction. Hence the ocean exerts an equal frictional couple on the Earth, and this couple tends to diminish the angular velocity of the Earth's diurnal rotation, and thus increase its period.

Therefore tidal friction tends to gradually lengthen the day.

But if the Moon exerts a couple on the Earth, tending to retard it, the Earth must exert an equal and opposite couple on the Moon, tending to accelerate it. That it really does so is manifest from Fig. 152. The portion of the ocean heaped up at A, being nearer the Moon, exerts a greater attraction than that at A', in addition to which the angle CMA is very slightly greater than CMA'. Hence the resultant of the attractions of equal masses of water at A and A' acts on M in a direction slightly in front of MC, and tends to pull the Moon forward. This tends to increase the Moon's areal velocity. (Compare § 413.) Since the areal velocity of a body revolving in a circle varies as the square root of the radius (§ 411, Cor.), the Moon's distance must be gradually increased by this means, and hence also its periodic time.*

Therefore tidal friction tends to increase the Moon's distance and to lengthen the month.

Still the final effect of tidal friction must be to equalize the lengths of the day and lunar month. The angular velocities of the Earth and Moon both decrease, but the effect of the couple, in producing retardation, is far more considerable on the Earth than on the Moon.

The student who has not read Rigid Dynamics may illustrate this statement by the comparative ease with which a small top can be spun with the fingers, and the great difficulty of imparting an equal angular velocity to the same body by whirling it round in a circle at the end of a string of considerable length. The top represents the Earth, and the body on the long string the Moon.

In Rigid Dynamics it is shown that when a system of bodies are revolving under their mutual reactions, their *angular momentum*, or *moment of momentum* about their centre of mass, remains constant. Hence the decrease in the Earth's angular momentum is equal to the increase in that of the Moon. Now the angular momentum of a particle revolving in an orbit is twice the product of its mass into its areal velocity, and this is also approximately true of the Moon. Hence, since the Moon's distance from the common centre of mass is far greater (about sixty times as great) than the distance of any point on the Earth from its axis of rotation, it is evident that the same change in angular momentum produces far more effect on the angular velocity of the Earth than on that of the Moon.

---

* This increase of the distance more than counterbalances the tendency to increase the Moon's actual velocity. For the actual velocity is *inversely* proportional to the square root of the distance (§ 409), and therefore diminishes as the distance increases. Similarly, the *angular* velocity is decreased.

It thus appears that, after the lapse of probably many millions of years, tidal friction will equalize the periods of rotation of the Earth and Moon, and the day and month will be of equal length, each being probably about 1,400 hours long. The Earth will then always turn the same face towards the Moon, just as the Moon now does towards the Earth; hence there will be no lunar tides, and the retardation due to lunar tidal friction will no longer exist.

The solar tides will, however, still continue to exist, provided that there is any water left on the Earth. The effect of solar tidal friction will be to retard the Earth's rotation, thus further lengthening the day; and this again will retard the Moon's orbital motion, and diminish its areal velocity. The Moon will, therefore, approach the Earth, and will ultimately fall into the Earth; and finally, the Earth will always turn the same face towards the Sun, so that there will always be day over one hemisphere and night over the other.

This theory of the probable future history of the Earth is due to Professor G. H. Darwin. It is certain that the effect of tidal friction on the Earth's rotation must be very small; hence a very long period must necessarily elapse before any perceptible increase in the length of the day can be detected. The records of history afford no data sufficiently accurate to furnish conclusive evidence of such a lengthening, but there are *some* grounds for believing that the sidereal day is increasing in length by about ·006 of a second in 1,000 years.

Moreover, the Earth is gradually cooling, and consequently is shrinking; and this shrinkage, by bringing the particles of the Earth nearer to the axis, causes an *increase* of the angular velocity of rotation.* It is quite possible that an increase of this nature is at the present time either wholly or partially counteracting the retardation due to tidal friction.

---

* For, according to the principles of Rigid Dynamics, the angular momentum of the Earth = (its angular velocity) × (its moment of inertia). And if the angular momentum remains constant, and the moment of inertia decreases through shrinkage, the angular velocity must increase.

454. The Moon's Form and Rotation.—The theory of tidal friction affords a simple explanation of how it is that the Moon always turns the same face to the Earth. Remembering that the Earth's mass is 81 times the Moon's, but that its radius is about four times as great, the Earth's tide-raising force at a point on the Moon would be about 81/4, or over twenty times as great as the Moon's on the Earth. Although there are now no oceans on the Moon, still we have some evidence that water may once have existed on its surface. Furthermore, the large volcanic craters with which its surface is dotted prove that the Moon was at one time filled with molten lava, and that it was probably wholly in a liquid or viscous state at an earlier period of its history. At that time the huge tides on the Moon, ever following the Earth, must, by their friction, have gradually equalized the Moon's period of rotation with its period of revolution about the Earth, in just the same way as if the Moon were surrounded by a friction belt attached to the Earth. This continued till the Moon always turned the same face to the Earth.

If the Moon was then not quite solid, the Earth's tide-raising force, which had then become constant, must have drawn it out into the form demanded by the equilibrium theory, namely, to a first approximation, to a prolate spheroid, with its longest diameter pointed towards the Earth.

It may easily be seen, from the expressions in § 440, that the tide-raising force of a body is slightly greater at the point just under it than at the opposite point (when we do not only consider *approximate* values). Hence the Moon is not quite spheroidal, but is more drawn out on the side toward the Earth than on the remote side. Its form is, therefore, that of an *egg*, the small end being towards the Earth. This result of theory cannot, of course, be confirmed by direct observation, the remote side being invisible; but Hansen, by the theory of perturbations, has shown that the Moon's centre of mass is further from the Earth than its centre of figure, thus furnishing independent evidence in favour of the theory.

*455. Application to Solar System.—Since the Sun's tide-raising force on different planets varies inversely as the cube of their distance, the solar tides are far greater on the nearer planets than on those more remote. It is, therefore, quite natural to suppose that the effects of tidal friction may have produced such a great retardation in the rotations of Mercury, and possibly also Venus, that one or both of these bodies already turn the same face towards the Sun, while the Earth, and the remoter planets, must necessarily take a much longer time to undergo the necessary retardation, and it would be very unnatural to expect Neptune, for example, always to turn the same face to the Sun. Thus Professor Schiaparelli's recent researches on the rotations of Mercury and Venus are in support of the theory of tidal friction.

SECTION III.—*Precession and Nutation.*

456. In § 141 we stated that the plane of the Earth's equator is not fixed in space, but that its intersections with the ecliptic have a slow retrograde motion. This phenomenon, which is known as Precession, is due to the fact that the Earth is not quite spherical, and that, in consequence of its spheroidal form, the Sun's and Moon's attractions exert a disturbing couple on it.

457. The Sun's and Moon's Disturbing Couples on the Earth.

Let the plane of the paper in Fig. 153 contain the Earth's polar axis PP', and the Moon's centre M, say at the time when the Moon's south declination is greatest.

Inside the Earth inscribe a sphere $PAP'A'$, touching its surface at the poles. Then we may (for the sake of illustration) regard the protuberant portion of the Earth outside this sphere as a kind of tide firmly fixed to the Earth, and the arguments of the last section (§ 453) show that the variations in the Moon's attraction at different points give rise to a distribution of disturbing force identical with the tide-raising force, tending to draw this protuberant part with its longest diameter QR pointing towards the Moon. The Moon's attraction on the matter inside the inscribed sphere passes exactly through the

Earth's centre C, and produces no such couple ; but the disturbing forces at A, A', which are represented by $AH, A'H'$, form a couple on the protuberant parts, AQ, $A'R$, tending to turn the diameter $A'A$ towards CM. The same is true of the disturbing forces at any other pair of opposite points of the Earth in the quadrants $HCK, H'CK'$. Of course there are couples in the two other quadrants tending in the reverse direction, but they have less matter to act on, and are therefore insufficient to balance the former couples.

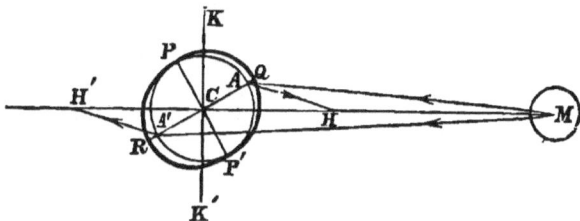

FIG. 153.

When the Moon is at the opposite point of its orbit, *i.e.*, at its greatest N. declination, it is again in the line CH', and again tends to draw the Earth's equatorial plane towards the line HH'. For any intermediate position of the Moon the couple is smaller, and it vanishes when the Moon is *on* the equator ; still, on the whole, *the Moon's disturbing force always tends to draw the plane of the Earth's equator towards the plane of the Moon's orbit.*

Similarly, *the Sun's disturbing force always tends to draw the plane of the Earth's equator towards the ecliptic.*

Since the Moon's nodes are rotating (§ 273), the plane of the Moon's orbit is not fixed ; but it is inclined to the ecliptic at a small angle (5°), while the plane of the equator is inclined to the ecliptic at a much larger angle ($23\frac{1}{2}°$). The *average* effect of the Moon's disturbing couple is thus to pull the Earth's equator towards the plane of the ecliptic. This tendency is increased by the Sun's disturbing couple ; and the two are proportional to the Sun's and Moon's tide-producing forces, *i.e.*, as 3 : 7 roughly. For this reason, the resulting phenomenon is sometimes called **luni-solar** precession.

*458. Effect of the Couple on the Earth's Axis.—
If the Earth were without rotation, the tendency of this
couple would be to bring the plane of the equator into coincidence with the ecliptic, with the result that the equator
would oscillate from side to side of the ecliptic, like a pendulum under gravity. But the rapid diurnal motion of the
Earth entirely alters the phenomena.

Let CR be a semi-diameter of the Earth, perpendicular to
CP and CM. The precessional couple would, alone, produce a
slow rotation in the direction PQM; *i.e.*, about CR. If
now the Earth's rotation be represented in magnitude and
direction by CP, measured along the Earth's axis, this additional rotation must be represented by a very short length
CR', measured along CR.

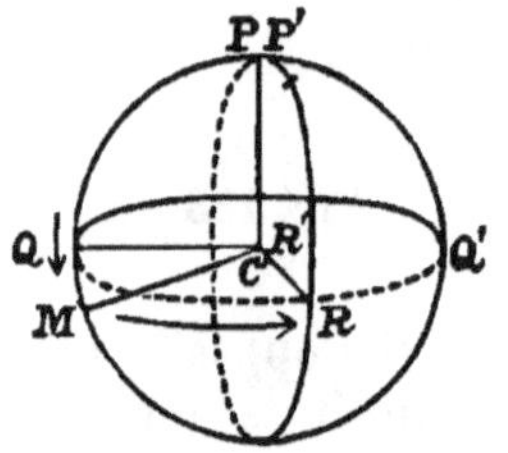

Fig. 154.

Take PP', equal and parallel to CR'; then, since PP' is
very small, CP' is of almost exactly the same length as CP.
But angular velocities, and momenta *about* lines which represent them in magnitude, are compounded by the same law as
forces, velocities, &c. [*cf.* § 387 (iii.)] *along* the same lines of
corresponding magnitudes.

Hence, the resultant axis of rotation is shifted from CP to
CP', *in a direction perpendicular to the plane of the acting
couple.*

A full explanation of what follows would be impossible
without a close acquaintance with Rigid Dynamics. But it is
evident that a body flattened at the poles will spin more
readily about the line CP than about any other line drawn in
its substance. Hence it is easy to understand that the polar
axis CP is *itself* deflected towards CP', and thus moves perpendicular to the acting couple.

This motion can be illustrated by that of a rapidly spinning top, or of a gyroscope, the phenomena of which can readily be investigated by experiment.

459. Precession of a Spinning Top.—*Experiment 1.*

Let a top be set spinning rapidly about its extremity, in the opposite direction to the hands of a watch, as seen from above, the top being supported at a point on its axis below its centre of gravity. The weight of the top, acting vertically through the centre of gravity, tends to upset the top by pulling its axis out of the vertical. But if the top is spinning sufficiently rapidly, we know that it will not fall, the only effect of gravity being to make it " reel," *i.e.*, to cause its axis of rotation to describe a cone about the vertical through the point of support, revolving slowly in the counter-clockwise direction. This slow revolution may be called the **precession** of the top, and the experiment shows that when a top is acted on by a couple (such as that due to its weight) tending to pull its axis away from the vertical, it precesses in the *same* direction in which it is spinning.

Experiment 2.—Now suppose the top *suspended* from its upper extremity, being thus supported above its centre of gravity. The couple due to the weight and the reaction of the support, now tends to draw the axis of the top *towards* the vertical. In this case the axis of the top will be found to slowly describe a cone in the *opposite* direction; that is, the top now precesses in the *opposite* direction to that in which it is spinning.

Experiment 3.—Suppose the top supported as in Experiment 1. If we give the top a push away from the vertical, its axis will not move in this direction, but its precessional motion will increase. If we give a push in the direction of precession, its axis will approach the vertical. If we push the axis in the direction of the vertical, it will not move towards the vertical, but its rate of precessional motion will be increased, *i.e.*, the top will acquire an additional increased precessional motion. If we push it in the direction opposite to that of precession, the axis will begin to move away from the vertical. In every case the axis of the top moves in a direction perpendicular to the direction of the force acting on it, and therefore a couple acting on a very rapidly spinning top produces displacement of the axis in a plane *perpendicular* to the plane of the couple.*

[If we push the top by pressing the side of a pencil against its axis, it thus always moves *in the direction in which the axis would roll along* the side of the pencil. Of course the displacement of the axis is *not due* to rolling, as may easily be shown by repeating the same experiment with a gyroscope, this time pushing one of the hoops carrying the top instead of touching the top itself; here no such rolling is possible.]

---

* These experiments may easily be performed by the reader with any good-sized top.

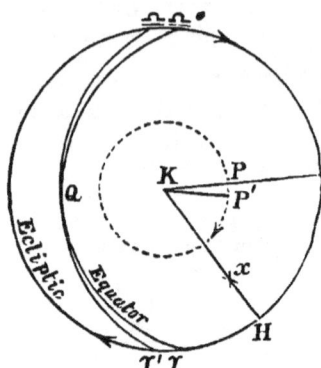

FIG. 155.

460. Precession of the Earth's Axis.—On the celestial sphere, let P, K be the poles of the equator and ecliptic respectively. The Sun's disturbing couple and the mean couple due to the Moon tend to pull the Earth's equator towards the ecliptic, or to pull the polar axis P towards the axis of the ecliptic K. Hence the Earth behaves like a top suspended from above its centre of gravity, and the polar axis slowly describes a cone about the axis of the ecliptic, revolving in the opposite direction to that of the Earth's rotation, *i.e.*, in the retrograde direction.* The pole P therefore slowly describes a small circle PP' about K, the pole of the ecliptic, with angular radius PK, equal to the obliquity of the ecliptic, *i.e.*, 23° 27'. As the pole revolves from P to P' it carries the equator from $\Upsilon Q \triangleq$ to $\Upsilon' Q \triangleq'$, thus carrying the equinoctial points Υ and $\triangleq$ slowly backwards along the ecliptic. The average angle $\Upsilon \Upsilon'$, or PKP'†, described in a year, is 50·2″, and P therefore performs a complete revolution about K in 25,800 years (§ 141).

---

* See also Fig. 154. If K be pole of ecliptic (CK nearly perpendicular to CM) it is evident that as P travels towards P' it moves in the retrograde direction about K.

† $P\Upsilon$ and $K\Upsilon$ are each 90°; ∴ Υ is pole of arc KP; ∴ ∠ ΥKP is a right angle. Similarly, $\Upsilon' KP'$ is a right angle;

∴ ∠ $PKP' = $ ∠ $\Upsilon K\Upsilon' = $ arc $\Upsilon \Upsilon'$,

since $\Upsilon \Upsilon' \triangleq$ is a *great circle*, whose pole is K.

The position of the *ecliptic* is not affected by precession. Hence the *celestial latitude xΠ of any star x remains constant, and its celestial longitude ϒ Π increases by the amount of precession ϒ ϒ', that is, at the rate of 50·2″ per year.*

A star's declination and right ascension are, however, continually changing. This change is, of course, due to the motion of the equator, and not of the star. Thus, as *P* moves to *P'*, the N.P.D. of the star *x* decreases from *Px* to *P'x*, and its R.A. changes from ϒ *Px* to ϒ'*P'x*. (The circles ϒ *P*, ϒ'*P'*, *xP*, *xP'* are not represented, in order not to complicate the figure unnecessarily. The reader should draw a figure, inserting them.)

The declinations of some stars are increasing, of others decreasing.

461. To apply the Corrections for Precession.— The changes in the decl. and R.A. of a star in one year are always small, except in the case of the Pole Star, which is so near the pole that a slight displacement of the pole produces a great change in the R.A. With this exception, the rates of change of the decl. and R.A. of a star remain sensibly constant for a considerable period. Hence, if the coordinates are observed on any given date, and their rates of variation are known, their values at any other date may be found by adding or subtracting corrections obtained by multiplying these rates of variation by the elapsed time.

The rates of variation may be regarded as constant so long as the interval of time is small compared with the period of rotation of the pole. They are therefore sensibly uniform for several years.

The most convenient plan, in correcting for precession, is to calculate the right ascensions and declinations of all stars for the same date or *epoch*.

For this purpose, the time of the vernal equinox in the year 1900 is now frequently chosen as the standard epoch of reference. When the R.A. and decl. of a star are known, their rates of variation can be calculated by Spherical Trigonometry in terms of the known rate of precession, and the correction can then be applied.

It would, of course, be possible to proceed somewhat differently, namely, from the decl. and R.A. to find the star's lat. and long. The long. could then be increased by the amount of the precession, namely, $50 \cdot 2'' \times$ (the number of years elapsed); and from the new lat. and long. the new decl. and R.A. could be found; but the calculations would be longer.

For the purpose of facilitating observations of time, latitude and longitude, and instrumental errors, the declinations and right ascensions of certain bright stars are calculated at intervals of ten days in the Nautical Almanack; these stars are the *clock stars* of § 54.

The effects of aberration, as well as of precession and nutation, are taken into account, the tabulated coordinates being those of the apparent and not the true positions of the star. Such stars can therefore be used to determine clock error and other errors, without applying any further correction.

462. Various Effects of Precession.

Since the R.A. and decl. of a star depend only on the *relative* positions of the star and equator, their variations due to precession are just the same as they would be if the equator and ecliptic were fixed, and the stars had a *direct* motion of rotation, of $50 \cdot 2''$ per annum, about the pole of the ecliptic.

If we make this supposition, the stars will describe circles about K in a period of 25,800 years.

(i.) If a star's distance Kx from the pole of the ecliptic is less than the obliquity i, or its latitude (l) greater than $90° - i$, it will describe a circle $ax_1a'x_2$ (Fig. 156), of radius $90° - l$, not enclosing the pole P, and its greatest and least N.P.D. will be

$$Pa' = i + (90° - l), \quad Pa = i - (90° - l).$$

Also the star's R.A. will fluctuate between the values ΥPx_1 and ΥPx_2. Now Υ is the pole of PK; hence $KP\Upsilon$ is a right angle, and $\Upsilon PK = 270°$; therefore the maximum and minimum R.A. are $270° + KPx_1$, and $270° - KPx_1$.

(ii.) If, on the other hand, the star's latitude is $< 90° - i$, it will describe a circle byb', enclosing the pole P. Its greatest and least N.P.D. will be

$$Pb' = (90° - l) + i, \quad Pb = (90° - l) - i.$$

The star's R.A. will continually increase from 0° to 360°.

In either case the star's N.P.D. will increase as its longitude increases from 90° (at a or b) to 270° (at a' or b'), and will decrease over the other half of the path.

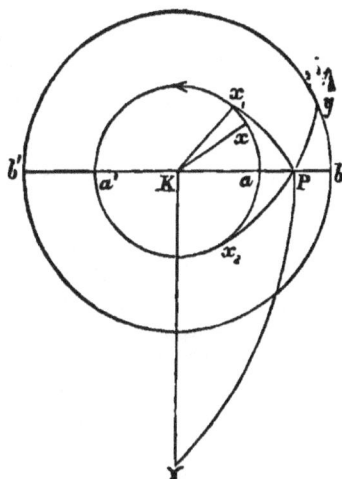

Fig. 156.

The Pole Star will, after a time, move away from the pole, and its place will be then occupied in succession by other stars whose latitude is very nearly $= 90° - i = 66° \ 33'$. If l, L be the latitude and longitude of such a star, it will be nearest the pole in an interval of $(90° - L) \div 50·2''$ years, and its N.P.D. will then be $(90° - l) \sim i$.

That precession has shifted the equinoctial points from the constellations Aries and Libra, into Pisces and Virgo, has already been mentioned. Since there are twelve signs of the zodiac, the equinoctial points shift from one "sign" into the next in 25,800/12 years, i.e., about 2,150 years,

463. Effects on the Climate of the Earth's Hemispheres.—We have seen (§ 132) that the fact of the Earth being in perihelion near the winter solstices renders the climate of the Earth's northern hemisphere more equable, but makes the seasons more marked in southern hemisphere. Owing, however, to precession, combined with the progressive motion of the apse line (§ 153), the reverse will be the case in $\dfrac{180 \times 60 \times 60}{50 \cdot 22 + 11 \cdot 25}$, or 10,545 years. The summer in the northern hemisphere will then be hotter, but shorter, and the winter colder and longer. On the whole, the climate will be colder, as the Earth's radiation will be more rapid during the heat of summer, and therefore a larger proportion of the heat received from the Sun will be lost before the winter.

In a recent paper, Sir Robert Ball has shown that the ice ages, of which we have geological evidence, can probably be accounted for in this manner. The eccentricity of the Earth's orbit is not constant, but is changing very slowly, and is decreasing at the present time. When the orbit had its greatest eccentricity and the winter solstice coincided with aphelion, the autumn and winter were 199 days long, spring and summer being only 166 days long. At this time the climate of the northern hemisphere must have been so exceedingly cold that the whole of northern Europe, including Germany and Switzerland, was ice-bound. When aphelion coincided with the summer solstice a similar effect took place in the southern hemisphere, but the northern hemisphere was warmer and more genial than it is now, spring and summer being 199 days long, and autumn and winter only 166 days long. Thus, at the time of greatest eccentricity there must have been long ages of arctic climate, oscillating from one hemisphere to the other and back in a period of 10,500 years, alternating with more equable, and, perhaps, almost tropical climates.

464. Nutation of the Earth's Axis.—In treating of precession, we have supposed the Earth's poles to describe small circles uniformly about the poles of the ecliptic. This they would do if the Sun's and Moon's disturbing couples on the Earth were always constant in magnitude, and always tended to pull the Earth's poles directly towards the poles of the ecliptic. But the couples, so far from being constant, are subject to periodic variations, in consequence of which the Earth's poles really describe a wavy curve (shown in Fig. 157), threading alternately in and out of the small circle which would be described under precession alone if the couple were constant. This phenomenon is called **Nutation,** because it causes the Earth's poles to *nod* to and from the pole of the ecliptic.

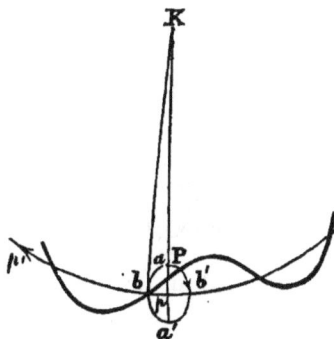

Fig. 157.

Nutation is really compounded of several independent periodic motions of the Earth's axis; the most important of these is known as **Lunar Nutation,** and has for its period the time of a sidereal revolution of the Moon's nodes, *i.e.*, about 18 years 220 days. The effect of lunar nutation may be represented by imagining the pole P to revolve in a small ellipse about its mean position p as centre, in the above period, in the retrograde direction, while p revolves about K, the pole of the ecliptic, with the uniform angular velocity of precession of 50·2″ per annum. The major and minor axes of the little ellipse are along and perpendicular to Kp respectively, their semi-lengths being $pa = 9″$ and $pb = 6·8″$ respectively. The angle $pKb = bp/\sin Kp$ (Sph. Geom. 17) $= 6·8″ \operatorname{cosec} 23° 27' = 17·1″$ nearly.

465. General Effects of Lunar Nutation.—In consequence of lunar nutation, the obliquity of the ecliptic is subject to periodic variations. For this obliquity is equal to the arc KP, and as P revolves about its mean position from one end to the other of the major axis of the little ellipse, the arc KP becomes alternately greater and less than its mean value Kp, by 9″. Thus the greatest and least values of the obliquity of the ecliptic differ by 18″, and the obliquity fluctuates between the values 23° 27′ 20″ and 23° 27′ 2″ once in about $18\frac{2}{3}$ years.

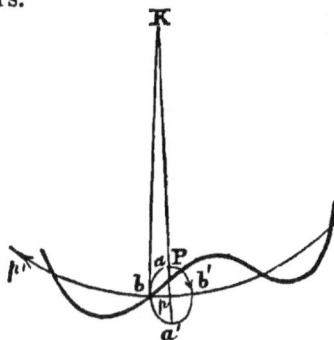

FIG. 158.

Again, when the pole is at an extremity of the minor axis b, it has regreded further than its mean position p by the angle pKb, which we have seen is about 17·1″. Hence, also, the first point of Aries has regreded 17·1″ further than it would have gone had its motion been uniform. Similarly, at b' it has regreded 17·1″ less than it would have done if moving uniformly. Hence the first point of Aries oscillates to and fro about its mean position through an arc of 34·2″ in the period of $18\frac{2}{3}$ years, while its mean position moves through an angle $18\frac{2}{3} \times 50·2″$, or about 15′ 37″.

The angular distance between the true and mean positions of the first point of Aries is called the *Equation of the Equinoxes*. It is, of course, equal to the angle pKP.

Nutation does not affect the position of the ecliptic; hence the *latitudes* of stars are unaltered by it. Their apparent *longitudes* are, however, increased by the equation of the equinoxes. Both this cause and the varying obliquity of the ecliptic produce variations in a star's R.A. and decl.

466. Discovery of Nutation.—Nutation was discovered by Bradley soon after his discovery of aberration, while continuing his observations on the star γ *Draconis* and on a small star in the constellation *Camelopardus*, by its effect on the declinations of these stars. The peculiarity which led him to separate nutation from aberration was their difference of period. The period of the former phenomenon is about 19 years, while that of the aberration displacement is only a year. Had the observed variations in declination been due to aberration alone, the declination would always have had the same apparent value at the same time of year, but such was not the case.

Newton had, sixty years previously (1687), proved the existence of nutation from theory, but had supposed that its effects would be inappreciable.

467. To correct for Nutation, the coordinates of a star are always referred to. the *mean* position of the ecliptic, *i.e.*, the position which the ecliptic would occupy if its pole were at p, the centre of the little ellipse. Hence, since the apparent decl. and R.A. of a star x are measured by $90° - Px$ and $\Upsilon Px\,(= 270° + KPx)$, the corrected decl. and R.A. are $90° - px$ and $270° + Kpx$. If the star's position is specified by its celestial latitude and longitude, the only correction required is to increase the longitude by the equation of the equinoxes.

*468. Bessel's Day Numbers.—If the declinations and right ascension of stars have been tabulated for a certain date, their apparent values for any other date, as affected by precession, nutation, and aberration, can be found by adding certain small corrections to the tabulated values, and it is found that these may be put into the form

$$\text{Change of R.A.} = Aa + Bb + Cc + Dd,$$

$$\text{Change of decl.} = Aa' + Bb' + Cc' + Dd',$$

where A, B, C, D are constants, whose values depend only on the date, and *are the same for all stars;* while $a, b, c, d,\ a',\ b',\ c',\ d'$ depend only on the coordinates of the star, being *always constant for the same star, and independent of the time of observation.*

The four quantities A, B, C, D are called *Bessel's Day Numbers,* and their logarithms are given in the Nautical Almanack for every day of the year. The logarithms of the eight constants $a, b, c, d,$ $a', b', c', d',$ have been tabulated for many thousands of stars in the star catalogues of the Royal Astronomical Society.

469. Physical Cause of Nutation.—If the Moon were
to move exactly in the ecliptic, the average couples exerted
by the Moon as well as the Sun would both tend to pull the
Earth's pole directly towards K, the pole of the ecliptic.
But the Moon's orbit is inclined to the ecliptic at an angle
of $5°$; hence, if L be its pole, $KL = 5°$, and the Moon's
average disturbing couple tends to pull the pole P towards
L instead of K. When we consider the Sun's action also,
the resultant of the two couples tends to pull the pole towards
a point H which is intermediate between K and L, but
nearer to L (because the Moon's disturbing couple is about
$2\frac{1}{3}$ times the Sun's). Hence the pole P moves off in a direction perpendicular to HP, and not to KP. In consequence
of the rotation of the Moon's nodes, L, and therefore also H,
revolves in a small circle about P in the period of $18\frac{2}{3}$ years
(see Fig. 159).

Let L_1, L_2, L_3, L_4, L_5 be the positions of L, and $P_1, P_2, P_3,$
P_4, P_5 the positions of P, when the angle PKL is $0°$, $90°$,
$180°$, $270°$, $360°$ respectively, H_2, H_4 the positions of H corresponding to L_2, L_4. Then at P_1 and P_3 the couple is
directed towards K, and therefore P is then moving perpendicular to KP. At P_2 the couple is directed towards H_2, and
the pole P_2 moves perpendicularly to H_2P_2, thus passing from the
inside to the outside of the small circle described by its mean
position. Similarly, at P_4 the pole, by moving perpendicularly
to H_4P_4, passes back from the outside to the inside of the
small circle which it would describe if the couple were
always directed towards K. Thus the wavy form of the
curve described by P is accounted for. And since the whole
space P_1KP_5 or L_1KL_5, traversed in a revolution of L, is very
small, the period of oscillation is almost exactly that of
revolution of the Moon's nodes.

Again, the Moon's couple depends on the angular distance
PL, and is greater the greater this distance (as may easily be
seen by § 457). Hence the resultant couple, and therefore
also the precessional motion, is least at P_1 and greatest at P_3.
This accounts for the variable rate of motion of P, which
gives rise to the equation of the equinoxes.

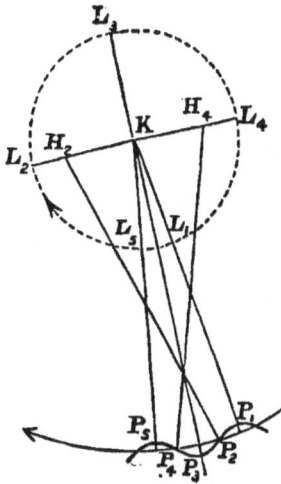

FIG. 159.

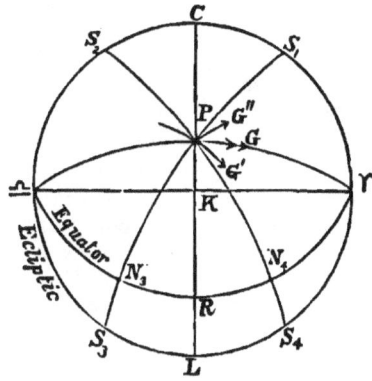

FIG. 160.

* 470. Solar and Monthly Nutations.—The variations in the intensity of the Sun's and Moon's disturbing couples during their orbital revolutions give rise to two other kinds of nutation. Let us first consider the variations in the Sun's disturbing couple, which produce Solar Nutation. It appears from § 457, that the couple vanishes when the Sun is on the equator, and that it is greater the greater the Sun's declination. Also it is readily evident from Fig. 153 that the couple in general acts in a plane through the Sun and the Earth's poles, tending to turn the poles more nearly perpendicular to the direction of the Sun. This shows that the couple is not really directed towards the pole of the ecliptic (though this is its average direction for the year) except at the solstices (Fig. 160).

Now at the vernal equinox, when the Sun is at Υ, the couple vanishes, and therefore the Earth's tendency to precession, due to the Sun, vanishes. Between the vernal equinox and the summer solstice, when the Sun is at S_1, the couple is along S_1P away from S_1, and this tends to make the pole precess along PG' perpendicularly to S_1P. At the summer solstice the couple along CP is a maximum, and tends to produce precession along PG perpendicular to KP. At S_2 the couple along S_2P tends to make the pole precess in the direction PG''. At the autumnal equinox, $\triangle$, the couple, and therefore the velocity of solar precession, vanishes. At S_3 the Sun's declination is negative, and the couple tends to draw P *towards* S_3; hence the Earth again tends to precess along PG'. At the winter solstice the direction of precession is again along PG, and the precessional velocity again a maximum. Finally, at S_4 the direction of precession is again along PG''.

Hence the variations in the Sun's declination cause the pole to thread its way in and out of the circle it would describe under uniform precession once every six months, and to cause the velocity of revolution about K to fluctuate in the same period. This gives rise to the nutation known as *Solar Nutation*, whose period is half a tropical year. In the case of the Moon the corresponding phenomenon is known as *Monthly Nutation*, and its period is half a month ; the explanation is exactly the same.

The variations in the obliquity of the ecliptic due to these two causes are small, because, owing to the comparatively small period in which they recur, the pole has not time to oscillate to and from K to any great extent. Moreover, the couple, and therefore the rate of motion of P, decreases as the inclination of PG' to PG increases. When the Sun is at ♈ or ♎ the displacement, if it existed, would be along PK, in the most advantageous direction for producing nutation, but at this instant the couple vanishes.

The solar nutation only displaces the pole about $1.2''$ to or from K, and the displacement due to monthly nutation is imperceptible. The effects on the equation of the equinoxes are more apparent. Under the Sun's action alone, the pole would come to rest *twice a year*, viz., at the equinoxes, and under the Moon's action its rate of motion would vanish *twice a month*, viz., when the Moon crossed the equator. At all other times the couples tend to produce *retrograde—never direct*—motion of the pole about K. Hence the precessional motion can never vanish unless the Sun and Moon should happen to cross the equator *simultaneously*.

Section IV.—*Lunar and Planetary Perturbations.*

471. In consequence of the universality of gravitation, every body in the solar system has its motion more or less disturbed by the attraction of every other body. Kepler's Laws (with the modification of the Third Law given in § 421) would only be strictly true if each planet were attracted solely by the Sun, and each satellite described its relative orbit solely under the attraction of its primary. Hence the fact that these laws very nearly agree with the results of observation shows that the mutual attractions of the planets are small compared with that which the Sun exerts on each of them, and that, in the orbital motion of a satellite, by far the greater part of the relative acceleration is due to the attraction of the primary.

472. Lunar Perturbations.—We have seen, in Section I., that the Moon's motion consists of two component parts, a monthly orbital motion relative to the Earth—or, more strictly, relative to the centre of mass of the Earth and Moon—and the annual orbital motion of this centre of mass in an ellipse about the Sun. If the acceleration of the Sun's attraction were the same in magnitude and direction at the Moon as at the Earth, it would be exactly the acceleration required to produce the latter component, and the relative orbit about the Earth would be determined by the Earth's attraction alone. This is very nearly the case, owing to the great distance of the Sun. But the small differences of the accelerations caused by the Sun's attraction on the Earth and Moon tend to modify the relative motion of these two bodies, by giving rise to **perturbations** (§ 272). The relative accelerations thus produced may be represented by a distribution of *disturbing force* due to the Sun, just in the same way that the relative accelerations of the oceans, which cause the tides, are determined by distributions of disturbing force due to the Sun and Moon. And since the Sun's distance is nearly 400 times the Moon's, the expressions for the disturbing force, corresponding to those investigated in § 441, are sufficiently approximate to account for the more important lunar perturbations.

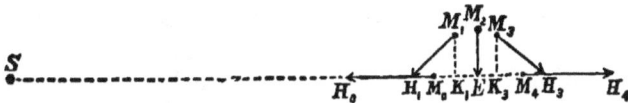

FIG. 161.

Let S, E, M denote the centres of the Sun, Earth, and Moon. Drop MK perpendicular on ES, and on EK produced take $KH = 2EK$. Then, if S denote the mass and r the distance of the Sun, the Sun's disturbing force produces at M a relative acceleration along MH of magnitude $kS \cdot MH/r^3$, its components being $k \cdot S \cdot MK/r^3$ along MK and $2k \cdot S \cdot EK/r^3$ parallel to EK.

This force tends to accelerate the Moon *towards* the Earth at quadrature (M_2), and *away* from the Earth at conjunction and opposition (M_0, M_4). At any other position it accelerates the Moon towards a point (H_1) in the line ES, and thus makes the Moon tend to approach the Sun, if its elongation ($M_1 ES$) is less than 90°; but it accelerates the Moon towards a point (H_3) away from the Sun if its angle of elongation from the Sun be obtuse.

473. The Rotation of the Moon's Nodes.—Let CL represent the ecliptic, $N_1 M_1 N_1'$ the great circle which the Moon would appear to describe on the celestial sphere if there were no disturbing force acting upon it, and let H, between N_1 and N_1' on the ecliptic, represent either the Sun's position on the celestial sphere or that of the point antipodal to it. Then the reasoning of the last paragraph shows that the disturbing force acts in the plane HEM_1, and therefore has a component at M_1 directed along the tangent to the great circle $M_1 H$.

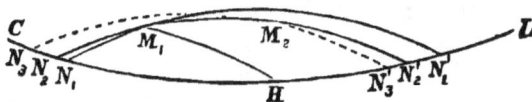

FIG. 162.

Now let us suppose that the Moon is revolving under the Earth's attraction alone, but that on arriving at M_1 it is acted on by a sudden impulse or blow directed towards H. Clearly the effect of such an impulse is to bend the direction of motion inward, from $M_1 N_1'$ to $M_1 N_2'$, and the Moon will then begin to describe a great circle $M_1 N_2'$, which, if produced both ways, will intercept the ecliptic at points N_2, N_2' *behind* N_1, N_1'. The inclination of the orbit to the ecliptic will also be diminished slightly if M_1 is within 90° of N_1 ; for the exterior angle $MN_1 H > MN_2 H$, *since the sides of the triangle* $M_1 N_1 N_2$ *are each less than* 90°. But when the Moon comes to M_2, let another impulse act towards H. This will deflect the direction of motion from $M_2 N_2'$ to $M_2 N_3'$, and the Moon will now begin to describe the great circle $N_3 M_2 N_3'$, whose nodes N_3, N_3' are still further behind their initial positions. The inclination of the orbit to the ecliptic will, however, be increased this time.

It is easy to see that the same general effect takes place when the Moon is acted on by a *continuous force*, always tending *towards* the ecliptic, instead of a series of impulses. Such a force continuously deflects the Moon's direction of motion, and draws the Moon down so that it returns to the ecliptic more quickly than it would otherwise. Hence the Moon, after leaving one node, arrives at the next before is has quite described 180°, and the result is an apparent *retrograde* (*never direct*) motion of the nodes, combined with periodic, but small, fluctuations in the inclination of the orbit.

*474. The retrograde motion of the Moon's nodes is, in some respects, analogous to the precession of the equinoxes, and, although the analogy is somewhat imperfect, the former phenomenon gives an illustration of the way in which the latter is produced. If the Earth had a string of satellites, like Saturn's rings, closely packed together in a circle in the plane of the equator, the Sun's disturbing force, ever ac celerating them towards the ecliptic, would, as in the case of the Moon, cause a retrograde motion of the points of intersection of all of their paths with the ecliptic, and this would give the appearance of a kind of retrograde *precession* of the plane of the rings. If the particles, instead of being separate, were united into a solid ring, the general phenomena would be the same. And it is not unnatural to expect that what occurs in a simple ring should also occur, to a greater or less degree, in the case of other bodies that are somewhat flattened out perpendicularly to their axis of rotation, such as the Earth, thus accounting for the precession of the equinoxes. (Of course this is only an *illustration*, not a rigorous proof; in fact, if the Earth were *quite* spherical it would behave very differently.)

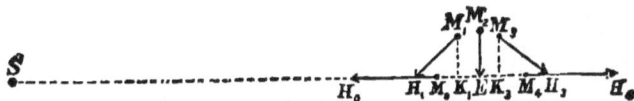

FIG. 163.

*475. **Perturbations due to Average Value of Radial Disturbing Force.**—Let d be the Moon's distance. Then, when the Moon is in conjunction or opposition, the Sun's disturbing force acts *away* from the Earth, and is of magnitude $2kSd/r^3$ (Fig. 163). When the Moon is in quadrature the disturbing force acts *towards* the Earth, but is only half as great. Hence, on the *average*, the disturbing force tends to pull the Moon *away* from the Earth.

In consequence, the Moon's average centrifugal force must be rather less than it would be at the same distance from the Earth if there were no disturbing force, and the effect of this is to *make the month a little longer* than it would be otherwise for the same distance of the Moon.

Moreover, the disturbing force increases as the Moon's distance increases, but the Earth's attraction diminishes, being proportional to the inverse square of the distance; this has the effect of making the whole *average* acceleration along the radius vector *decrease more rapidly as the distance increases* than it would according to the law of inverse squares.   The result of this cause is *the progressive motion of the apse line.*   It is difficult to explain this in a simple manner, but the following arguments may give some idea of how the effect takes place.   At apogee the Moon's average acceleration is less, and at perigee it is greater than if it followed the law of inverse squares and had the same mean value.   Hence, when the Moon's distance is greatest, as at apogee, the Earth does not pull the Moon back so quickly, and it takes longer to come back to its least distance, so that it does not reach perigee till it has revolved through *a little more than* 180°.   Similarly, at perigee the greater average acceleration to the Earth does not allow the Moon to fly out again quite so quickly, and it does not reach apogee till it has described *rather more than* 180°.   Hence, in each case, the line of apsides moves forward on the whole.

*476. **Variation, Evection, Annual Equation, Parallactic Inequality.**—When the Moon is nearer than the Earth to the Sun (M_1, Fig. 162), the Moon is more attracted than the Earth, and therefore the disturbing force is towards the Sun (§ 472).   Its effect is, therefore, to accelerate the Moon from last quarter to conjunction, and to retard it from conjunction to first quarter.   When the Moon is more distant than the Earth from the Sun (M_3, Fig. 163), it is less attracted than the Earth, and therefore the disturbing force is away from the Sun.   Thus the Moon is accelerated from first quarter to full Moon, and retarded from full Moon to last quarter.†   Hence we see that the Moon's motion in each case must be swiftest at conjunction and opposition, and slowest at the quadratures.   This phenomenon is known as the Variation.

The force *towards the Earth* is greatest at the quadratures, and least at the conjunction and opposition, since at the former the Sun pulls the Moon towards, and at the latter away from the Earth.   Either cause tends to make the orbit more curved at the quadratures and less curved at the syzygies.   For, if v is the velocity, R the radius of curvature, then v^2/R = normal acceleration.   Hence R is greatest, and the orbit therefore least curved, when v is greatest, and the normal acceleration is least.   The effect of this cause would be to distort the orbit, if it were a circle, into a slightly oval curve, which would be most flattened, and therefore narrowest (compare

---

† These retardations and accelerations are closely analogous to those of the water in an equatorial canal (§ 445).

arguments of §§ 114, 115), at the points towards and opposite the Sun; most rounded, and therefore broadest, at the points distant 90° from the Sun.

Of course the Moon's undisturbed orbit is not really circular, but elliptic, and far more elliptic than the oval into which a circular orbit would be thus distorted. But a distortion still takes place, and gives rise to periodical changes in the eccentricity, depending on the position of the apse line, and known as evection.

The Sun's disturbing force is greatest when the Sun is nearest, and least when the Sun is furthest. These fluctuations, between perihelion and aphelion, give rise to another perturbation, called the annual equation, whose most noticeable effect consists in the consequent variations in the length of the month (§ 475).

If, instead of resorting to a first approximation, we employ more accurate expressions for the Sun's disturbing force on the Moon, it is evident that this force is greater when the Moon is near conjunction than at the corresponding position near opposition; just as the disturbing force which produces the tides is really greater under the Moon than at the opposite point. Hence the Moon is more disturbed from last quarter through new Moon to first quarter than from first quarter through full Moon to last quarter. Hence the time of first quarter is slightly accelerated, and that of last quarter retarded. This is called the Moon's Parallactic Inequality. Its amount is proportional to kSd^2/r^4, instead of kSd/r^3 (like the other perturbations). For many reasons this perturbation is of considerable use in determinations of the Sun's mass and distance.

477. Planetary Perturbations.

The Sun's mass is so great, compared with the masses of the planets, that the orbital motion of one planet about the Sun is but slightly affected by the attraction of any other planet. The mutual attractions of the planets, and their actions on the Sun, give rise to small **planetary perturbations**, which cause each planet to diverge slowly from its elliptical orbit, besides accelerating or retarding its motion.

Since the orbital motions of the planets are all usually referred to the Sun as their common centre or " origin," and not to the centre of mass of the solar system, the perturbations of one planet, due to a second, depend, not on the actual acceleration produced by the latter, but on the differences of the accelerations which it produces on the former planet and on the Sun.

As in the case of the Moon, the force which produces this difference of accelerations is called the *disturbing force*.

ASTRON.   2 E

#478. Geometrical Construction for the Disturbing Force.—The approximate expressions, investigated in § 472, for the Sun's disturbing force on the Moon, are inapplicable to the disturbing force of one planet on another, because the distance of the *disturbing* body from the Sun is no longer very large, compared with that of the *disturbed* body. We must, therefore, adopt the following construction (Fig. 164) :—

Let P, Q be two planets, of masses M, M'; S the Sun. Then the planet P produces an acceleration kM/PQ^2 on Q along QP, and an acceleration kM/PS^2 on S along SP. To find the acceleration of Q, relative to S, due to this cause, take a point T on PQ such that $PT : PS = PS^2 : PQ^2$. Then the accelerations of S, Q, due to P, are $kM . SP/SP^3$ and $kM . TP/SP^3$ respectively. Hence, by the triangle of accelerations, the acceleration of Q, relative to S, is represented in magnitude and direction by $kM . TS/SP^3$. Therefore the disturbing force per unit mass on Q, due to P, is parallel to TS, and of magnitude $kM . TS/SP^3$.

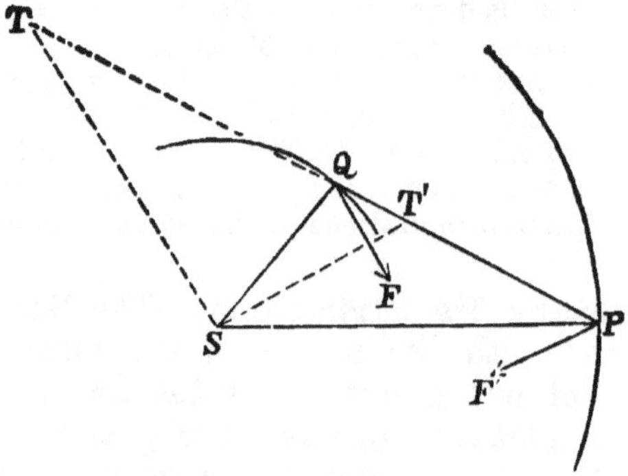

FIG. 164.

Similarly, if we take a point T' on QP such that $QT' : QS = QS^2 : QP^2$, the disturbing force per unit mass on P, due to Q, is parallel to $T'S$, and is of magnitude $kM' . T'S/SQ^3$.

The disturbing force on Q, due to P, and that on P, due to Q, are not equal and opposite, because they depend on the planets' attractions on S, as well as on their mutual attractions.

When $PQ = PS$, the points Q, T evidently coincide, and the disturbing force on Q is along the radius vector QS. When $PQ < PS$, $PT > PQ$, so that the disturbing force on Q tends to pull Q about S (as in Fig. 164) *towards* P, and when $PQ > PS$, the disturbing force tends to push Q about S *away* from P.

Similarly, when $QP = QS$, the disturbing force on P is along PS. When $QP < QS$ it tends to pull P about S *towards* Q, and when $QP > QS$, it tends to push P about S *away* from Q.

*479. Periodic Perturbations on an Interior Planet.—Let us consider, in the first place, the perturbations produced by one planet E on another planet V, whose orbit is nearer the Sun; as, for example, the perturbations produced by the Earth on Venus, by Jupiter or Mars on the Earth, or by Neptune on Uranus.

Let A, B be the positions of the planet, relative to E, when in heliocentric conjunction and opposition respectively; U, U' points on the relative orbit such that $EU = EU' = ES$. (These points are near, but not quite coincident with the positions of greatest elongation.)   Then, if we only consider the component relative acceleration of V perpendicular to the radius vector VS, this vanishes when the planet is at U or U', as shown in the last paragraph.

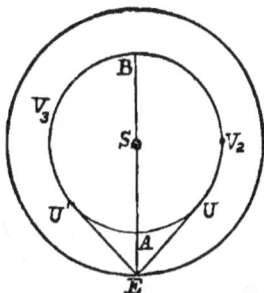

FIG. 165.

The tangential acceleration also vanishes at A and B.   Over the arc $U'AU$ the relative acceleration is *towards* E, therefore the planet's orbital velocity is accelerated from U' to A; similarly it is retarded from A to U.

Again, at a point V_2 on the arc UBU', the relative acceleration is away from the Earth, and this accelerates the planet's orbital velocity between U and B, and retards it between B and U'.

It follows that V is moving most swiftly at A and B, and most slowly at U and U'.   Hence, if we neglect the eccentricity of the orbit, we see that the planet, after passing A, will shoot ahead of the position it would occupy if moving uniformly; thus the disturbing force displaces the planet *forwards* during its path from A to near U.   Somewhere near U, when the planet is moving with its least velocity, it begins to lag behind the position it would occupy if moving uniformly; thus from near U to B the disturbing force displaces the planet *backwards*.   Similarly, it may be seen that from B to near U' the planet is displaced *forwards*, and from near U' to A it is displaced *backwards*.

The principal effect of the component of the disturbing force along the radius vector, is to cause rotation of the planet's apsides, as in the case of the Moon. The direction of their rotation depends on the direction of the force, and is not always direct. The eccentricity of the orbit is also affected by this cause, as in the phenomenon of lunar *evection*, and the periodic time is slightly changed.

Owing to the inclination of the planes of the orbits of E, V, the attraction of E, in general, gives rise to a small component perpendicular to the plane of V's orbit, which is *always directed towards* the plane of E's orbit. This component produces *rotation of the line of nodes*, or line of intersection of the planes of the two orbits. This rotation is *always in the retrograde direction*, and is to be explained in exactly the same way as the rotation of the Moon's nodes.

It is thus a remarkable fact that since all the bodies in the solar system (except the satellites of Uranus and Neptune) rotate in the direct direction, all the planes of rotation and revolution, and all their lines of intersection (*i.e.*, the lines of nodes, and the lines of equinoxes) in the whole solar system, with the above exceptions, have a *retrograde* motion.

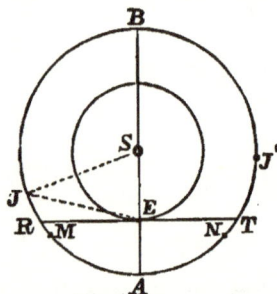

FIG. 166.

*480. Periodic Perturbations of an Exterior Planet.—The accelerations and retardations produced by a planet E on one J, whose orbit is more remote from the Sun, during the course of a synodic period, may be investigated in a similar manner to the corresponding perturbations of an interior planet, assuming the orbits to be nearly circular.

If SJ is less than $2SE$ there are two points M, N on the relative orbit at which $EM = EN = ES$. At these points the disturbing force is purely radial, and it appears, as before, that the planet J is accelerated from heliocentric conjunction A to M, and from heliocentric opposition B to N; retarded from N to A, and from M to B.

If $SJ > 2SE$, then $ES < EA$; hence the attraction of E is greater on the Sun than on J, and the disturbing force therefore always accelerates the planet J towards B. Thus the planet's orbital velocity increases from A to B, and decreases from B to A, and it is greatest at B and least at A. Therefore from B to A the planet is displaced in advance of its mean position, and from A to B falls behind its mean position.

The effects of the radial and orthogonal components of the disturbing force in altering the period and causing rotation of the apse line, and regression of the nodes, can be investigated in the same way for a superior as for an inferior planet.

*481. Inequalities of Long Period.—If the orbits of the planets were circular (except for the effects of perturbations), and in the same plane, their mutual perturbations would be strictly periodic, and would recur once in every synodic period. Owing, however, to the inclinations and eccentricities of the orbits, this is not the case. The mutual attractions of the planets produce small changes in the eccentricities and inclinations, and even in their periodic times, which depend on the positions of conjunction and opposition relative to the lines of nodes and apses. Neglecting the motion of these latter lines, the perturbations would only be strictly periodic if the periodic times of two planets were commensurable; the period of recurrence being the least common multiple of the periods of the two planets. But when the periodic times of two planets are *nearly but not quite* in the proportion of two small whole numbers, **inequalities of long period** are produced, whose effects may, in the course of time, become considerable.

Thus, for example, the periodic times of Jupiter and Saturn are *very nearly but not quite* in the proportion of 2 to 5. If the proportionality were exact, then 5 revolutions of Jupiter would take the same time as 2 revolutions of Saturn; and, since Jupiter would thus gain three revolutions on Saturn, the interval would contain 3 synodic periods. Thus, after 3 synodic periods had elapsed from conjunction, another conjunction would occur at exactly the same place in the two orbits, and the perturbations would be strictly periodical.

But, in reality, the proportionality of periods is not exact; the positions of every third conjunction are very slowly revolving in the direct direction. They perform a complete revolution in 2,640 years. But there are three points on the orbits at which conjunctions occur, and these are distant very nearly 120° from one another. It follows that when the positions of conjunction have revolved through 120°, they will again occur at the same points on the orbits, and the perturbations will again be of the same kind as initially. The time required for this is one-third the above period, or 880 years, and consequently Jupiter and Saturn are subject to long-period inequalities which recur only once in 880 years.

Again, the periodic times of Venus and the Earth are nearly in the proportion of 8 to 13; consequently 5 conjunctions of Venus occur in almost exactly 8 years, thus giving rise to perturbations having a period of 8 years. But the proportion is not exact, and, consequently, there are other mutual perturbations having a very long period.

One of the most important secular perturbations is the alternate increase and decrease in the eccentricity of the Earth's orbit. This, at the present time, is becoming gradually more and more circular, but in about 24,000 years the eccentricity will be a minimum, and will then once more begin to increase. The effects of this cause on the climate of the Earth's two hemispheres have already been considered (§ 463).

482. Gravitational Methods of Finding the Sun's Distance.—The Earth's perturbations on Mars and Venus furnish a good method of finding the Sun's distance. For the magnitude of these perturbations depends on the ratio of the Earth's mass, or rather the sum of the masses of the Earth and Moon (since both are instrumental in producing the perturbations), to the Sun's mass. Hence, if S, M, m denote the masses of the Sun, Earth, and Moon, it is possible, from observations of these perturbations, to find the ratio of $(M+m) : S$.

But, if r, d be the distances of the Sun and Moon from the Earth, T and Y the length of the sidereal lunar month and year, we have, by Kepler's corrected Third Law,

$$(M+m)\, T^2 : (S+M+m)\, Y^2 = d^3 : r^3 ;$$

whence the ratio of r to d is known. If, now, the Moon's distance d be determined by observation in any of the ways described in Chapter VIII., or by the gravitational method of § 423, the Sun's distance r may be immediately found.

This method was used by Leverrier in 1872. From observations of certain perturbations of Venus he found the values $8 \cdot 853''$ and $8 \cdot 859''$ for the Sun's parallax, while the rotation of the apse line of Mars gave the value $8 \cdot 866''$.

The perturbations of Encke's comet were used in a similar way by Von Asten, in 1876, to find the Sun's parallax, the value thus obtained being rather greater, viz., $9 \cdot 009''$.

The lunar perturbations also furnish data for determining the Sun's distance, the principal of these being the parallactic inequality of the Moon (§ 476). Several computations of the Sun's parallax have thus been made, the results being $8 \cdot 6''$ by Laplace in 1804, $8 \cdot 95''$ by Leverrier in 1858, $8 \cdot 838''$ by Newcomb in 1867. See also § 437 for the determination of the parallax from the apparent monthly displacement of the Sun.

483. Determination of Masses. — The mass of any planet which is not furnished with a satellite can be determined in terms of the Sun's mass by means of the perturbations it produces on the orbits of other planets. The amount of these perturbations is always proportional to the disturbing force, and this again is proportional to the *mass* of the disturbing planet. In this manner the mass of Venus has been found to be about 1/400,000 of the Sun's mass, and that of Mercury about 1/5,000,000.

484. The Discovery of Neptune. — The narrative of the discovery of Neptune is one of the most striking and remarkable in the annals of theoretical astronomy, and forms a fitting conclusion to this chapter.

In 1795, or about 14 years after its discovery, the planet Uranus was observed to deviate slightly from its predicted position, the observed longitude becoming slightly greater than that given by theory. The discrepancy increased till 1822, when Uranus appeared to undergo a retardation, and to again approach its predicted position. About 1830 the observed and computed longitudes of the planet were equal, but the retardation still continued, and by 1845 Uranus had fallen behind its computed position by nearly 2'.

As early as 1821, Alexis Bouvard pointed out that these discrepancies indicated the existence of a planet exterior to Uranus, but the matter remained in abeyance until 1846, when the late Mr. (afterwards Prof.) Adams, in Cambridge, and M. Leverrier, in Paris, independently and almost simultaneously, undertook the problem of determining the position, orbit, and mass of an unknown planet which would give rise to the observed perturbations. Adams was undoubtedly the first by a few months in performing the computations, but the actual search for the planet at the observatory of Cambridge was delayed from pressure of other work. Meanwhile Leverrier sent the results of his calculations to Dr. Galle, of Berlin, who, within a few hours of receiving them, turned his telescope towards the place predicted for the planet, and found it within about 52' of that place. Subsequent examination of star charts showed that the planet had been previously observed on several occasions, but had always been mistaken for a fixed star.

It will be seen from § 479 that the acceleration of Uranus up to 1822, and its subsequent retardation, are at once accounted for by supposing an exterior planet to be in helio-centric conjunction with the Sun about the year 1822. But Adams and Leverrier sought for far more accurate details concerning the planet. At the same time the data afforded by the observed perturbations of Uranus were insufficient to determine all the unknown elements of the new planet's orbit, and therefore the problem admitted of any number of possible solutions. In other words, any number of different planets could have produced the observed perturbations. To render the problem less indeterminate, however, both astronomers assumed that the disturbing body moved nearly in the plane of the ecliptic and in a nearly circular orbit, that its distance and period were connected by Kepler's Third Law, and that its distance from the Sun followed Bode's Law. The latter assumption led to considerable errors, including an erroneous estimation of the planet's period by Kepler's Third Law. For when Neptune was observed, its distance was found to be only 30·04 times the Earth's distance, instead of 38·8 times, as it would have been according to Bode's Law. Nevertheless, the actual planet was subsequently found to fully account for all the observed perturbations of Uranus.

The discovery of Neptune affords most powerful evidence of the truth of the Law of Gravitation, and so indeed does the theory of perturbations generally. The fact that the planetary motions are observed to agree closely with theory, that computations of astronomical constants (such as the Sun's and Moon's distances), based upon gravitational methods, agree so closely with those obtained by other methods, when possible errors of observation are taken into account, affords an indisputable proof that the resultant acceleration of any body in the solar system can always be resolved into com-ponents directed to the various other bodies, each component being proportional directly to the mass and inversely to the square of the distance of the corresponding body. Such a truth cannot be regarded as a fortuitous coincidence; it can only be explained by supposing every body in the universe to attract every other body in accordance with Newton's Law of Universal Gravitation.

EXAMPLES.—XIV.

1. If the Sun's parallax be 8·80″, and the Sun's displacement at first quarter of Moon 6·52″, calculate the mass of the Moon, the Earth's radius being taken as 3,963 miles.

2. Supposing the Moon's distance to be 60 of the Earth's radii, and the Sun's distance to be 400 times that of the Moon, while his mass is 25,600,000 times the Moon's mass, compare the effects of the Sun and Moon in creating a tide at the equator, in the event of a total eclipse occurring at the equinox.

3. If the Earth and Moon were only half their present distance from the Sun, what difference would this make to the tides? Calculate roughly what the proportion between the Sun's tide-raisng power and the Moon's would then be, assuming the Moon's distance from the Earth remained the same as at present.

4. Taking the Moon's mass as $\frac{1}{80}$ of the Earth's, and its distance as 60 times the Earth's radius, show that the Moon's tide-raising force increases the intensity of gravity by 1/17,280,000 when the Moon is on the horizon, and that it decreases the intensity of gravity by 1/8,640,000 when the Moon is in the zenith.

5. Compare the heights of the solar tides on the Earth and on Mercury, taking the density of Mercury to be twice that of the Earth, its diameter ·38 of the Earth's diameter, and its solar distance ·38 of the Earth's solar distance.

6. Explain how the pushing forward of the Moon by the tidal wave enlarges the Moon's orbit.

7. Show that, owing to precession, the right ascension of a star at a greater distance than $23\frac{1}{2}°$ from the pole of the ecliptic will undergo all possible changes, but that a star at a less distance than $23\frac{1}{2}°$ will always have a right ascension greater than twelve hours.

8. Prove that for a short time precession does not alter the declinations of stars whose right ascensions are 6h., or 18h.

9. Exhibit in a diagram the position of the pole star (R.A. = 1h. 20m., decl. = 88° 40′) relative to the poles of the equator and ecliptic, and hence show that owing to precession its R.A. is increasing rapidly, but that its polar distance is decreasing.

10. Describe the disturbing effects of Neptune on Uranus for a short time before and after heliocentric conjunction, pointing out when Uranus is displaced in the direct, and when in the retrograde direction.

EXAMINATION PAPER.—XIV.

1. Show that the Moon's orbit is everywhere concave to the Sun.

2. Show that the tide-raising force of a heavenly body is nearly proportional to its (mass) ÷ (distance)³.

3. How is it that we have tides on opposite sides of the Earth at once?

4. Explain the production of the tides on the equilibrium theory.

5. Define the terms *spring tide*, *neap tide*, *priming* and *lagging*, *establishment of the port, lunar time*.

6. What is meant by the expression " Luni-solar Precession" ? Describe the action of the Sun and of the Moon in causing the Precession.

7. Give a general description of Precession. Does precession change the position of (*a*) the equator, (*b*) the ecliptic among the stars?

8. Describe *nutation*. What is the cause of Lunar Nutation? What is meant by the *equation of the equinoxes* ?

9. Give a brief account of the discovery of Neptune.

10. Explain how the retrograde motion of the Moon's nodes is caused by the Sun's attraction on the Earth and Moon.

NOTE I.

DIAGRAM FOR SOUTH LATITUDES.

In order to familiarize the student with astronomical diagrams drawn under different conditions, we subjoin a figure showing the principal circles of the celestial sphere of an observer in South latitude 45° at about 19h. of sidereal time ($Q W R \Upsilon = 270° + 15° = 19h.$). The figure shows also the Sun's daily paths at the solstices; also the arcs $\Upsilon R \triangleq Q M$, and Mx, which measure the R.A. and N. decl. of the star x.

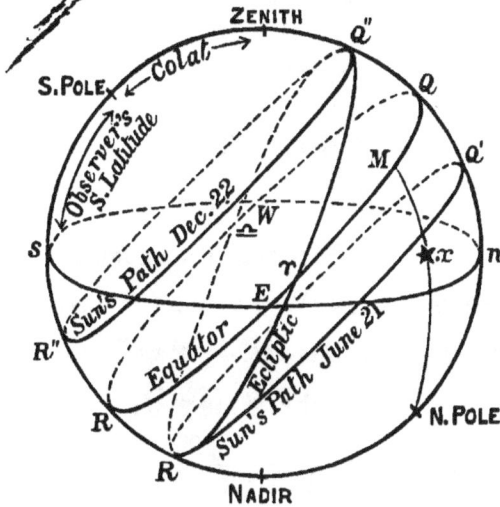

FIG. 169.

NOTE II.

THE PHOTOCHRONOGRAPH.

Quite recently photography has been applied to recording transits, as an alternative for the methods explained in Chap. II., §§ 49, 50. The image of the observed star is projected on a sensitized plate placed in front of the transit circle, and, owing to the diurnal motion, it moves *horizontally* across the plate. The plate is made to oscillate slightly in a vertical direction, by means of clockwork, say once in a second, and this motion, combined with the horizontal motion of the image, causes it to describe a zigzag or wavy streak on the plate. The star's position at each second is indicated by the undulations, and the position of these is capable of being measured with great exactness.

NOTE III.

NOTE ON § 104.

It may be proved, by Spherical Trigonometry, that

$$\sin nP = \sin xP \sin nxP, \quad \text{or} \quad \sin l = \cos d \sin nxP \; ;$$

$$\therefore \; \cos^2 d \cos^2 nxP = \cos^2 d - \cos^2 d \sin^2 nxP = \cos^2 d - \sin^2 l$$

$$= \cos(d+l)\cos(d-l) \; ;$$

$$\therefore \; \text{acceleration } t = \frac{D''}{15 \, \sqrt{(\cos^2 d - \sin^2 l)}}$$

$$= \frac{D''}{15 \, \sqrt{\{\cos(d+l)\cos(d-l)\}}} \text{ secs.}$$

The same formula is applicable to §§ 135, 190.

APPENDIX.

PROPERTIES OF THE ELLIPSE.

For the benefit of those readers who have not studied Conic Sections, we subjoin a list of those properties of the ellipse which are of astronomical importance. The proofs are given in books on Conic Sections.

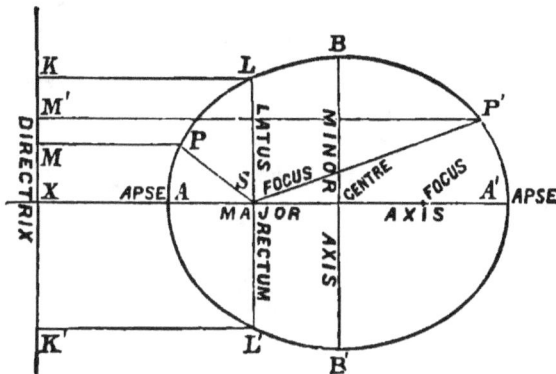

FIG. 168.

1. DEFINITION.—A conic section is a curve such that the distance of every point on it from a certain fixed point is proportional to its perpendicular distance from a certain fixed straight line.

The fixed point is called the focus, the fixed line is called the directrix, and the constant ratio of distances is called the eccentricity.

If this constant ratio or eccentricity is less than unity, the curve is called an ellipse. In this case the curve assumes the form of a closed oval, as shown in the figure.

If S is the focus, and if from A, P, L, P', A', &c., any points on the curve, perpendiculars AX, PM, &c., be drawn on the directrix, and if the eccentricity be e, the definition requires that

$$e = \frac{SA}{AX} = \frac{SP}{PM} = \frac{SL}{LK} = \frac{SP'}{P'M'} = \frac{SA'}{A'X} = \&c.,$$

and that e is less than unity.

The other conic sections, the *parabola* and *hyperbola*, are defined by the same property, save that in the former $e = 1$, and in the latter $e > 1$; but they are of little astronomical importance, except as representing the paths described by non-periodic comets.

2. An ellipse has two foci (each focus having a corresponding directrix), and the sum of the distances of any point from the two foci is constant.

Thus in Fig 169, S, H are the two foci, and the sum $SP + PH$ is the same for all positions of P on the curve.

From this property an ellipse may easily be drawn. For, let two small pins be fixed at S and H, and let a loop of string SPH be passed over them and round a pencil-point P; then, if the pencil be moved so as to keep the string tight, its point P will trace out an ellipse. For $SP + PH + HS =$ constant, and $\therefore$ $SP + PH =$ constant.

3. For all positions of P on the ellipse, SP is inversely proportional to $1 + e \cos ASP$, so that

$$SP (1 + e \cos ASP) = l = \text{constant,}$$

e being the eccentricity and SA the line through S perpendicular to the directrix.

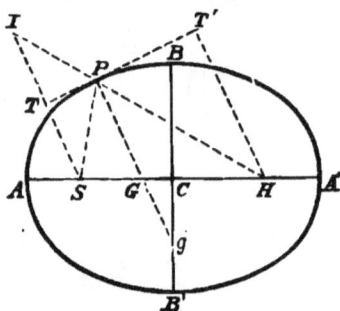

FIG. 169.

4. The line joining the two foci is perpendicular to the directrices.

The portion of this line (AA'), bounded by the curve, is called the major axis or axis major. Its middle point C is called the centre, and the curve is symmetrical about this point.

The line BCB', drawn through the centre perpendicular to ACA' and terminated by the curve, is called the minor axis or axis minor. The lengths of the major and minor axes are usually denoted by $2a$ and $2b$ respectively.

5. The extremities A, A' of the major axis are called the apses or apsides. Since, by (2), $SP + HP$ is constant, therefore, taking P at A or A', $SP + HP = SA + HA = SA' + HA'$

$$= \tfrac{1}{2} (SA + HA + SA' + HA') \text{ evidently}$$
$$= AA' = 2a.$$

Taking P at B, $SB + HB = 2a$;

$$\therefore SB \text{ (evidently)} = HB = a = CA.$$

6. The eccentricity $e = CS/CA$; $\therefore$ $CS = e \cdot CA$, and
$$b^2 = CB^2 = SB^2 - CS^2 \text{ (Euc. I. 47)} = a^2 - a^2 e^2 = a^2 (1 - e^2);$$
$$\therefore e^2 = (a^2 - b^2)/a^2.$$
Hence also
$$SA = CA - CS = a(1 - e) \quad \text{and} \quad SA' = CA' + CS = a(1 + e).$$

7. The **latus rectum** is the chord LSL' drawn through the focus perpendicular to the major axis AA'. Its length is $2l$, where $l = a (1 - e^2)$. Also l is the constant of (3), for when P coincides with L, $ASP = 90°$; $\therefore$ cos $ASL = 0$, and $SL = l$. [Fig. 168.]

8. The **tangent** $T'PT$ and **normal** PGg, at P, bisect respectively the exterior and interior angles (SPI, SPH) formed by the lines SP, HP.

9. If the normal meets the major and minor axes in G, g,
$$PG : Pg = CB^2 : CA^2 \quad (\equiv b^2 : a^2).$$

10. If ST, drawn perpendicular on the tangent at P, meets HP produced in I, then evidently $SP = IP$;
$$\therefore HI = SP + HP = 2a \quad \text{[by (2)]}.$$

If HT' is the other focal perpendicular on the tangent, it is known that rectangle $ST \cdot HT' = \text{constant} = b^2$.

11. **Relation between the focal radius SP and the focal perpendicular on the tangent ST.**

Let $\qquad\qquad\qquad SP = r, \; ST = p.$
Then $\qquad\qquad\qquad$ cos $TIP = $ cos $TSP = p/r.$
By Trigonometry,
$$SH^2 = IS^2 + IH^2 - 2 \cdot IS \cdot IH \cdot \cos SIH;$$
$$\therefore 4a^2 e^2 = 4p^2 + 4a^2 - 8pa \times p/r;$$
$$\therefore \frac{a^2 (1 - e^2)}{p^2} = \frac{2a}{r} - 1,$$
or by (6) $\qquad\qquad \frac{b^2}{p^2} = \frac{2a}{r} - 1 = \frac{2a - r}{r} = \frac{HP}{SP}.$

This may also be proved from the similarity of the triangles SPT, HPT', which gives $\quad ST : HT' = SP : HP$;
$$\therefore ST^2 : ST \cdot HT' = SP : HP \text{ and } ST \cdot HT' = b^2 \quad (10);$$
$$\therefore p^2 : b^2 = r : 2a - r.$$

12. If a circular cone (*i.e.*, either a right or oblique cone on a circular base) is cut in two by a plane not intersecting its base, the curve of section is an *ellipse*. More generally, the form of a circle represented in perspective, or the oval shadow cast by a spherical globe or a circular disc on any plane, are *ellipses*. A circle is a particular form of ellipse for the case where $b = a$ and $\therefore e = 0$.

13. The *area of the ellipse* is $\pi a b$.

TABLE OF ASTRONOMICAL CONSTANTS.

(Approximate values, calculated, when variable, for the Spring
Equinox, A.D. 1900.)

THE CELESTIAL SPHERE.

Latitude of London (Greenwich Observatory),   51° 28′ 31″,
     ,,   Cambridge Observatory,   52° 12′ 51″.
Obliquity of Ecliptic,   23° 27′ 8″.

OPTICAL CONSTANTS.

Coefficient of Astronomical Refraction,   57″.
Horizontal Refraction,   33′.
Coefficient of Aberration,   20·493″.
Velocity of Light in miles per second,   186,330.
    ,,   ,,   ,,   metres   ,,   299,860,000.
Equation of Light,   8m. 18s

TIME CONSTANTS.

Sidereal Day in mean solar units $= 1 - 1/366\frac{1}{4}$ days $=$ 23h. 56m. 4·1s.
Mean Solar Day in sidereal units $= 1 + 1/365\frac{1}{4}$ days $=$ 24h. 3m. 56·5s.
Year, Tropical, in mean time,   365d. 5h. 48m. 45·51s.
  ,,   Sidereal,   ,,   365d. 6h. 9m. 8·97s.
  ,,   Anomalistic, ,,   365d. 6h. 13m. 48·09s.
  ,,   Civil, if the number of the year is not divisible by 4,
       or if it be divisible by 100, but not by 400,   365 days.
       In other cases,   366   ,,
Month, Sidereal,   27·32166d. $=$ 27d. 7h. 43m. 11·4s.
  ,,   Synodic,   29·53059d. $=$ 29d. 12h. 44m. 30s.
Metonic Cycle,   235 Synodic Months $=$ 6939·69d
              $=$ 19 tropical years (all but 2 hours).
Period of Rotation of Moon's Nodes (Sidereal), 6793·391d. $=$ 18·60 yr.
  ,,   ,,   ,,   ,, (Synodic), 346·644d.
                               $=$ 346d. 14½h.
  ,,   ,,   ,, Apsides (Sidereal), 3232·575d. $=$ 8·85 yr.
  ,,   ,,   ,,   ,, (Synodic),   411·74d.
Saros   223 Synodic Months $=$ 6585·29d. $=$ 18·0906 yr.
                          $=$ 18 yr. 10 or 11 days.
              $=$ 19 Synodic periods of Moon's Nodes (very nearly,
              $=$ 16   ,,   ,,   ,,   Apsides (nearly).
Equation of Time, Maximum due to Eccentricity,   7m.
     ,,   ,,   ,,   Obliquity,   10m.

THE EARTH.

Equatorial Radius,	3963·296 miles.
Polar „	3949·791 „
Mean „	3959·1 „

Equatorial Circumference, $\left\{ \begin{array}{l} 22,902 \quad\text{„} \\ 360 \times 60 \ = 21,600 \text{ geographical miles.} \\ 4 \times 10^7 = 40,000,000 \text{ metres.} \end{array} \right.$

Ellipticity or Compression,	1 ÷ 293.
Eccentricity,	·0826.
Density (Water = 1),	5·58.
Mass,	6067 × 10¹⁸ tons.
Mean Acceleration of Gravity in ft. per sec. per sec.,	32·18.
Ratio of Centrifugal Force to Gravity at Equator,	1 ÷ 289.
Eccentricity of its Orbit,	1 ÷ 60.
Annual Progressive Motion of Apse Line,	11·25″.
„ Retrograde Motion of Equinoxes (Precession),	50·22″.
Period of Precession,	25,695 years.
„ Nutation,	18·6 „
Greatest change in Obliquity due to Nutation,	9·23″.
Equation of Equinoxes,	15′ 37″.

THE SUN.

Mean Parallax,	8·80″.
„ Angular Semi-diameter,	16′ 1″.
„ Distance in miles,	92,800,000.
Diameter in miles,	866,400.
„ in Earth's radii,	109.
Density in terms of Earth's,	¼.
„ (taking water as 1),	1·4.
Mass in terms of Earth's,	324,439.
Period of Axial Rotation,	25d. 5h. 37m.

THE MOON.

Mean Parallax,	57′ 2·707″.
„ Angular Semi-diameter,	15′ 34″.
„ Distance in miles,	238,840.
„ „ in Earth's radii,	60·27.
„ „ in terms of Sun's distance,	1/389.
Diameter in miles,	2,162.
„ in terms of Earth's,	3/11.
Density in terms of Earth's,	·61.
„ (taking water as 1),	3·4.
Mass, in terms of Earth's,	1/81.
Eccentricity of Orbit,	1/18.
Inclination of Orbit to Ecliptic,	5° 8′.
Ecliptic Limits, Lunar,	12° 5′ and 9° 30′.
„ Solar,	18° 31′ and 15° 21′.
Tide-raising force in terms of Sun's,	7/3.

ASTRON.   2 F

ANSWERS.

NOTE.—Where only rough values of the astronomical data are given in the questions, the answers can only be regarded as rough approximations, not as highly accurate results. It is impossible to calculate results correctly to a greater number of significant figures than are given in the data employed, and any extra figures so calculated will necessarily be incorrect. As the use of working examples is to learn astronomy rather than arithmetic, it is advisable to supply from memory the rough values of such astronomical constants as are not given in the questions. These values will thus be remembered more easily than if the more accurate values were taken from the tables on pages 426, 427, though reference to the latter should be made until the student is familiar with them.

I. EXAMPLES (p. 33).

1. Only their *relative* positions are stated; these do not completely fix them.

2. 6 P.M., 6 A.M.; on the meridian.   **8.** On September 19.

9. (i.) Early in July; (ii.) middle of June—the Sun passes it about June 26.

10. 304° = 20h. 16m.; at 8h. 13m. P.M.

11. Near the S. horizon about 10 P.M. early in October.

12. 38° 27′, 51° 33′, 28° 5′, or if Sun transits N. of zenith 8° 27′, 81° 33′, 58° 5′.

I. EXAMINATION PAPER (p. 34).

7. 30°.   **8.** 61° 58′ 37″, 15° 4′ 21″.   **9.** 6h. 43m. 16s. (roughly).

10. The figure should make *Capella* slightly W. of N., altitude about 15°; *a Lyræ* a little S.E. of zenith, altitude about 75°; *a Scorpii* slightly W. of S., altitude about 12°; *a Ursæ Majoris* N.W., altitude about 60°.

ANSWERS.

II. Examples (p. 61).

6. Direct. **7.** Interval = 12 sidereal hours. **9.** 2° 29′ 58·5″.
11. 12° 39′ 9″. **12.** 17h. 29m. 52·42s.

II. Examination Paper (p. 62).

6. Positive. **10.** 1m. 2·52s., + 0·71s.

III. Examples (p. 84).

2. 4,267 ft.
3. $a°$ N., $L° - 90°$ W. and $a°$ S., $L° + 90°$ W., if $L° =$ W. longitude
given place.
5. 13m. **6.** 39·8 miles. **7.** 3960.
8. 6084 ft. **10.** 49′ 6″ per hour.

Miscellaneous Questions (p. 85).

2. N.P.D. = 85°, hour angle = 30° W.
3. Because declination circle has not been defined.
5. 22h. 40m., 9h. 20m., 14h. 0m., 19h. 36m.. **10.** 52″.

III. Examination Paper (p. 86).

1. 24,840 miles, 3,953 miles.
2. 3·285 ft., 6,084 ft., 1·69 ft. per second. **3.** 50·7 ft.
5. 3,437,700 fathoms, 6,366,200 metres (roughly), 1,851·851 metres.
9. See § 97, cor.

IV. Examples (p. 113).

5. 45°. **7.** Star, 6h. 15m. 26·35s.; Sun, 0h. 13m. 51·90s.
10. 3481 : 3721, or 29 : 31 nearly.

IV. Examination Paper (p. 114).

3. See §§ 130, 151.
8. 0h. 36m. 21·26s. (Note that the clock has a losing rate of
3m. 22·05s. on sidereal time; it gives solar time approximately.)

V. Examples (p. 137).

1. Retrograde. **3.** −3·9m. **6.** 347 centuries exactly.
7. Star's hour angle = 4h. 11m. 3s., N.P.D. = 53°.
8. October 28, 15h. 39m. 27·32s.
10. 12h. 27m. 13·26s. at Louisville = 18h. 9m. 13·26 at Greenwich.

Miscellaneous Questions (p. 138).

3. Eastward.   **5.** Use Figs. 47, 50.   **6.** See § 439.
7. See § 161.   **8.** 11h. 59m. 15·9s.; − 1m. 7·4s.
9. 366·25 : 365·25 or 1465 : 1461.

V. Examination Paper (p. 139).

4. −10m.; morning 20m. longer.   **5.** See § 172.
8. (i.) 7h. 13m. 5s.; (ii.) 7h. 12m. 48s.   **9.** June 26.
10. 1824, 1852, 1880, 1920.

VI. Examples (p. 151).

3. 3,963 miles.
4. From 50° 9′ 47″ to 49° 59′ 55″ (refraction at altitude 5° = 9′ 47″
     by tables).
5. 44° 53′ 28″.   **8.** 84° 33′; 377 miles or 327 nautical miles.

VI. Examination Paper (p. 152).

4. 462″.   **7.** 44° 58′ 54″.   **10.** 1h. 12m.

VII. Examples (p. 188).

1. 37° 49′.   **2.** 51° 44′ 26·09″.
4. 50° 54′ 58·6″ or 60° 43′ 23·6″ according as star transits N. or S.
     of zenith.
5. 44° 55′, or, if corrected for refraction (cf. Ex. 2, p. 168),
     44° 53′ 54″.
6. 51° 33′, 38° 27′, 61° 54′.   **8.** −10m., i.e., 10m. fast.
9. 12° 30′.   **10.** 1h. 0m.   **11.** 2° 32′.   **12.** 27′.
13. See § 237.   **18.** Lat. = $\cos^{-1}\frac{1}{37}$ = 87° 54′ nearly.

VIII. Examples (p. 217).

2. 92,819,000 (see Ex. 2, p. 195).
3. At 6 p.m.; about same length as Midsummer Sun, i.e., 16½h.
4. See § 264.   **5.** 8′ 48″.   **6.** Use § 266.
7. 10d. 4½h. at noon.
8. Gibbous, bright limb turned slightly below direction of W.
     Hour angle = 30°, decl. = 0.
10. (i.) No harvest moon; (ii.) Phenomena practically unaltered.

VIII. Examination Paper (p. 218).

4. See § 260.   **7.** 71° 33″.   **9.** When we have a solar eclipse.

IX. Examples (p. 236).

1. $23\frac{1}{2}$° S.

2. Favourable if moon passes from N. to S. at ecliptic on March 21.

4. 4m. 38s.   **5.** Length = (Earth's radius) ÷ sin $(S - P)$.

7. 6h. 32m. if month unaltered; or, by § 329, a lunation = about 10 days, and then time = 2h. 10m.

8. 40 Earth's radii = 158,000 miles (roughly).

9. Total Solar.   **10.** 1° 28′ (*cf.* § 291).

IX. Examination Paper (p. 237).

6. 850,000, 230,000, and 5,800 miles (roughly).

7. See §§ 292, 295-297.   **9.** No.

10. In Fig. 93 take M on xm produced, such that $\sin xM = xm/(p - P)$.

X. Examples (p. 265).

1. 291·96 days, or, if conjunctions are of the same kind, 583·92 days.

2. 40°.   **3.** 19 : 6, or nearly 3 : 1.   **4.** $10\frac{1}{9}$h., 120h.

5. $p + P - s$ with notation of § 290.   **6.** 888 million miles, 164 yrs.

7. 6 months; $\sqrt[3]{\frac{1}{4}}$ or ·63 of Earth's mean distance.   **8.** 398 days.

9. $\frac{3}{8}$ of a year = 137 days.

10. Stationary at heliocentric conjunction only, never retrograde.

X. Examination Paper (p. 266).

3. $1\frac{1}{30}$ years = 378 days.

4. See §§ 323, 324.   The alterations in Venus's brightening are really not inconsiderable (see Ex. 3, p. 205).

6. Most rapid approach at quadrature; velocity that with which the *Earth* would describe its orbit in synodic period.

9. 287 days.

10. Draw the circular orbits about ☉, radii 4, 7, 10, 16, 52 (§ 304). The *heliocentric* longitudes (measured from ☉ ♈) are roughly as follows: ☿ 153°, ♀ 175°, ⊕ 220°, ♂ 20°, ♃ 211°. The ☾ should be drawn close to ⊕ at an elongation ☉ ⊕ ☾ = 90° at first quarter.

XI. EXAMPLES (p. 311).

2. 432,000 miles.   **3.** 2,250 miles.

6. 9,282,000 and 92,820,000 million miles respectively.

7. 37·8 billion miles $= 378 \times 10^{11}$ miles.

8. $5 : \pi = 1·6 : 1$ roughly.

10. It will always appear half-way between its actual direction and a point on the ecliptic 90° behind Sun. Path is roughly a small circle of angular radius 45°.

11. 4° 35'.

13. (i.) On ecliptic 90° from Sun. (ii.) In same or opposite direction to Sun. Effects greatest along great circles distant 90° from these points.

14. (i.) At either pole of ecliptic. (ii.) In ecliptic.

16. Jan. 21, 10·25" Eastwards; Feb., 17·75" E.; Mar., 20·50" E.; April, 17·75" E.; May, 10·25" E.; June, 0"; July, 10·25" Westwards; Aug., 17·75" W.; Sept., 20·50" W.; Oct., 17·75" W.; Nov., 10·25" W.; Dec., 0".

18. 973,800 miles.

MISCELLANEOUS QUESTIONS (p. 313).

5. 15° E.   **6.** In the autumn.

7. 17d. 5h.; star is on equator, hour angle 60° E.   **8.** $1 : \sqrt{7} : 7$.

9. 24h. 50m. 30s. mean units = 24h. 54m. 35s. sidereal units.

10. At the equinoxes.   **11.** See § 376.

XII. EXAMPLES (p. 335).

1. $12\sqrt{e}$ sidereal hours = 16h. 58m. 5s. sidereal time.

2. Pendulum revolving in direction of hands of watch will have less velocity in S. hemisphere.

7. Increased (i.) 59° 54' 51"; (ii.) 60° 15' 27".   **12.** 109 lbs.

XII. EXAMINATION PAPER (p. 336).

3. By observing deviation of a projectile (§ 390), or by § 387 or § 389.

4. $16\sqrt{3} = 27·7157$ sidereal hours = 1d. 3h. 33m. mean time.

5. 3·368 cm. per sec. per sec.; $\frac{1}{110}$.   **9.** See § 390.

XIII. Examples (p. 368).

1. 2·97 miles per sec.   **2.** 15½ ft., or, if $g = 32·2$, 15·576 ft.
3. 5h. 35m.   **4.** 5·39 days.   **5.** 2,959,000.   **11.** 8·98″.
13. The distances from the centre of the Sun are 457,579 miles,
457,579 + 278 miles, and 457,579 − 281 miles ; but these results
can only be considered as approximate.
14. 32·155 greater, owing to attraction of mountain.
17. ·253 of Earth's density ; 1·415, taking water = 1.
18. 894 poundals.
20. At first a hyperbola under the Earth's attraction. After going
some distance this attraction would become insensible, and
the Moon would describe an ellipse about the Sun rather
more eccentric than the Earth's present orbit.

XIV. Examples (p. 419).

1. $\dfrac{1}{80·34}$.   **2.** 2 : 5.

3. 24 : 7, by Ex. 1, § 442 Cor., or 16 : 5, using result of last example.
5. Tide on Mercury is higher in proportion 1 : ·2888, or 45 : 13, or
7 : 2 nearly.
10. Direct shortly before, retrograde shortly after.

XIV. Examination Paper (p. 420).

7. (a) Yes; (b) No.

INDEX.

(The numbers refer to the pages throughout.)

PRINTED AT THE BURLINGTON PRESS, CAMBRIDGE.